NEUTRINO PHYSICS

Series in High Energy Physics, Cosmology and Gravitation

Other books in the series

The Mathematical Theory of Cosmic Strings
M R Anderson

Geometry and Physics of Branes
Edited by U Bruzzo, V Gorini and U Moschella

Modern Cosmology
Edited by S Bonometto, V Gorini and U Moschella

Gravitation and Gauge Symmetries
M Blagojevic

Gravitational Waves
Edited by I Ciufolini, V Gorini, U Moschella and P Fré

Classical and Quantum Black Holes
Edited by P Fré, V Gorini, G Magli and U Moschella

Pulsars as Astrophysical Laboratories for Nuclear and Particle Physics
F Weber

The World in Eleven Dimensions
Supergravity, Supermembranes and M-Theory
Edited by M J Duff

Particle Astrophysics
Revised paperback edition
H V Klapdor-Kleingrothaus and K Zuber

Electron–Positron Physics at the Z
M G Green, S L Lloyd, P N Ratoff and D R Ward

Non-Accelerator Particle Physics
Revised edition
H V Klapdor-Kleingrothaus and A Staudt

Idea and Methods of Supersymmetry and Supergravity
or **A Walk Through Superspace**
Revised edition
I L Buchbinder and S M Kuzenko

NEUTRINO PHYSICS

K Zuber

Denys Wilkinson Laboratory, University of Oxford, UK

I₀P

INSTITUTE OF PHYSICS PUBLISHING
BRISTOL AND PHILADELPHIA

British Library Cataloguing-in-Publication Data

A catalogue record for this book is available from the British Library.

ISBN 0 7503 0750 1

Library of Congress Cataloging-in-Publication Data are available

Commissioning Editor: John Navas
Production Editor: Simon Laurenson
Production Control: Sarah Plenty
Cover Design: Victoria Le Billon
Marketing: Nicola Newey and Verity Cooke

Published by Institute of Physics Publishing, wholly owned by The Institute of Physics, London

Institute of Physics Publishing, Dirac House, Temple Back, Bristol BS1 6BE, UK

US Office: Institute of Physics Publishing, The Public Ledger Building, Suite 929, 150 South Independence Mall West, Philadelphia, PA 19106, USA

Typeset in LaTeX 2_ε by Text 2 Text, Torquay, Devon
Printed in the UK by MPG Books Ltd, Bodmin, Cornwall

Dedicated to
Tom Ypsilantis and Ray Davis Jr

Contents

Preface   **xiii**

Notation   **xv**

1   Important historical experiments   **1**
  1.1   'The birth of the neutrino'   1
  1.2   Nuclear recoil experiment by Rodeback and Allen   4
  1.3   Discovery of the neutrino by Cowan and Reines   4
  1.4   Difference between ν_e and $\bar{\nu}_e$ and solar neutrino detection   5
  1.5   Discovery of parity violation in weak interactions   7
  1.6   Direct measurement of the helicity of the neutrino   10
  1.7   Experimental proof that ν_μ is different from ν_e 11
  1.8   Discovery of weak neutral currents   12
  1.9   Discovery of the weak gauge bosons W and Z   14
  1.10 Observation of neutrinos from SN 1987A   16
  1.11 Number of neutrino flavours from the width of the Z^0 16

2   Properties of neutrinos   **19**
  2.1   Helicity and chirality   19
  2.2   Charge conjugation   22
  2.3   Parity transformation   23
  2.4   Dirac and Majorana mass terms   24
    2.4.1   Generalization to n flavours   28
  2.5   Lepton number   29
    2.5.1   Experimental status of lepton number violation   30

3   The standard model of particle physics   **33**
  3.1   The V–A theory of the weak interaction   33
  3.2   Gauge theories   35
    3.2.1   The gauge principle   36
    3.2.2   Global symmetries   38
    3.2.3   Local (= gauge) symmetries   39
    3.2.4   Non-Abelian gauge theories (= Yang–Mills theories)   40
  3.3   The Glashow–Weinberg–Salam model   41
    3.3.1   Spontaneous symmetry breaking and the Higgs mechanism   44

		3.3.2	The CKM mass matrix	48
		3.3.3	*CP* violation	49
	3.4	Experimental determination of fundamental parameters		51
		3.4.1	Measurement of the Fermi constant G_F	51
		3.4.2	Neutrino–electron scattering and the coupling constants g_V and g_A	52
		3.4.3	Measurement of the Weinberg angle	58
		3.4.4	Measurement of the gauge boson masses m_W and m_Z	59
		3.4.5	Search for the Higgs boson	61
4	**Neutrinos as a probe of nuclear structure**			**64**
	4.1	Neutrino beams		64
		4.1.1	Conventional beams	64
		4.1.2	ν_τ beams	69
		4.1.3	Neutrino beams from muon decay	69
	4.2	Neutrino detectors		70
		4.2.1	CDHS	70
		4.2.2	NOMAD	71
		4.2.3	CHORUS	72
	4.3	Total cross section for neutrino–nucleon scattering		73
	4.4	Kinematics of deep inelastic scattering		75
	4.5	Quasi-elastic neutrino–nucleon scattering		77
		4.5.1	Quasi-elastic CC reactions	78
		4.5.2	(Quasi-)elastic NC reactions	79
	4.6	Coherent, resonant and diffractive production		81
	4.7	Structure function of nucleons		83
	4.8	The quark–parton model, parton distribution functions		84
		4.8.1	Deep inelastic neutrino proton scattering	85
	4.9	*y* distributions and quark content from total cross sections		90
		4.9.1	Sum rules	93
	4.10	Charm physics		95
	4.11	Neutral current reactions		98
	4.12	Neutrino cross section on nuclei		101
5	**Neutrino masses and physics beyond the standard model**			**105**
	5.1	Running coupling constants		106
	5.2	The minimal SU(5) model		107
		5.2.1	Proton decay	110
	5.3	The SO(10) model		111
		5.3.1	Left–right symmetric models	113
	5.4	Supersymmetry		114
		5.4.1	The minimal supersymmetric standard model (MSSM)	115
		5.4.2	*R*-parity	117
		5.4.3	Experimental search for supersymmetry	117
	5.5	Neutrino masses		121

		5.5.1	Neutrino masses in the electroweak theory	121
		5.5.2	Neutrino masses in the minimal SU(5) model	122
		5.5.3	Neutrino masses in the SO(10) model and the seesaw mechanism	123
		5.5.4	Neutrino masses in SUSY and beyond	124
	5.6		Neutrino mixing	124

6   Direct neutrino mass searches   **127**

	6.1		Fundamentals of β-decay	127
		6.1.1	Matrix elements	129
		6.1.2	Phase space calculation	131
		6.1.3	Kurie plot and ft values	133
	6.2		Searches for $m_{\bar{\nu}_e}$	136
		6.2.1	General considerations	136
		6.2.2	Searches using spectrometers	137
		6.2.3	Cryogenic searches	141
		6.2.4	Kinks in β-decay	142
	6.3		Searches for m_{ν_e}	144
	6.4		m_{ν_μ} determination from pion-decay	145
	6.5		Mass of the ν_τ from tau-decay	146
	6.6		Electromagnetic properties of neutrinos	147
		6.6.1	Electric dipole moments	148
		6.6.2	Magnetic dipole moments	149
	6.7		Neutrino decay	151
		6.7.1	Radiative decay $\nu_H \rightarrow \nu_L + \gamma$	152
		6.7.2	The decay $\nu_H \rightarrow \nu_L + e^+ + e^-$	154
		6.7.3	The decay $\nu_H \rightarrow \nu_L + \chi$	154

7   Double β-decay   **156**

	7.1		Introduction	156
	7.2		Decay rates	161
		7.2.1	The $2\nu\beta\beta$ decay rates	161
		7.2.2	The $0\nu\beta\beta$ decay rates	164
		7.2.3	Majoron accompanied double β-decay	166
	7.3		Nuclear structure effects on matrix elements	167
	7.4		Experiments	170
		7.4.1	Practical considerations in low-level counting	172
		7.4.2	Direct counting experiments	172
		7.4.3	Geochemical experiments	177
		7.4.4	Radiochemical experiments	179
	7.5		Interpretation of the obtained results	179
		7.5.1	Effects of MeV neutrinos	181
		7.5.2	Transitions to excited states	182
		7.5.3	Majoron accompanied decays	182
		7.5.4	Decay rates for SUSY-induced $0\nu\beta\beta$-decay	182

	7.6	The future	183
	7.7	$\beta^+\beta^+$-decay	184
	7.8	CP phases and double β-decay	185
	7.9	Generalization to three flavours	186
		7.9.1 General considerations	186
8	**Neutrino oscillations**		**190**
	8.1	General formalism	190
	8.2	CP and T violation in neutrino oscillations	193
	8.3	Oscillations with two neutrino flavours	194
	8.4	The case for three flavours	195
	8.5	Experimental considerations	197
	8.6	Nuclear reactor experiments	199
		8.6.1 Experimental status	200
		8.6.2 Future	206
	8.7	Accelerator-based oscillation experiments	206
		8.7.1 LSND	207
		8.7.2 KARMEN	209
		8.7.3 Future test of the LSND evidence—MiniBooNE	209
	8.8	Searches at higher neutrino energy	210
		8.8.1 CHORUS and NOMAD	211
	8.9	Neutrino oscillations in matter	215
	8.10	CP and T violation in matter	218
	8.11	Possible future beams	218
		8.11.1 Off-axis beams and experiments	219
		8.11.2 Beta beams	220
		8.11.3 Superbeams	220
		8.11.4 Muon storage rings—neutrino factories	220
9	**Atmospheric neutrinos**		**222**
	9.1	Cosmic rays	222
	9.2	Interactions within the atmosphere	224
	9.3	Experimental status	230
		9.3.1 Super-Kamiokande	230
		9.3.2 Soudan-2	239
		9.3.3 MACRO	239
	9.4	Future activities—long-baseline experiments	242
		9.4.1 K2K	242
		9.4.2 MINOS	244
		9.4.3 CERN–Gran Sasso	245
		9.4.4 MONOLITH	247
		9.4.5 Very large water Cerenkov detectors	248
		9.4.6 AQUA-RICH	248

10 Solar neutrinos **250**
  10.1 The standard solar model 250
    10.1.1 Energy production processes in the Sun 250
    10.1.2 Reaction rates 254
    10.1.3 The solar neutrino spectrum 255
  10.2 Solar neutrino experiments 260
    10.2.1 The chlorine experiment 262
    10.2.2 Super-Kamiokande 264
    10.2.3 The gallium experiments 266
    10.2.4 The Sudbury Neutrino Observatory (SNO) 269
  10.3 Attempts at theoretical explanation 271
    10.3.1 Neutrino oscillations as a solution to the solar neutrino
           problem 271
    10.3.2 Neutrino oscillations in matter and the MSW effect 272
    10.3.3 Experimental signatures and results 280
    10.3.4 The magnetic moment of the neutrino 281
  10.4 Future experiments 286
    10.4.1 Measuring ^{7}Be neutrinos with the Borexino experiment 287
    10.4.2 Real-time measurement of pp neutrinos using
           coincidence techniques 288

11 Neutrinos from supernovae **290**
  11.1 Supernovae 290
    11.1.1 The evolution of massive stars 291
    11.1.2 The actual collapse phase 294
  11.2 Neutrino emission in supernova explosions 299
  11.3 Detection methods for supernova neutrinos 301
  11.4 Supernova 1987A 302
    11.4.1 Characteristics of supernova 1987A 304
    11.4.2 Neutrinos from SN 1987A 308
    11.4.3 Neutrino properties from SN 1987A 310
  11.5 Supernova rates and future experiments 315
    11.5.1 Cosmic supernova relic neutrino background 316
  11.6 Neutrino oscillations and supernova signals 316
    11.6.1 Effects on the prompt ν_e burst 318
    11.6.2 Cooling phase neutrinos 319
    11.6.3 Production of r-process isotopes 319
    11.6.4 Neutrino mass hierarchies from supernove signals 320
    11.6.5 Resonant spin flavour precession in supernovae 325

12 Ultra-high energetic cosmic neutrinos **327**
  12.1 Sources of high-energy cosmic neutrinos 327
    12.1.1 Neutrinos produced in acceleration processes 328
    12.1.2 Neutrinos produced in annihilation or decay of heavy
           particles 332

12.1.3 Event rates   334
12.1.4 v from AGNs   334
12.1.5 v from GRBs   337
12.1.6 Cross sections   340
12.2 Detection   344
12.2.1 Water Cerenkov detectors   349
12.2.2 Ice Cerenkov detectors—AMANDA, ICECUBE   353
12.2.3 Alternative techniques—acoustic and radio detection   355
12.2.4 Horizontal air showers—the AUGER experiment   356

13 Neutrinos in cosmology   **361**
13.1 Cosmological models   361
13.1.1 The cosmological constant Λ 365
13.1.2 The inflationary phase   367
13.1.3 The density in the universe   368
13.2 The evolution of the universe   370
13.2.1 The standard model of cosmology   370
13.3 The cosmic microwave background (CMB)   376
13.3.1 Spectrum and temperature   376
13.3.2 Measurement of the spectral form and temperature of the
        CMB   377
13.3.3 Anisotropies in the 3 K radiation   378
13.4 Neutrinos as dark matter   382
13.5 Candidates for dark matter   382
13.5.1 Non-baryonic dark matter   382
13.5.2 Direct and indirect experiments   386
13.6 Neutrinos and large-scale structure   386
13.7 The cosmic neutrino background   390
13.8 Primordial nucleosynthesis   391
13.8.1 The process of nucleosynthesis   392
13.8.2 The relativistic degrees of freedom g_{eff} and the number of
        neutrino flavours   396
13.9 Baryogenesis via leptogenesis   397

14 Summary and outlook   **401**

References   **406**

Index   **432**

Preface

The last decade has seen a revolution in neutrino physics. The establishment of a non-vanishing neutrino mass in neutrino oscillation experiments is one of the major new achievements. In this context the problem of missing solar neutrinos could be solved. In addition, limits on the absolute neutrino mass could be improved by more than an order of magnitude by beta decay and double beta decay experiments. Massive neutrinos have a wide impact on particle physics, astrophysics and cosmology. Their properties might guide us to theories Beyond the Standard Model of Particle Physics in form of grand unified theories (GUTs). The precise determination of the mixing matrix like the one in the quark sector lies ahead of us with new machines, opening the exciting possibility to search for CP-violation in the lepton sector. Improved absolute mass measurements are on their way. Astrophysical neutrino sources like the Sun and supernovae still offer a unique tool to investigate neutrino properties. A completely new window in high astrophysics using neutrino telescopes has just opened and very exciting results can be expected soon. Major new important observations in cosmology sharpen our view of the universe and its evolution, where neutrinos take their part as well.

The aim of this book is to give an outline of the essential ideas and basic lines of developments. It tries to cover the full range of neutrino physics, being as comprehensive and self-contained as possible. In contrast to some recent, excellent books containing a collection of articles by experts, this book tries to address a larger circle of readers. This monograph developed out of lectures given at the University of Dortmund, and is therefore well suited as an introduction for students and a valuable source of information for people working in the field. The book contains extensive references for additional reading. In order to be as up-to-date as possible many preprints have been included, which can be easily accessed electronically via preprint servers on the World Wide Web.

It is a pleasure to thank my students M Althaus, H Kiel, M Mass and D Münstermann for critical reading of the manuscript and suggestions for improvement. I am indebted to my colleagues S M Bilenky, C P Burgess, L diLella, K Eitel, T K Gaisser, F Halzen, D H Perkins, L Okun, G G Raffelt, W Rhodejohann, J Silk, P J F Soler, C Weinheimer and P Vogel for valuable comments and discussions.

Many thanks to Mrs S Helbich for the excellent translation of the manuscript and to J Revill, S Plenty and J Navas of Institute of Physics Publishing for their faithful and efficient collaboration in getting the manuscript published. Last, but not least, I want to thank my wife for her patience and support.

K Zuber Oxford, August 2003

Notation

Covering the scales from particle physics to cosmology, various units are used. A system quite often used is that of natural units ($c = \hbar = k_B = 1$) which is used throughout this book. Deviations are used if they aid understanding. The table overleaf gives useful conversion factors in natural units.

In addition, here are some useful relations:

$$\hbar c = 197.33 \, \text{MeV fm}$$
$$1 \, \text{erg} = 10^7 \, \text{J}$$
$$1 \, M_\odot = 1.988 \times 10^{30} \, \text{kg}$$
$$1 \, \text{pc} = 3.262 \, \text{light years} = 3.0857 \times 10^{16} \, \text{m}.$$

Among the infinite amount of Web pages from which to obtain useful information, the following URLs should be mentioned:

- http://xxx.lanl.gov (Los Alamos preprint server)
- http://adsabs.harvard.edu (Search for astrophysical papers)
- http://www.slac.stanford.edu/spires/hep (SLAC Spires—Search for High Energy Physics papers)
- http://neutrinooscillation.org (The Neutrino Oscillation Industry)

Conversion factors for natural units.

	s^{-1}	cm^{-1}	K	eV	amu	erg	g
s^{-1}	1	0.334×10^{-10}	0.764×10^{-11}	0.658×10^{-15}	0.707×10^{-24}	1.055×10^{-27}	1.173×10^{-48}
cm^{-1}	2.998×10^{10}	1	0.229	1.973×10^{-5}	2.118×10^{-14}	3.161×10^{-17}	0.352×10^{-37}
K	1.310×10^{11}	4.369	1	0.862×10^{-4}	0.962×10^{-13}	1.381×10^{-16}	1.537×10^{-37}
eV	1.519×10^{15}	0.507×10^{5}	1.160×10^{4}	1	1.074×10^{-9}	1.602×10^{-12}	1.783×10^{-33}
amu	1.415×10^{24}	0.472×10^{14}	1.081×10^{13}	0.931×10^{9}	1	1.492×10^{-3}	1.661×10^{-24}
erg	0.948×10^{27}	0.316×10^{17}	0.724×10^{16}	0.624×10^{12}	0.670×10^{3}	1	1.113×10^{-21}
g	0.852×10^{48}	2.843×10^{37}	0.651×10^{37}	0.561×10^{33}	0.602×10^{24}	0.899×10^{21}	1

Chapter 1

Important historical experiments

With the discovery of the electron in 1897 by J J Thomson a new era of physics—today called elementary particle physics—started. By destroying the atom as the fundamental building block of matter the question arose as to what other particles could be inside the atom. Probing smaller and smaller length scales equivalent to going to higher and higher energies by using high-energy accelerators a complete 'zoo' of new particles was discovered, which finally led to the currently accepted standard model (SM) of particle physics (see chapter 3). Here, the building blocks of matter consist of six quarks and six leptons shown in table 1.1, all of them being spin-$\frac{1}{2}$ fermions. They interact with each other through four fundamental forces: gravitation, electromagnetism and the strong and weak interactions.

In quantum field theory these forces are described by the exchange of the bosons shown in table 1.2. Among the fermions there is one species—neutrinos—where our knowledge today is still very limited. Being leptons (they do not participate in strong interactions) and having zero charge (hence no electromagnetic interactions) they interact only via weak interactions (unless they have a non-vanishing mass, in which case electromagnetic and gravitational interactions might be possible), making experimental investigations extremely difficult. However, neutrinos are the obvious tool with which to explore weak processes and the history of neutrino physics and weak interactions is strongly connected.

The following chapters will depict some of the historic milestones. For more detailed discussions on the history, see [Sie68, Pau77].

1.1 'The birth of the neutrino'

Ever since its discovery the neutrino's behaviour has been out of the ordinary. In contrast to the common way of discovering new particles, i.e. in experiments, the neutrino was first postulated theoretically. The history of the neutrino began with the investigation of β-decay (see chapter 6).

Table 1.1. (*a*) Properties of the quarks: I, isospin; S, strangeness; C, charm; Q, charge: B, baryon number; B^*, bottom; T, top. (*b*) Properties of leptons: L_i, flavour-related lepton number, $L = \sum_{i=e,\mu,\tau} L_i$.

(*a*) Flavour	Spin	B	I	I_3	S	C	B^*	T	$Q[e]$
u	1/2	1/3	1/2	1/2	0	0	0	0	2/3
d	1/2	1/3	1/2	−1/2	0	0	0	0	−1/3
c	1/2	1/3	0	0	0	1	0	0	2/3
s	1/2	1/3	0	0	−1	0	0	0	−1/3
b	1/2	1/3	0	0	0	0	−1	0	−1/3
t	1/2	1/3	0	0	0	0	0	1	2/3

(*b*) Lepton	$Q[e]$	L_e	L_μ	L_τ	L
e^-	−1	1	0	0	1
ν_e	0	1	0	0	1
μ^-	−1	0	1	0	1
ν_μ	0	0	1	0	1
τ^-	−1	0	0	1	1
ν_τ	0	0	0	1	1

Table 1.2. Phenomenology of the four fundamental forces and the hypothetical GUT interaction. Natural units $\hbar = c = 1$ are used.

Interaction	Strength	Range R	Exchange particle	Example
Gravitation	$G_N \simeq 5.9 \times 10^{-39}$	∞	Graviton?	Mass attraction
Weak	$G_F \simeq 1.02 \times 10^{-5} m_p^{-2}$	$\approx m_W^{-1}$ $\simeq 10^{-3}$ fm	$W^\pm, Z^0$	β-decay
Electro-magnetic	$\alpha \simeq 1/137$	∞	γ	Force between electric charges
Strong (nuclear)	$g_\pi^2/4\pi \approx 14$	$\approx m_\pi^{-1}$ ≈ 1.5 fm	Gluons	Nuclear forces
Strong (colour)	$\alpha_s \simeq 1$	confinement	Gluons	Forces between the quarks
GUT	$M_X^{-2} \approx 10^{-30} m_p^{-2}$ $M_X \approx 10^{16}$ GeV	$\approx M_X^{-1}$ $\approx 10^{-16}$ fm	X, Y	p-decay

After the observation of discrete lines in the α-and γ-decay of atomic nuclei, it came as a surprise when J Chadwick discovered a continuous energy spectrum of electrons emitted in β-decay [Cha14]. The interpretation followed two lines; one assumed primary electrons with a continuous energy distribution (followed mainly by C D Ellis) and the other assumed secondary processes, which broaden an initially discrete electron energy (followed mainly by L Meitner). To resolve the question, a calorimetric measurement which should result in either the average electron energy (if C D Ellis was right) or the maximal energy (if L Meitner was correct) was done. This can be understood in the following way: β-decay is nowadays described by the three-body decay

$$M(A, Z) \to D(A, Z + 1) + e^- + \bar{\nu}_e \tag{1.1}$$

where $M(A, Z)$ describes the mother nucleus and $D(A, Z + 1)$ its daughter. The actual decay is that of a neutron into a proton, electron and antineutrino. For decay at rest of $M(A, Z)$ the electron energy should be between

$$E_{min} = m_e \tag{1.2}$$

and using energy conservation

$$E_{max} = m_M - m_D. \tag{1.3}$$

In (1.3) the small kinetic recoil energy T_D of the daughter nucleus was neglected and $m_M - m_D = T_D + E_e + E_\nu = 0$ (assumption: $m_\nu = 0$). Hence, if there are only electrons in the final state the calorimetric measurement should always result in $E_{max} = m_M - m_D$.

The experiment was done using the β-decay (see chapter 6) of the isotope RaE (today known as ^{210}Bi) with a nuclear transition Q-value of 1161 keV. The measurement resulted in a value of 344 000 eV $\pm$ 10% ($\equiv$344 $\pm$ 10% keV) [Ell27] clearly supporting the first explanation. L Meitner, still not convinced, repeated the experiment ending up with 337 000 eV $\pm$ 6% confirming the primary origin of the continuous electron spectrum [Mei30]. To explain this observation only two solutions seemed to be possible: either the energy conservation law is only valid statistically in such a process (preferred by N Bohr) or an additional undetectable new particle (later called the neutrino by E Fermi) carrying away the additional energy and spin (preferred by W Pauli) is emitted. There was a second reason for Pauli's proposal of a further particle, namely angular momentum conservation. It was observed in β-decay that if the mother atom carries integer/fractional spin then the daughter also does, which cannot be explained by the emission of only one spin-$\frac{1}{2}$ electron. In a famous letter dated 4 December 1930 W Pauli proposed his solution to the problem; a new spin-$\frac{1}{2}$ particle (which we nowadays call the neutrino) produced together with the electron but escaping detection. In this way the continous spectrum can be understood: both electron and neutrino share the transition energy in a way that the sum of both always corresponds to the full transition energy. Shortly afterwards the neutron was discovered [Cha32], the understanding of β-decay changed rapidly and this led E Fermi to develop his successful theory of β-decay [Fer34]. The first experiments to support the notion of the neutrino were to come about 20 years later.

1.2   Nuclear recoil experiment by Rodeback and Allen

The first experimental evidence for neutrinos was found in the electron capture (EC) of ^{37}Ar:

$$^{37}\text{Ar} + \text{e}^- \rightarrow {}^{37}\text{Cl} + \nu_e + Q \tag{1.4}$$

with a Q-value of 816 keV. Because the process has only two particles in the final state the recoil energy of the nucleus is fixed. Using energy and momentum conservation, the recoil energy T_{Cl} is given by

$$T_{Cl} = \frac{E_\nu^2}{2m_{Cl}} \approx \frac{Q^2}{2m_{Cl}} = 9.67\,\text{eV} \tag{1.5}$$

because the rest mass of the ^{37}Cl atom is much larger than $Q \approx E_\nu$. This energy corresponds to a velocity for the ^{37}Cl nucleus of 0.71 cm μs^{-1}. Therefore, the recoil velocity could be measured by a delayed coincidence measurement. It is started by the Auger electrons emitted after electron capture and stopped by detecting the recoiling nucleus. In using a variable time delay line a signal should be observed if the delay time coincides with the time of flight of the recoil ions. With a flight length of 6 cm, a time delay of 8.5 μs was expected. Indeed, the expected recoil signal could be observed at about 7 μs. After several necessary experimental corrections (e.g. thermal motion caused a 7% effect in the velocity distribution), both numbers were in good agreement [Rod52].

Soon afterwards the measurement was repeated with an improved spectrometer and a recoil energy of $T_{Cl} = (9.63\pm0.03)$ eV was measured [Sne55] in good agreement with (1.5).

1.3   Discovery of the neutrino by Cowan and Reines

The discovery finally took place at nuclear reactors, which were the strongest neutrino sources available. The basic detection reaction was

$$\bar{\nu}_e + \text{p} \rightarrow \text{e}^+ + \text{n}. \tag{1.6}$$

The detection principle was a coincident measurement of the 511 keV photons associated with positron annihilation and a neutron capture reaction a few μs later. Cowan and Reines used a water tank with dissolved CdCl$_2$ surrounded by two liquid scintillators (figure 1.1). The liquid scintillators detect the photons from positron annihilation as well as the ones from the ^{113}Cd(n, γ) ^{114}Cd reaction after neutron capture. The detector is shown in figure 1.2. The experiment was performed in different configurations and at different reactors and finally resulted in the discovery of the neutrino.

In 1953, at the Hanford reactor (USA) using about 300 l of a liquid scintillator and rather poor shielding against background, a vague signal was observed. The experiment was repeated in 1956 at the Savannah River reactor

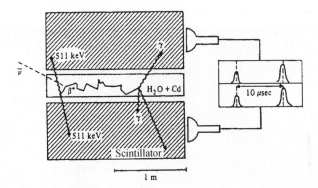

Figure 1.1. Schematic illustration of the experimental set-up for neutrino detections used by Cowan and Reines. A CdCl$_2$ loaded water tank is surrounded by liquid scintillators. They are used for a coincidence measurement of the 511 keV annihilation photons and the γ-rays emitted by the neutron capture on Cd (from [Rei58]).

(USA) with 4200 l of scintillator, finally proving the existence of neutrinos. For more historical information on this experiment see [Los97]. The obtained energy averaged cross section for reaction (1.6) was [Rei53, Rei56]

$$\bar{\sigma} = (11 \pm 2.6) \times 10^{-44} \text{ cm}^2 \qquad (1.7)$$

which, when fully revised, agreed with the V–A theory.

1.4 Difference between ν_e and $\bar{\nu}_e$ and solar neutrino detection

The aim of the experiment was to find out whether neutrinos and antineutrinos are identical particles. If so, the reactions

$$\nu_e + p \rightarrow e^- + n \qquad (1.8)$$
$$\bar{\nu}_e + p \rightarrow e^- + n \qquad (1.9)$$

should occur with the same cross section. In the real experiment Davis was looking for

$$\bar{\nu}_e + {}^{37}\text{Cl} \rightarrow e^- + {}^{37}\text{Ar} \qquad (1.10)$$

by using the Brookhaven reactor (USA). He was using 4000 l of liquid CCl$_4$. The produced Ar atoms were extracted by flooding He through the liquid and then freezing out the Ar atoms in a cooled charcoal trap. By not observing the process (1.9) he could set an upper limit of

$$\bar{\sigma}(\bar{\nu}_e + {}^{37}\text{Cl} \rightarrow e^- + {}^{37}\text{Ar}) < 0.9 \times 10^{-45} \text{ cm}^2 \qquad (1.11)$$

where the theoretical prediction was $\bar{\sigma} \approx 2.6 \times 10^{-45} \text{ cm}^2$ [Dav55].

(*a*)

(*b*)

Figure 1.2. (*a*) The experimental group of Clyde Cowan (left) and Fred Reines (right) of 'Unternehmen Poltergeist' (Project 'Poltergeist') to search for neutrinos. (*b*) The detector called 'Herr Auge' (Mr Eye) (with kind permission of Los Alamos Science).

This detection principle was used years later in a larger scale version in the successful detection of solar neutrinos. This showed that ν_e do indeed cause the reaction (1.8). This pioneering effort marks the birth of neutrino astrophysics and will be discussed in detail in section 10.4.2.

Later it was found at CERN that the same applies to muon neutrinos because in ν_μ interactions only μ^-s in the final state were ever detected but never a μ^+ [Bie64].

1.5 Discovery of parity violation in weak interactions

Parity is defined as a symmetry transformation by an inversion at the origin resulting in $x \rightarrow -x$. It was assumed that parity is conserved in all interactions. At the beginning of the 1950s, however, people were irritated by observations in kaon decays (the so called 'τ–θ' puzzle). Lee and Yang [Lee56], when investigating this problem, found that parity conservation had never been tested for weak interactions and this would provide a solution to this problem.

Parity conservation implies that any process and its mirrored one run with the same probability. Therefore, to establish parity violation, an observable quantity which is different for both processes must be found. This is exactly what pseudo-scalars do. Pseudoscalars are defined in such a way that they change sign under parity transformations. They are a product of a polar and an axial vector e.g. $p_e \cdot I_{\text{nuc}}$, $p_e \cdot s_e$ with I_{nuc} as the spin of the nucleus and p_e and s_e as momentum and spin of the electron. Any expectation value for a pseudo-scalar different from zero would show parity violation. Another example of a pseudo-scalar is provided by possible angular distributions like

$$\Delta\theta = \lambda(\theta) - \lambda(180° - \theta) \tag{1.12}$$

where λ is the probability for an electron to be emitted under an angle θ with respect to the spin direction of the nucleus. Under parity transformation the emission angle changes according to $\theta \rightarrow \pi - \theta$ which leads to $\Delta\theta \rightarrow -\Delta\theta$. In the classical experiment of Wu *et al* , polarized ^{60}Co atoms were used [Wu57]. To get a significant polarization, the ^{60}Co was implemented in a paramagnetic salt and kept at 0.01 K. The polarization was measured via the angular anisotropy of the emitted γ-rays from ^{60}Ni using two NaI detectors. The decay of ^{60}Co is given by

$$^{60}\text{Co} \rightarrow {}^{60}\text{Ni}^* + e^- + \bar{\nu}_e. \tag{1.13}$$

The emitted electrons were detected by an anthracene detector producing scintillation light. The mirror configuration was created by reversing the applied magnetic field. A schematic view of the experiment is shown in figure 1.3, the obtained data in figure 1.4. It shows that electrons are preferably emitted in the opposite spin direction to that of the mother nucleus. This could be described by an angular distribution

$$W(\cos\theta) \propto 1 + \alpha\cos\theta \tag{1.14}$$

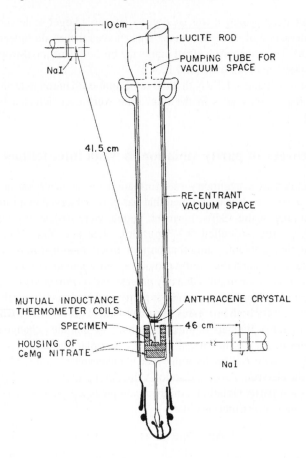

Figure 1.3. Schematic diagram showing the demagnetisation cryostat used in the measurement of the angular distribution of the electrons from the β-decay of ^{60}Co nuclei (from [Wu57]).

with a measured $\alpha \approx -0.4$. This was clear evidence that $\Delta\theta \neq 0$ and β-decay does indeed violate parity. The reason is that α is given by $\alpha = -P_{Co}\frac{\langle v_e \rangle}{c}$ where P_{Co} is the polarisation of the ^{60}Co nuclei and $\langle v_e \rangle$ the electron velocity averaged over the electron spectrum. With the given parameters of $P_{Co} \simeq 0.6$ and $\langle v_e \rangle / c = 0.6$ a value of $\alpha = 0.4$ results showing that parity is not only violated but is maximally violated in weak interactions. Another example is pion decay at rest [Gar57]. The positive pion decays via

$$\pi^+ \rightarrow \mu^+ + \nu_\mu. \tag{1.15}$$

Considering the fact that the pion carries spin-0 and decays at rest, this implies that the spins of the muon and neutrino are opposed to each other (figure 1.5).

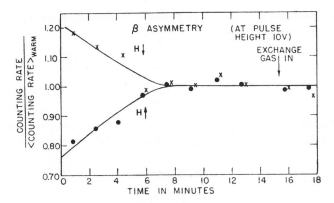

Figure 1.4. Observed β-decay counting rate as a function of time normalized to a warmed-up state. A typical run with a reasonable polarization of ^{60}Co lasted only about 8 min. But in this interval a clear difference for the two magnetic field configurations emerges, showing the effect of parity violation (from [Wu57]).

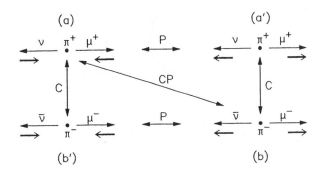

Figure 1.5. Schematic illustration of π^+ decay at rest. The spin and momentum alignment is also shown after applying parity transformation P(a'), change conjugation C(b') and the CP operation (b) (long thin arrows: flight directions, short thick arrows: spin directions).

Defining the helicity $\mathcal{H}$ as

$$\mathcal{H} = \frac{\sigma \cdot p}{|p|} \quad (1.16)$$

this results in $\mathcal{H}(\mu^+) = \mathcal{H}(\nu_\mu) = -1$. Applying a parity transformation, $\mathcal{H}(\mu^+)$ and $\mathcal{H}(\nu_\mu)$ both become $+1$. Parity invariance would imply that both helicities should have the same probability and no longitudinal polarization of the muon should be observed. Parity violation would already be established if there were some polarization. By measuring only $\mathcal{H}(\mu^+) = +1$ it turned out that parity is maximally violated. These observations finally led to the V–A theory of weak interaction (see chapter 6).

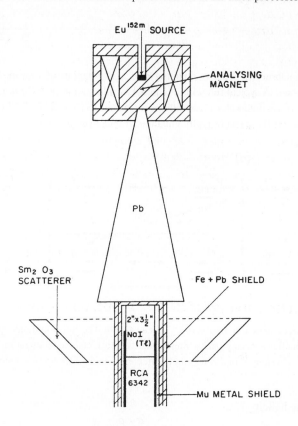

K-Capture			γ – Emission		Resonance-Scattering			
ν_e	Eu	Sm*	Sm	γ	Sm	Sm*	Sm	γ

$$-\vec{p}_\nu \;=\; \vec{p}_{Sm*} \;=\; \vec{p}_\gamma \;=\; \vec{p}_{Sm*} \;=\; \vec{p}_\gamma$$

Figure 1.6. Neutrino helicity in the Goldhaber experiment. Long thin arrows are the momenta and short thick arrows are the spin directions in the three processes.

Figure 1.7. Experimental set-up of the Goldhaber experiment to observe the longitudinal polarization of neutrinos in EC reactions. For details see text (from [Gol58]).

1.6   Direct measurement of the helicity of the neutrino

The principle idea of this experiment was that the neutrino helicity could be measured under special circumstances by a measurement of the polarization of photons in electron capture reactions. In the classical experiment by Goldhaber *et al* the electron capture of ^{152}Eu was used [Gol58]. The decay is given by

$$^{152}\text{Eu} + e^- \rightarrow \nu_e + {}^{152}\text{Sm}^* \rightarrow {}^{152}\text{Sm} + \gamma. \qquad (1.17)$$

The experimental set-up is shown in figure 1.7. The decay at rest of ^{152}Eu results from momentum conservation in $\boldsymbol{p}_{^{152}\text{Sm}^*} = -\boldsymbol{p}_\nu$. The emission of forward photons (961 keV) will stop the Sm nucleus implying $\boldsymbol{p}_\gamma = -\boldsymbol{p}_\nu$ (figure 1.6). Such photons also carry the small recoil energy of the ^{152}Sm* essential for resonant absorption (to account for the Doppler effect) which is used for detection. The resonant absorption is done in a ring of Sm$_2$O$_3$ and the re-emitted photons are detected under large angles by a well-shielded NaI detector. The momentum of these photons is still antiparallel to the neutrino momentum. Concerning the spin, the initial state is characterized by the spin of the electron $J_z = \pm 1/2$ (defining the emission direction of the photon as the z-axis, using the fact that $J(^{152}\text{Eu}) = 0$ and that the K-shell electron has angular momentum $l = 0$) the final state can be described by two combinations $J_z = J_z(\nu) + J_z(\gamma) = (+1/2, -1)$ or $(-1/2, +1)$. Only these result in $J_z = \pm \frac{1}{2}$. This implies, however, that the spins of the neutrino and photon are opposed to each other. Combining this with the momentum arrangement implies that the helicity of the neutrino and photon are the same: $\mathcal{H}(\nu) = \mathcal{H}(\gamma)$. Therefore, the measurement of $\mathcal{H}(\nu)$ is equivalent to a measurement of $\mathcal{H}(\gamma)$. The helicity of the photon is nothing else than its circular polarization, which was measured by Compton scattering in a magnetized iron block before the absorption process. After several measurements a polarization of $67\pm10\%$ was observed in agreement with the assumed 84% [Gol58]. Applying several experimental corrections the outcome of the experiment was that neutrinos do indeed have a helicity of $\mathcal{H}(\nu) = -1$.

1.7 Experimental proof that ν_μ is different from ν_e

In 1959, Pontecorvo investigated whether the neutrino emitted together with an electron in β-decay is the same as the one emitted in pion decay [Pon60]. The idea was that if ν_μ and ν_e are identical particles, then the reactions

$$\nu_\mu + n \rightarrow \mu^- + p \qquad (1.18)$$
$$\bar{\nu}_\mu + p \rightarrow \mu^+ + n \qquad (1.19)$$

and

$$\nu_\mu + n \rightarrow e^- + p \qquad (1.20)$$
$$\bar{\nu}_\mu + p \rightarrow e^+ + n \qquad (1.21)$$

should result in the same rate, because the latter could be done by ν_e and $\bar{\nu}_e$, otherwise the last two should not be observed at all. At the same time, the use of high-energy accelerators as neutrino sources was discussed by Schwarz [Sch60].

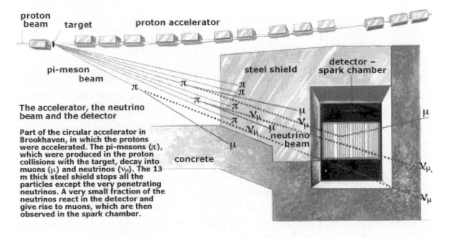

Part of the circular accelerator in Brookhaven, in which the protons were accelerated. The pi-mesons (π), which were produced in the proton collisions with the target, decay into muons (μ) and neutrinos (ν_μ). The 13 m thick steel shield stops all the particles except the very penetrating neutrinos. A very small fraction of the neutrinos react in the detector and give rise to muons, which are then observed in the spark chamber.

Figure 1.8. Plan view of the AGS neutrino experiment (with kind permission of the Royal Swedish Academy of Sciences).

Thus, the experiment was done at the Brookhaven AGS using a 15 GeV proton beam hitting a beryllium-target (figure 1.8) [Dan62]. The created secondary pions and kaons produced an almost pure ν_μ beam. Behind a shielding of 13.5 m iron to absorb all the hadrons and most of the muons, 10 modules of spark chambers with a mass of 1 t each were installed. Muons and electrons were discriminated by their tracking properties, meaning muons produce straight lines, while electrons form an electromagnetic shower. In total, 29 muon-like and six electron-like events were observed clearly showing that $\nu_\mu \neq \nu_e$. Some electron events were expected from ν_e beam contaminations due to K-decays (e.g. $K^+ \rightarrow e^+ \nu_e \pi^0$). The experiment was repeated shortly afterwards at CERN with higher statistics and the result confirmed [Bie64].

1.8 Discovery of weak neutral currents

The development of the electroweak theory by Glashow, Weinberg and Salam, which will be discussed in more detail in chapter 3, predicted the existence of new gauge bosons called W and Z. Associated with the proposed existence of the Z-boson, weak neutral currents (NC) should exist in nature. They were discovered in a bubble chamber experiment (Gargamelle using the proton synchrotron (PS) $\nu_\mu/\bar{\nu}_\mu$ beam at CERN [Has73, Has74]. The bubble chamber was filled with high-density fluid freon (CF_3Br, $\rho = 1.5$ g cm^{-3}) and it had a volume of 14 m^3, with a fiducial volume of 6.2 m^3. The search relied on pure hadronic events without a charged lepton (neutral current events, NC) in the final state which is described

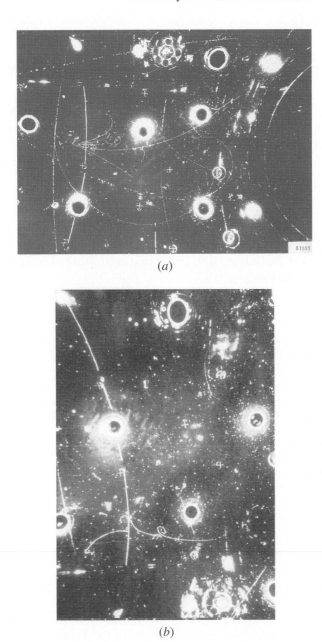

(a)

(b)

Figure 1.9. (a) A hadronic NC event with charged hadrons in the final state as observed by the Gargamelle bubble chamber. (b) A leptonic NC event $\bar{\nu}_\mu e \to \bar{\nu}_\mu e$ as obtained by Gargamelle (with kind permission of CERN).

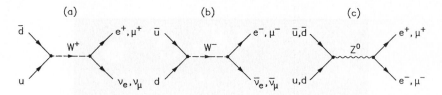

Figure 1.10. Lowest-order Feynman diagrams for $W^{\pm}$- and Z-boson production in $p\bar{p}$ collisions and their leptonic decays.

by

$$\nu_{\mu} + N \rightarrow \nu_{\mu} + X \tag{1.22}$$
$$\bar{\nu}_{\mu} + N \rightarrow \bar{\nu}_{\mu} + X \tag{1.23}$$

where X denotes the hadronic final state (see chapter 4). In addition, the charged current (CC) interactions

$$\nu_{\mu} + N \rightarrow \mu^{-} + X \tag{1.24}$$
$$\bar{\nu}_{\mu} + N \rightarrow \mu^{+} + X \tag{1.25}$$

were detected. In total, 102 NC and 428 CC events were observed in the ν_{μ} beam and 64 NC and 148 CC events in the $\bar{\nu}_{\mu}$ run (figure 1.9). The total number of pictures taken was of the order 83 000 in the ν_{μ} beam and 207 000 in the $\bar{\nu}_{\mu}$ run. After background subtraction, due to the produced neutrons and K_{L}^{0} which could mimic NC events, the ratios for NC/CC turned out to be (see also chapter 4)

$$R_{\nu} = \frac{\sigma(\text{NC})}{\sigma(\text{CC})} = 0.21 \pm 0.03 \tag{1.26}$$

$$R_{\bar{\nu}} = \frac{\sigma(\text{NC})}{\sigma(\text{CC})} = 0.45 \pm 0.09. \tag{1.27}$$

Purely leptonic NC events resulting from $\bar{\nu}_{\mu} + e \rightarrow \bar{\nu}_{\mu} + e$ were also discovered [Has73a] (figure 1.9). Soon afterwards, these observations were confirmed by several other experiments [Cno78, Fai78, Hei80].

1.9   Discovery of the weak gauge bosons W and Z

The weak gauge bosons predicted by the Glashow–Weinberg–Salam (GWS) model were finally discovered at CERN in 1983 by the two experiments UA1 and UA2 [Arn83, Bag83, Ban83]. They used the SPS as a $p\bar{p}$-collider with a centre-of-mass energy of $\sqrt{s} = 540$ GeV. The production processes were weak charged and neutral currents given at the quark level by (figure 1.10)

$$\bar{d} + u \rightarrow W^{+} \rightarrow e^{+} + \nu_{e}(\mu^{+} + \nu_{\mu})$$

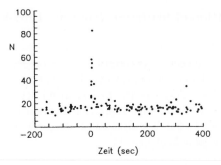

Figure 1.11. Number of struck photomultipliers in KamiokandeII on 23 February 1987. The zero on the time axis marks 7:35 UT. The increase in count rate is clearly visible and attributed to SN 1987A (from [Sut92]).

$$\bar{u} + d \rightarrow W^- \rightarrow e^- + \bar{\nu}_e (\mu^- + \bar{\nu}_\mu)$$
$$\bar{d} + d \rightarrow Z^0 \rightarrow e^+ + e^- (\mu^+ + \mu^-)$$
$$\bar{u} + u \rightarrow Z^0 \rightarrow e^+ + e^- (\mu^+ + \mu^-). \tag{1.28}$$

These were difficult experiments because the cross sections for W and Z production at that energy are rather small. They are including the branching ratio (BR)

$$\sigma(p\bar{p} \rightarrow W^\pm X) \times BR(W \rightarrow l\nu) \approx 1 \text{ nb} = 10^{-33} \text{ cm}^2 \tag{1.29}$$
$$\sigma(p\bar{p} \rightarrow Z^0 X) \times BR(Z^0 \rightarrow l^+l^-) \approx 0.1 \text{ nb} = 10^{-34} \text{ cm}^2 \tag{1.30}$$

while the total cross section $\sigma(p\bar{p})$ is 40 mb![1] The signature was for W detection an isolated lepton ℓ with high transverse momentum p_T balanced by a large missing transverse momentum and for Z detection two high p_T leptons with an invariant mass around the Z-boson mass. With regard to the latter, the Z-boson mass could be determined to be (neglecting the lepton mass)

$$m_Z^2 = 2E^+E^-(1 - \cos\theta) \tag{1.31}$$

with $\cos\theta$ being the angle between the two leptons $\ell^\pm$ of energy E^+ and E^-. Both experiments came up with a total of about 25 W or Z events which were later increased. With the start of the e^+e^--collider LEP at CERN in 1989 and the SLC at SLAC the number of produced Z-bosons is now several million and its properties are well determined. The W properties are investigated at LEP and at the Tevatron at Fermilab. Both gauge bosons are discussed in more detail in chapter 3.

---

[1] 1 barn $= 10^{-24}$ cm^2.

1.10   Observation of neutrinos from SN 1987A

The observation of neutrinos from a supernova type-II explosion by large underground neutrino detectors was one of the great observations in last century's astrophysics (figure 1.11). About 25 neutrino events were observed within a time interval of 12 s. This was the first neutrino detection originating from an astrophysical source beside the Sun. The supernova SN1987A occurred in the Large Magellanic Cloud at a distance of about 50 kpc. This event will be discussed in greater detail in chapter 11.

1.11   Number of neutrino flavours from the width of the Z^0

The number N_ν of light ($m_\nu < m_Z/2$) neutrinos was determined at LEP by measuring the total decay width Γ_Z of the Z^0 resonance. Calling the hadronic decay width Γ_{had} (consisting of $Z^0 \to q\bar{q}$) and assuming lepton universality (implying that there is a common partial width Γ_l for the decay into charged lepton pairs $\ell^+\ell^-$), the invisible width Γ_{inv} is given by

$$\Gamma_{inv} = \Gamma_Z - \Gamma_{had} - 3\Gamma_l. \tag{1.32}$$

As the invisible width corresponds to

$$\Gamma_{inv} = N_\nu \cdot \Gamma_\nu \tag{1.33}$$

the number of neutrino flavours N_ν can be determined. The partial widths of decays in fermions $Z \to f\bar{f}$ are given in electroweak theory (see chapter 3) by

$$\Gamma_f = \frac{G_F m_Z^3}{6\sqrt{2}\pi} c_f [(g_V)^2 + (g_A)^2] = \Gamma_0 c_f [(g_V)^2 + (g_A)^2] \tag{1.34}$$

with

$$\Gamma_0 = \frac{G_F m_Z^3}{6\sqrt{2}\pi} = 0.332 \text{ GeV}. \tag{1.35}$$

In this equation c_f corresponds to a colour factor ($c_f = 1$ for leptons, $c_f = 3$ for quarks) and g_V and g_A are the vector and axial vector coupling constants respectively. They are closely related to the Weinberg angle $\sin^2 \theta_W$ and the third component of weak isospin I_3 (see chapter 3) via

$$g_V = I_3 - 2Q \sin^2 \theta_W \tag{1.36}$$

$$g_A = I_3 \tag{1.37}$$

with Q being the charge of the particle. Therefore, the different branching ratios are

$$\Gamma(Z^0 \to u\bar{u}, c\bar{c}) = (\tfrac{3}{2} - 4\sin^2 \theta_W + \tfrac{16}{3}\sin^4 \theta_W)\Gamma_0 = 0.286 \text{ GeV}$$

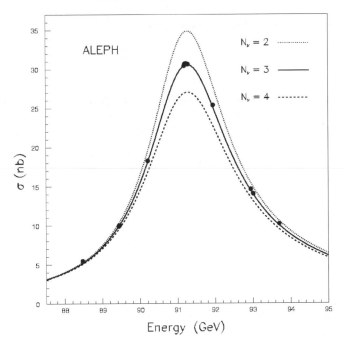

Figure 1.12. Cross section as a function of $\sqrt{s}$ for the reaction $e^+e^- \to$ hadrons as obtained by the ALEPH detector at LEP. The different curves show the standard model predictions for two, three and four light neutrino flavours (with kind permission of the ALEPH collaboration).

$$\Gamma(Z^0 \to d\bar{d}, s\bar{s}, b\bar{b}) = (\tfrac{3}{2} - 2\sin^2\theta_W + \tfrac{4}{3}\sin^4\theta_W)\Gamma_0 = 0.369 \text{ GeV}$$
$$\Gamma(Z^0 \to e^+e^-, \mu^+\mu^-, \tau^+\tau^-) = (\tfrac{1}{2} - 2\sin^2\theta_W + 4\sin^4\theta_W)\Gamma_0 = 0.084 \text{ GeV}$$
$$\Gamma(Z^0 \to \nu\bar{\nu}) = \tfrac{1}{2}\Gamma_0 = 0.166 \text{ GeV}. \tag{1.38}$$

Summing all decay channels into quarks results in a total hadronic width $\Gamma_{\text{had}} = 1.678$ GeV. The different decay widths are determined from the reaction $e^+e^- \to f\bar{f}$ for $f \neq e$ whose cross section as a function of the centre-of-mass energy $\sqrt{s}$ is measured ($\sqrt{s} \approx m_Z$) and is dominated by the Z^0 pole. The cross section at the resonance is described in the Born approximation by a Breit–Wigner formula:

$$\sigma(s) = \sigma^0 \frac{s\Gamma_Z^2}{(s - m_Z^2)^2 + s^2\Gamma_Z^2/m_Z^2} \qquad \text{with } \sigma^0 = \frac{12\pi}{m_Z^2} \frac{\Gamma_e\Gamma_f}{\Gamma_Z^2} \tag{1.39}$$

with σ^0 being the maximum of the resonance. Γ_Z can be determined from the width and $\Gamma_e\Gamma_f$ from the maximum of the observed resonance (figure 1.12).

Experimentally, the Z^0 resonance is fitted with four different parameters which have small correlations with each other:

$$m_Z, \Gamma_Z, \sigma_{\text{had}}^0 = \frac{12\pi}{m_Z^2} \frac{\Gamma_e \Gamma_{\text{had}}}{\Gamma_Z^2} \qquad \text{and} \qquad R_l = \frac{\Gamma_{\text{had}}}{\Gamma_l} \tag{1.40}$$

σ_{had}^0 is determined from the maximum of the resonance in $e^+e^- \rightarrow$ hadrons. Assuming again lepton-universality, which is justified by the equality of the measured leptonic decay width, the number of neutrino flavours can be determined as

$$N_\nu = \frac{\Gamma_{\text{inv}}}{\Gamma_l} \left(\frac{\Gamma_l}{\Gamma_\nu}\right)_{SM} = \left[\sqrt{\frac{12\pi R_l}{m_Z^2 \sigma_{\text{had}}^0}} - R_l - 3\right] \left(\frac{\Gamma_l}{\Gamma_\nu}\right)_{SM}. \tag{1.41}$$

This form is chosen because in this way radiative corrections are already included in the Standard Model (SM) prediction. Using the most recent fit to the data of the four LEP experiments a number of

$$N_\nu = 2.9841 \pm 0.0083 \tag{1.42}$$

can be deduced [PDG02], in excellent agreement with the theoretical expectation of three.

Chapter 2

Properties of neutrinos

In quantum field theory spin-$\frac{1}{2}$ particles are described by four-component wavefunctions $\psi(x)$ (spinors) which obey the Dirac equation. The four independent components of $\psi(x)$ correspond to particles and antiparticles with the two possible spin projections $J_Z = \pm 1/2$ equivalent to the two helicities $\mathcal{H} = \pm 1$. Neutrinos as fundamental leptons are spin-$\frac{1}{2}$ particles like other fermions; however, it is an experimental fact that only left-handed neutrinos ($\mathcal{H} = -1$) and right-handed antineutrinos ($\mathcal{H} = +1$) are observed. Therefore, a two-component spinor description should, in principle, be sufficient (Weyl spinors). In a four-component theory they are obtained by projecting out of a general spinor $\psi(x)$ the components with $\mathcal{H} = +1$ for particles and $\mathcal{H} = -1$ for antiparticles with the help of the operators $P_{L,R} = \frac{1}{2}(1 \mp \gamma_5)$. The two-component theory of the neutrinos will be discussed in detail later. Our discussion will be quite general, for a more extensive discussion see [Bjo64, Bil87, Kay89, Kim93, Sch97].

2.1 Helicity and chirality

The Dirac equation is the relativistic wave equation for spin-$\frac{1}{2}$ particles and given by (using Einstein conventions)

$$\left(i\gamma^\mu \frac{\partial}{\partial x^\mu} - m \right) \psi = 0. \tag{2.1}$$

Here ψ denotes a four-component spinor and the 4×4 γ-matrices are given in the form[1]

$$\gamma_0 = \begin{pmatrix} 1 & 0 \\ 0 & -1 \end{pmatrix} \qquad \gamma_i = \begin{pmatrix} 0 & \sigma_i \\ -\sigma_i & 0 \end{pmatrix} \tag{2.2}$$

---

[1] Other conventions of the γ-matrices are also commonly used in the literature, which leads to slightly different forms for the following expressions.

where σ_i correspond to the 2×2 Pauli matrices. Detailed introductions and treatments can be found in [Bjo64]. The matrix γ_5 is given by

$$\gamma_5 = i\gamma_0\gamma_1\gamma_2\gamma_3 = \begin{pmatrix} 0 & 1 \\ 1 & 0 \end{pmatrix} \tag{2.3}$$

and the following anticommutator relations hold:

$$\{\gamma^\alpha, \gamma^\beta\} = 2g_{\alpha\beta} \tag{2.4}$$
$$\{\gamma^\alpha, \gamma_5\} = 0 \tag{2.5}$$

with $g_{\alpha\beta}$ as the metric $(+1, -1, -1, -1)$. Multiplying the Dirac equation from the left with γ_0 and using $\gamma_i = \gamma_0\gamma_5\sigma_i$ results in

$$\left(i\gamma_0^2 \frac{\partial}{\partial x^0} - i\gamma_0\gamma_5\sigma_i \frac{\partial}{\partial x_i} - m\gamma_0\right)\psi = 0 \qquad i = 1, \dots, 3. \tag{2.6}$$

Another multiplication of (2.6) from the left with γ_5 and using $\gamma_5\sigma_i = \sigma_i\gamma_5$ (which follows from (2.5)) leads to ($\gamma_0^2 = 1$, $\gamma_5^2 = 1$)

$$\left(i \frac{\partial}{\partial x^0}\gamma_5 - i\sigma_i \frac{\partial}{\partial x_i} - m\gamma_0\gamma_5\right)\psi = 0. \tag{2.7}$$

Subtraction and addition of the last two equations results in the following system of coupled equations:

$$\left(i \frac{\partial}{\partial x^0}(1 + \gamma_5) - i\sigma_i \frac{\partial}{\partial x_i}(1 + \gamma_5) - m\gamma_0(1 - \gamma_5)\right)\psi = 0 \tag{2.8}$$
$$\left(i \frac{\partial}{\partial x^0}(1 - \gamma_5) - i\sigma_i \frac{\partial}{\partial x_i}(1 - \gamma_5) - m\gamma_0(1 + \gamma_5)\right)\psi = 0. \tag{2.9}$$

Now let us introduce left- and right-handed components by defining two projection operators P_L and P_R given by

$$P_L = \tfrac{1}{2}(1 - \gamma_5) \qquad \text{and} \qquad P_R = \tfrac{1}{2}(1 + \gamma_5) \tag{2.10}$$

Because they are projectors, the following relations hold:

$$P_L P_R = 0 \qquad P_L + P_R = 1 \qquad P_L^2 = P_L \qquad P_R^2 = P_R. \tag{2.11}$$

With the definition

$$\psi_L = P_L\psi \qquad \text{and} \qquad \psi_R = P_R\psi \tag{2.12}$$

it is obviously valid that

$$P_L\psi_R = P_R\psi_L = 0. \tag{2.13}$$

Then the following eigenequation holds:

$$\gamma_5 \psi_{L,R} = \mp \psi_{L,R}. \tag{2.14}$$

The eigenvalues ± 1 to γ_5 are called chirality and $\psi_{L,R}$ are called chiral projections of ψ. Any spinor ψ can be rewritten in chiral projections as

$$\psi = (P_L + P_R)\psi = P_L \psi + P_R \psi = \psi_L + \psi_R. \tag{2.15}$$

The equations (2.8) and (2.9) can now be expressed in these projections as

$$\left(i \frac{\partial}{\partial x^0} - i\sigma_i \frac{\partial}{\partial x_i} \right) \psi_R = m\gamma_0 \psi_L \tag{2.16}$$

$$\left(i \frac{\partial}{\partial x^0} + i\sigma_i \frac{\partial}{\partial x_i} \right) \psi_L = m\gamma_0 \psi_R. \tag{2.17}$$

Both equations decouple in the case of a vanishing mass $m = 0$ and can then be depicted as

$$i \frac{\partial}{\partial x^0} \psi_R = i\sigma_i \frac{\partial}{\partial x_i} \psi_R \tag{2.18}$$

$$i \frac{\partial}{\partial x^0} \psi_L = -i\sigma_i \frac{\partial}{\partial x_i} \psi_L. \tag{2.19}$$

But this is identical to the Schrödinger equation ($x_0 = t, \hbar = 1$)

$$i \frac{\partial}{\partial t} \psi_{L,R} = \mp i\sigma_i \frac{\partial}{\partial x_i} \psi_{L,R} \tag{2.20}$$

or in momentum space ($i\frac{\partial}{\partial t} = E, -i\frac{\partial}{\partial x_i} = p_i$)

$$E\psi_{L,R} = \pm \sigma_i p_i \psi_{L,R}. \tag{2.21}$$

The latter implies that the $\psi_{L,R}$ are also eigenfunctions to the helicity operator $\mathcal{H}$ given by (see chapter 1)

$$\mathcal{H} = \frac{\boldsymbol{\sigma} \cdot \boldsymbol{p}}{|\boldsymbol{p}|} \tag{2.22}$$

ψ_L is an eigenspinor with helicity eigenvalues $\mathcal{H} = +1$ for particles and $\mathcal{H} = -1$ for antiparticles. Correspondingly ψ_R is the eigenspinor to the helicity eigenvalues $\mathcal{H} = -1$ for particles and $\mathcal{H} = +1$ for antiparticles. Therefore, in the case of massless particles, chirality and helicity are identical.[2] For $m > 0$ the decoupling of (2.16) and (2.17) is no longer possible. This means that the chirality eigenspinors ψ_L and ψ_R no longer describe particles with fixed helicity and helicity is no longer a good conserved quantum number.

[2] May be of opposite sign depending on the representation used for the γ-matrices.

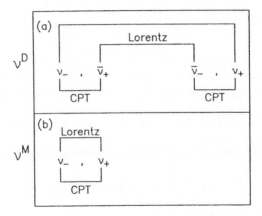

Figure 2.1. Schematic drawing of the difference between massive Dirac and Majorana neutrinos. (*a*) The Dirac case: ν_L is converted via CPT into a $\bar{\nu}_R$ and via a Lorentz boost into a ν_R. An application of CPT on the latter results in $\bar{\nu}_L$ which is different from the one obtained by applying CPT on ν_L. The result is four different states. (*b*) The Majorana case: Both operations CPT and a Lorentz boost result in the same state ν_R, there is no difference between particle and antiparticle. Only two states emerge.

The two-component theory now states that the neutrino spinor ψ_ν in weak interactions always reads as

$$\psi_\nu = \tfrac{1}{2}(1 - \gamma_5) = \psi_L \tag{2.23}$$

meaning that the interacting neutrino is always left-handed and the antineutrino always right-handed. For $m = 0$, this further implies that ν always has $\mathcal{H} = +1$ and $\bar{\nu}$ always $\mathcal{H} = -1$. The proof that indeed the Dirac spinors ψ_L and ψ_R can be written as the sum of two independent 2-component Weyl spinors can be found in [Sch97].

2.2   Charge conjugation

While for all fundamental fermions of the Standard Model (see chapter 3) a clear discrimination between particle and antiparticle can be made by their electric charge, for neutrinos it is not so obvious. If particle and antiparticle are not identical, we call such a fermion a Dirac particle which has four independent components. If particle and antiparticle are identical, they are called Majorana particles (figure 2.1). The latter requires that all additive quantum numbers (charge, strangeness, baryon number, lepton number etc) have to vanish. Consequently, the lepton number is violated if neutrinos are Majorana particles.

The operator connecting particle $f(x, t)$ and antiparticle $\bar{f}(x, t)$ is charge conjugation C:

$$C|f(x,t)\rangle = \eta_c|\bar{f}(x,t)\rangle. \tag{2.24}$$

If $\psi(x)$ is a spinor field of a free neutrino then the corresponding charge conjugated field ψ^c is defined by

$$\psi \overset{C}{\to} \psi^c \equiv C\psi C^{-1} = \eta_c C \bar{\psi}^T \tag{2.25}$$

with η_c as a phase factor with $|\eta_c| = 1$. The 4×4 unitary charge conjugation matrix C obeys the following general transformations:

$$C^{-1}\gamma_\mu C = -\gamma_\mu^T \qquad C^{-1}\gamma_5 C = \gamma_5^T \qquad C^\dagger = C^{-1} = C^T = -C. \tag{2.26}$$

A possible representation is given as $C = i\gamma_0\gamma_2$. Using the projection operators $P_{L,R}$, it follows that

$$P_{L,R}\psi = \psi_{L,R} \overset{C}{\to} P_{L,R}\psi^c = (\psi^c)_{L,R} = (\psi_{R,L})^c. \tag{2.27}$$

It is easy to show that if ψ is an eigenstate of chirality, ψ^c is an eigenstate too but it has an eigenvalue of opposite sign. Furthermore, from (2.27) it follows that the charge conjugation C transforms a right(left)-handed particle into a right(left)-handed antiparticle, leaving the helicity (chirality) untouched. Only the additional application of a parity transformation changes the helicity as well. However, the operation of (2.25) converts a right(left)-handed particle into a left(right)-handed antiparticle. Here helicity and chirality are converted as well.

To include the fact that $\psi_{L,R}$ and $\psi_{L,R}^c$ have opposite helicity, one avoids calling $\psi_{L,R}^c$ the charge conjugate of $\psi_{L,R}$. Instead it is more frequently called the CP (or CPT) conjugate with respect to $\psi_{L,R}$ [Lan88]. In the following sections we refer to ψ^c as the CP or CPT conjugate of the spinor ψ, assuming CP or CPT conservation correspondingly.

2.3   Parity transformation

A parity transformation P operation is defined as

$$\psi(x,t) \overset{P}{\to} P\psi(x,t)P^{-1} = \eta_P\gamma_0\psi(-x,t). \tag{2.28}$$

The phase factor η_P with $|\eta_P| = 1$ corresponds for real $\eta_P = \pm 1$ with the inner parity. Using (2.25) for the charge conjugated field, it follows that

$$\psi^c = \eta_C C\bar{\psi}^T \overset{P}{\to} \eta_C \eta_P^* C\gamma_0^T \bar{\psi}^T = -\eta_P^*\gamma_0\psi^c. \tag{2.29}$$

This implies that a fermion and its corresponding antifermion have opposite inner parity, i.e. for a Majorana particle $\psi^c = \pm\psi$ holds which results in $\eta_P = -\eta_P^*$.

Therefore, an interesting point with respect to the inner parity occurs for Majorana neutrinos. A Majorana field can be written as

$$\psi_M = \frac{1}{\sqrt{2}}(\psi + \eta_C \psi^c) \qquad \text{with } \eta_C = \lambda_C e^{2i\phi}, \ \lambda_C = \pm 1 \qquad (2.30)$$

where λ_C is sometimes called creation phase. By applying a phase transformation

$$\psi_M \rightarrow \psi_M e^{-i\phi} = \frac{1}{\sqrt{2}}(\psi e^{-i\phi} + \lambda_C \psi^c e^{i\phi}) = \frac{1}{\sqrt{2}}(\psi + \lambda_C \psi^c) \equiv \psi_M \qquad (2.31)$$

it can be achieved that the field ψ_M is an eigenstate with respect to charge conjugation C

$$\psi_M^c = \frac{1}{\sqrt{2}}(\psi^c + \lambda_C \psi) = \lambda_C \psi_M \qquad (2.32)$$

with eigenvalues $\lambda_C = \pm 1$. This means the Majorana particle is identical to its antiparticle, i.e. ψ_M and ψ_M^c cannot be distinguished. With respect to CP, one obtains

$$\psi_M(\boldsymbol{x}, t) \xrightarrow{C} \psi_M^c = \lambda_C \psi_M \xrightarrow{P} \frac{\lambda_C}{\sqrt{2}}(\eta_P \gamma_0 \psi - \lambda_C \eta_P^* \gamma_0 \psi^c)$$
$$= \lambda_C \eta_P \gamma_0 \psi_M = \pm i \gamma_0 \psi_M(-\boldsymbol{x}, t) \qquad (2.33)$$

because $\eta_P^* = -\eta_P$. This means that the inner parity of a Majorana particle is imaginary, $\eta_P = \pm i$ if $\lambda_C = \pm 1$. Finally, from (2.31) it follows that

$$(\gamma_5 \psi_M)^c = \eta_C C \gamma_5 \bar{\psi}_M^T = -\eta_C C \gamma_5^T \bar{\psi}_M^T = -\gamma_5 \psi_M^c = -\lambda_C \gamma_5 \psi_M \qquad (2.34)$$

because $\gamma_5 \bar{\psi}_M = (\gamma_5 \psi_M)^\dagger \gamma_0 = \psi_M^\dagger \gamma_5 \gamma_0 = -\bar{\psi}_M \gamma_5$. Using this together with (2.27) one concludes that an eigenstate to C cannot be at the same time an eigenstate to chirality. A Majorana neutrino, therefore, has no fixed chirality. However, because ψ and ψ^c obey the Dirac equation, ψ_M will also do so.

For a discussion of T transformation and C, CP and CPT properties, see [Kay89, Kim93].

2.4   Dirac and Majorana mass terms

Consider the case of free fields without interactions and start with the Dirac mass. The Dirac equation can then be deduced with the help of the Euler–Lagrange equation from a Lagrangian [Bjo64]:

$$\mathcal{L} = \bar{\psi} \left(i\gamma^\mu \frac{\partial}{\partial x^\mu} - m_D \right) \psi \qquad (2.35)$$

where the first term corresponds to the kinetic energy and the second is the mass term. The Dirac mass term is, therefore,

$$\mathcal{L} = m_D \bar{\psi} \psi \qquad (2.36)$$

where the combination $\bar{\psi}\psi$ has to be Lorentz invariant and Hermitian. Requiring $\mathcal{L}$ to be Hermitian as well, m_D must be real ($m_D^* = m_D$). Using the following relations valid for two arbitrary spinors ψ and ϕ (which follow from (2.10) and (2.11))

$$\bar{\psi}_L \phi_L = \bar{\psi} P_R P_L \phi = 0 \qquad \bar{\psi}_R \phi_R = 0 \qquad (2.37)$$

it follows that

$$\bar{\psi}\phi = (\bar{\psi}_L + \bar{\psi}_R)(\phi_L + \phi_R) = \bar{\psi}_L \phi_R + \bar{\psi}_R \phi_L. \qquad (2.38)$$

In this way the Dirac mass term can be written in its chiral components (Weyl spinors) as

$$\mathcal{L} = m_D(\bar{\psi}_L \psi_R + \bar{\psi}_R \psi_L) \qquad \text{with } \bar{\psi}_R \psi_L = (\bar{\psi}_L \psi_R)^\dagger. \qquad (2.39)$$

Applying this to neutrinos, it requires both a left- and a right-handed Dirac neutrino to produce such a mass term. In the Standard Model of particle physics only left-handed neutrinos exist, that is the reason why neutrinos remain massless as will be discussed in chapter 3.

In a more general treatment including ψ^c one might ask which other combinations of spinors behaving like Lorentz scalars can be produced. Three more are possible: $\bar{\psi}^c \psi^c$, $\bar{\psi}\psi^c$ and $\bar{\psi}^c \psi$. $\bar{\psi}^c \psi^c$ is also hermitian and equivalent to $\bar{\psi}\psi$; $\bar{\psi}\psi^c$ and $\bar{\psi}^c \psi$ are hermitian conjugates, which can be shown for arbitrary spinors

$$(\bar{\psi}\phi)^\dagger = (\psi^\dagger \gamma_0 \phi)^\dagger = \phi^\dagger \gamma_0 \psi = \bar{\phi}\psi. \qquad (2.40)$$

With this we have an additional hermitian mass term, called the Majorana mass term and given by

$$\mathcal{L} = \frac{1}{2}(m_M \bar{\psi}\psi^c + m_M^* \bar{\psi}^c \psi) = \frac{1}{2}m_M \bar{\psi}\psi^c + h.c.^3 \qquad (2.41)$$

m_M is called the Majorana mass. Now using again the chiral projections with the notation

$$\psi_{L,R}^c = (\psi^c)_{R,L} = (\psi_{R,L})^c \qquad (2.42)$$

one gets two hermitian mass terms:

$$\mathcal{L}^L = \frac{1}{2}m_L(\bar{\psi}_L \psi_R^c + \bar{\psi}_R^c \psi_L) = \frac{1}{2}m_L \bar{\psi}_L \psi_R^c + h.c. \qquad (2.43)$$

$$\mathcal{L}^R = \frac{1}{2}m_R(\bar{\psi}_L^c \psi_R + \bar{\psi}_R \psi_L^c) = \frac{1}{2}m_R \bar{\psi}_L^c \psi_R + h.c. \qquad (2.44)$$

with $m_{L,R}$ as real Majorana masses because of (2.40). Let us define two Majorana fields (see (2.30) with $\lambda_C = 1$)

$$\phi_1 = \psi_L + \psi_R^c \qquad \phi_2 = \psi_R + \psi_L^c \qquad (2.45)$$

---

[3] *h.c.* throughout the book signifies Hermitian conjugate.

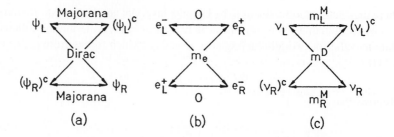

Figure 2.2. Coupling schemes for fermion fields via Dirac and Majorana masses: (*a*) general scheme for left- and right-handed fields and the charge conjugate fields; (*b*) the case for electrons (because of its electric charge only Dirac-mass terms are possible) and (*c*) coupling scheme for neutrinos. It is the only fundamental fermion that allows all possible couplings (after [Mut88]).

which allows (2.43) to be rewritten as

$$\mathcal{L}^L = \frac{1}{2} m_L \bar{\phi}_1 \phi_1 \qquad \mathcal{L}^R = \frac{1}{2} m_R \bar{\phi}_2 \phi_2. \tag{2.46}$$

While $\psi_{L,R}$ are interaction eigenstates, $\phi_{1,2}$ are mass eigenstates to $m_{L,R}$.

The most general mass term (the Dirac–Majorana mass term) is a combination of (2.39) and (2.43) (figure 2.2):

$$2\mathcal{L} = m_D(\bar{\psi}_L \psi_R + \bar{\psi}_L^c \psi_R^c) + m_L \bar{\psi}_L \psi_R^c + m_R \bar{\psi}_L^c \psi_R + h.c.$$

$$= (\bar{\psi}_L, \bar{\psi}_L^c) \begin{pmatrix} m_L & m_D \\ m_D & m_R \end{pmatrix} \begin{pmatrix} \psi_R^c \\ \psi_R \end{pmatrix} + h.c. \tag{2.47}$$

$$= \bar{\Psi}_L M \Psi_R^c + \bar{\Psi}_R^c M \Psi_L$$

where, in the last step, the following was used:

$$M = \begin{pmatrix} m_L & m_D \\ m_D & m_R \end{pmatrix} \qquad \Psi_L = \begin{pmatrix} \psi_L \\ \psi_L^c \end{pmatrix} = \begin{pmatrix} \psi_L \\ (\psi_R)^c \end{pmatrix} \tag{2.48}$$

implying

$$(\Psi_L)^c = \begin{pmatrix} (\psi_L)^c \\ \psi_R \end{pmatrix} = \begin{pmatrix} \psi_R^c \\ \psi_R \end{pmatrix} = \Psi_R^c.$$

In the case of *CP* conservation the elements of the mass matrix M are real. Coming back to neutrinos, in the known neutrino interactions only ψ_L and ψ_R^c are present (active neutrinos) and not the fields ψ_R and ψ_L^c (called sterile neutrinos), it is quite common to distinguish between both types in the notation: $\psi_L = \nu_L$, $\psi_R^c = \nu_R^c$, $\psi_R = N_R$, $\psi_L^c = N_L^c$. With this notation, (2.47) becomes

$$2\mathcal{L} = m_D(\bar{\nu}_L N_R + \bar{N}_L^c \nu_R^c) + m_L \bar{\nu}_L \nu_R^c + m_R \bar{N}_L^c N_R + h.c.$$

$$= (\bar{\nu}_L, \bar{N}_L^c) \begin{pmatrix} m_L & m_D \\ m_D & m_R \end{pmatrix} \begin{pmatrix} \nu_R^c \\ N_R \end{pmatrix} + h.c. \tag{2.49}$$

The mass eigenstates are obtained by diagonalizing M and are given as

$$\psi_{1L} = \cos\theta\,\psi_L - \sin\theta\,\psi_L^c \qquad \psi_{1R}^c = \cos\theta\,\psi_R^c - \sin\theta\,\psi_R \qquad (2.50)$$

$$\psi_{2L} = \sin\theta\,\psi_L + \cos\theta\,\psi_L^c \qquad \psi_{2R}^c = \sin\theta\,\psi_R^c + \cos\theta\,\psi_R \qquad (2.51)$$

while the mixing angle θ is given by

$$\tan 2\theta = \frac{2m_D}{m_R - m_L}. \qquad (2.52)$$

The corresponding mass eigenvalues are

$$\tilde{m}_{1,2} = \frac{1}{2}\left[(m_L + m_R) \pm \sqrt{(m_L - m_R)^2 + 4m_D^2}\right]. \qquad (2.53)$$

To get positive masses,[4] we use [Lan88, Gro90]

$$\tilde{m}_k = \epsilon_k m_k \qquad \text{with } m_k = |\tilde{m}_k| \text{ and } \epsilon_k = \pm 1 \ (k = 1, 2). \qquad (2.54)$$

To get a similar expression as (2.45), two independent Majorana fields with masses m_1 and m_2 (with $m_k \geq 0$) are introduced via $\phi_k = \psi_{kL} + \epsilon_k\psi_{kR}^c$ or, explicitly,

$$\phi_1 = \psi_{1L} + \epsilon_1\psi_{1R}^c = \cos\theta(\psi_L + \epsilon_1\psi_R^c) - \sin\theta(\psi_L^c + \epsilon_1\psi_R) \qquad (2.55)$$

$$\phi_2 = \psi_{2L} + \epsilon_2\psi_{2R}^c = \sin\theta(\psi_L + \epsilon_2\psi_R^c) + \cos\theta(\psi_L^c + \epsilon_2\psi_R) \qquad (2.56)$$

and, as required for Majorana fields,

$$\phi_k^c = (\psi_{kL})^c + \epsilon_k\psi_{kL} = \epsilon_k(\epsilon_k\psi_{kR}^c + \psi_{kL}) = \epsilon_k\phi_k \qquad (2.57)$$

ϵ_k is the CP eigenvalue of the Majorana neutrino ϕ_k. So we finally get the analogous expression to (2.45):

$$2\mathcal{L} = m_1\bar{\phi}_1\phi_1 + m_2\bar{\phi}_2\phi_2. \qquad (2.58)$$

From this general discussion one can take some interesting special aspects:

(1) $m_L = m_R = 0$ ($\theta = 45°$), resulting in $m_{1,2} = m_D$ and $\epsilon_{1,2} = \mp 1$. As Majorana eigenstates, two degenerated states emerge:

$$\phi_1 = \frac{1}{\sqrt{2}}(\psi_L - \psi_R^c - \psi_L^c + \psi_R) = \frac{1}{\sqrt{2}}(\psi - \psi^c) \qquad (2.59)$$

$$\phi_2 = \frac{1}{\sqrt{2}}(\psi_L + \psi_R^c + \psi_L^c + \psi_R) = \frac{1}{\sqrt{2}}(\psi + \psi^c). \qquad (2.60)$$

---

[4] An equivalent procedure for $\tilde{m}_k < 0$ would be a phase transformation $\psi_k \to i\psi_k$ resulting in a change of sign of the $\bar{\psi}^c\psi$ terms in (2.43). With $m_k = -\tilde{m}_k > 0$, positive m_k terms in (2.43) result.

These can be used to construct a Dirac field ψ:

$$\frac{1}{\sqrt{2}}(\phi_1 + \phi_2) = \psi_L + \psi_R = \psi. \tag{2.61}$$

The corresponding mass term (2.58) is (because $\bar{\phi}_1\phi_2 + \bar{\phi}_2\phi_1 = 0$)

$$\mathcal{L} = \frac{1}{2}m_D(\bar{\phi}_1 + \bar{\phi}_2)(\phi_1 + \phi_2) = m_D\bar{\psi}\psi. \tag{2.62}$$

We are left with a pure Dirac field. As a result, a Dirac field can be seen, using (2.61), to be composed of two degenerated Majorana fields, i.e. a Dirac ν can be seen as a pair of degenerated Majorana ν. The Dirac case is, therefore, a special solution of the more general Majorana case.

(2) $m_D \gg m_L, m_R$ ($\theta \approx 45°$): In this case the states $\phi_{1,2}$ are, almost degenerated with $m_{1,2} \approx m_D$ and such an object is called a pseudo-Dirac neutrino.

(3) $m_D = 0$ ($\theta = 0$): In this case $m_{1,2} = m_{L,R}$ and $\epsilon_{1,2} = 1$. So $\phi_1 = \psi_L + \psi_R^c$ and $\phi_2 = \psi_R + \psi_L^c$. This is the pure Majorana case.

(4) $m_R \gg m_D, m_L = 0$ ($\theta = (m_D/m_R) \ll 1$): One obtains two mass eigenvalues:

$$m_\nu = m_1 = \frac{m_D^2}{m_R} \qquad m_N = m_2 = m_R\left(1 + \frac{m_D^2}{m_R^2}\right) \approx m_R \tag{2.63}$$

and

$$\epsilon_{1,2} = \mp 1.$$

The corresponding Majorana fields are

$$\phi_1 \approx \psi_L - \psi_R^c \qquad \phi_2 \approx \psi_L^c + \psi_R. \tag{2.64}$$

The last scenario is especially popular within the seesaw model of neutrino mass generation and will be discussed in more detail in chapter 5.

2.4.1   Generalization to *n* flavours

The discussion so far has only related to one neutrino flavour. The generalization to *n* flavours will not be discussed in greater detail, only some general statements are made—see [Bil87, Kim93] for a more complete discussion. A Weyl spinor is now an *n*-component vector in flavour space, given, for example, as

$$\nu_L = \begin{pmatrix} \nu_{1L} \\ \cdot \\ \cdot \\ \cdot \\ \nu_{nL} \end{pmatrix} \qquad N_R = \begin{pmatrix} N_{1R} \\ \cdot \\ \cdot \\ \cdot \\ N_{nR} \end{pmatrix} \tag{2.65}$$

where every ν_{iL} and N_{iR} are normal Weyl spinors with flavour i. Correspondingly, the masses m_D, m_L, m_R are now $n \times n$ matrices M_D, M_L and M_R with complex elements and $M_L = M_L^T, M_R = M_R^T$. The general symmetric $2n \times 2n$ matrix is then, in analogy to (2.48),

$$M = \begin{pmatrix} M_L & M_D \\ M_D^T & M_R \end{pmatrix}. \tag{2.66}$$

The most general mass term (2.47) is now

$$2\mathcal{L} = \bar{\Psi}_L M \Psi_R^c + \bar{\Psi}_R^c M^\dagger \Psi_L \tag{2.67}$$
$$= \bar{\nu}_L M_D N_R + \bar{N}_L^c M_D^T \nu_R^c + \bar{\nu}_L M_L \nu_R^c + \bar{N}_L^c M_R N_R + h.c. \tag{2.68}$$

where

$$\Psi_L = \begin{pmatrix} \nu_L \\ N_L^c \end{pmatrix} \quad \text{and} \quad \Psi_R^c = \begin{pmatrix} \nu_R^c \\ N_R \end{pmatrix}. \tag{2.69}$$

Diagonalization of M results in $2n$ Majorana mass eigenstates with associated mass eigenvalues $\epsilon_i m_i (\epsilon_i = \pm 1, m_i \geq 0)$. In the previous discussion, an equal number of active and sterile flavours ($n_a = n_s = n$) is assumed. In the most general case with $n_a \neq n_s$, M_D is an $n_a \times n_s$, M_L an $n_a \times n_a$ and M_R an $n_s \times n_s$ matrix. So the full matrix M is an $(n_a + n_s) \times (n_a + n_s)$ matrix whose diagonalization results in $(n_a + n_s)$ mass eigenstates and eigenvalues.

In seesaw models light neutrinos are given by the mass matrix (still to be diagonalized)

$$M_\nu = M_D M_R^{-1} M_D^T \tag{2.70}$$

in analogy to m_ν in (2.63).

Having discussed the formal description of neutrinos in some detail, we now take a look at the concept of lepton number.

2.5 Lepton number

Conserved quantum numbers arise from the invariance of the equation of motion under certain symmetry transformations. Continuous symmetries (e.g. translation) can be described by real numbers and lead to additive quantum numbers, while discrete symmetries (e.g. spatial reflections through the origin) are described by integers and lead to multiplicative quantum numbers. For some of them the underlying symmetry operations are known, as discussed in more detail in chapter 3. Some quantum numbers, however, have not yet been associated with a fundamental symmetry such as baryon number B or lepton number L and their conservation is only motivated by experimental observation. The quantum numbers conserved in the individual interactions are shown in table 2.1. Lepton number was introduced to characterize experimental observations of weak interactions. Each lepton is defined as having a lepton number $L = +1$, each

Table 2.1. Summary of conservation laws. B corresponds to baryon number and L to total lepton number.

Conservation law	Strong	Electromagnetic	Weak
Energy	yes	yes	yes
Momentum	yes	yes	yes
Angular momentum	yes	yes	yes
B, L	yes	yes	yes
P	yes	yes	no
C	yes	yes	no
CP	yes	yes	no
T	yes	yes	no
CPT	yes	yes	yes

antilepton $L = -1$. Moreover, each generation of leptons has its own lepton number L_e, L_μ, L_τ with $L = L_e + L_\mu + L_\tau$. Individual lepton number is not conserved, as has been established with the observation of neutrino oscillations (see chapter 8).

Consider the four Lorentz scalars discussed under a global phase transformation $e^{i\alpha}$:

$$\psi \to e^{i\alpha}\psi \qquad \bar{\psi} \to e^{-i\alpha}\bar{\psi} \qquad \text{so that} \qquad \bar{\psi}\psi \to \bar{\psi}\psi \qquad (2.71)$$

$$\psi^c \to (e^{i\alpha}\psi)^c = \eta_C C e^{\bar{i}\alpha}\bar{\psi}^T = e^{-i\alpha}\psi^c \qquad \bar{\psi}^c \to e^{i\alpha}\bar{\psi}^c. \qquad (2.72)$$

As can be seen, $\bar{\psi}\psi$ and $\bar{\psi}^c\psi^c$ are invariant under this transformation and are connected to a conserved quantum number, namely lepton number: ψ annihilates a lepton or creates an antilepton, $\bar{\psi}$ acts oppositely. $\bar{\psi}\psi$ and $\bar{\psi}^c\psi^c$ result in transitions $\ell \to \ell$ or $\bar{\ell} \to \bar{\ell}$ with $\Delta L = 0$. This does not relate to the other two Lorentz scalars $\bar{\psi}\psi^c$ and $\bar{\psi}^c\psi$ which force transitions of the form $\ell \to \bar{\ell}$ or $\bar{\ell} \to \ell$ corresponding to $\Delta L = \pm 2$ according to the assignment made earlier. For charged leptons such lepton-number-violating transitions are forbidden (i.e. $e^- \to e^+$) and they have to be Dirac particles. But if one associates a mass to neutrinos both types of transitions are, in principle, possible.

If the lepton number is related to a global symmetry which has to be broken spontaneously, a Goldstone boson is associated with the symmetry breaking. In this case it is called a majoron (see [Moh86, 92, Kim93] for more details).

2.5.1 Experimental status of lepton number violation

As no underlying fundamental symmetry is known to conserve lepton number, one might think about observing lepton flavour violation (LFV) at some level. Several searches for LFV are associated with muons. A classic test for the conservation of individual lepton numbers is the muon conversion on nuclei:

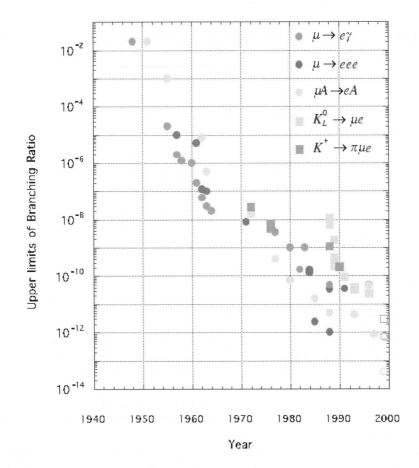

Figure 2.3. Time evolution of experimental limits of branching ratios on some rare LFV muon and kaon decays (from [Kun01]).

	μ^-	$+\,{}^A_Z X$	$\rightarrow {}^A_Z X$	$+e^-$
L_e	0	$+\,0$	$\rightarrow 0$	$+\,1$
L_μ	1	$+\,0$	$\rightarrow 0$	$+\,0$

This would violate both L_e and L_μ conservation but would leave the total lepton number unchanged. It has not yet been observed and the current experimental limit for this decay is [Win98]

$$BR(\mu^- + \mathrm{Ti} \rightarrow e^- + \mathrm{Ti}) < 6.1 \times 10^{-13} \qquad (90\% \text{ CL}). \qquad (2.73)$$

New proposals exist (MECO and PRISM) to go down to 10^{-16} or even 10^{-18} [Kun01]. Other processes studied intensively with muons are the radiative decay $\mu \rightarrow e\gamma$, $\mu \rightarrow 3e$, muon–positron conversion on nuclei ($\mu^-(A, Z) \rightarrow e^+(A, Z - 2)$) and muonium–antimuonium conversion ($\mu^+e^- \rightarrow \mu^-e^+$). The

Table 2.2. Some selected experimental limits on lepton-number-violating processes. The values are taken from [PDG00] and [Kun01].

Process	Exp. limit on BR
$\mu \to e\gamma$	$< 1.2 \times 10^{-11}$
$\mu \to 3e$	$< 1.0 \times 10^{-12}$
$\mu(A, Z) \to e^-(A, Z)$	$< 6.1 \times 10^{-13}$
$\mu(A, Z) \to e^+(A, Z)$	$< 1.7 \times 10^{-12}$
$\tau \to \mu\gamma$	$< 1.1 \times 10^{-6}$
$\tau \to e\gamma$	$< 2.7 \times 10^{-6}$
$\tau \to 3e$	$< 2.9 \times 10^{-6}$
$\tau \to 3\mu$	$< 1.9 \times 10^{-6}$
$K^+ \to \pi^- e^+ e^+$	$< 6.4 \times 10^{-10}$
$K^+ \to \pi^- e^+ \mu^+$	$< 5.0 \times 10^{-10}$
$K^+ \to \pi^+ e^+ \mu^-$	$< 5.2 \times 10^{-10}$

evolution over time of experimental progress of some of the searches is shown in figure 2.3. Searches involving τ-leptons e.g. $\tau \to \mu\gamma$ are also performed but are not as sensitive. A compilation of obtained limits on some selected searches is given in table 2.2. Another LFV process is neutrino oscillation, discussed in chapter 8. For a comprehensive list see [PDG00].

The 'gold-plated' reaction to distinguish between Majorana and Dirac neutrino and therefore establish total lepton number violation is the process of neutrinoless double β-decay

$$(A, Z) \to (A, Z + 2) + 2e^-. \tag{2.74}$$

This process is only possible if neutrinos are massive Majorana particles and it is discussed in detail in chapter 7. A compilation of searches for $\Delta L = 2$ processes is given in table 7

Chapter 3

The standard model of particle physics

In this chapter the basic features of the current standard model of elementary particle physics are discussed. As the main interest lies in neutrinos, the focus is on the weak or the more general electroweak interaction. For a more extensive introduction, see [Hal84, Nac86, Kan87, Ait89, Don92, Mar92, Lea96, Per00].

3.1 The V–A theory of the weak interaction

Historically, the first theoretical description of the weak interaction as an explanation for β-decay (see chapter 6) was given in the classical paper by Fermi [Fer34]. Nowadays, we rate this as a low-energy limit of the Glashow–Weinberg–Salam (GWS) model (see section 3.3) but it is still valid to describe most of the weak processes. Fermi chose a local coupling of four spin-$\frac{1}{2}$ fields (a four-point interaction) and took an ansatz quite similar to that in quantum electrodynamics (QED). In QED, the interaction of a proton with an electromagnetic field A_μ is described by a Hamiltonian

$$H_{em} = e \int d^3x \, \bar{p}(x)\gamma^\mu p(x)A_\mu(x) \qquad (3.1)$$

where $p(x)$ is the Dirac field-operator of the proton. In analogy, Fermi introduced an interaction Hamiltonian for β-decay:

$$H_\beta = \frac{G_F}{\sqrt{2}} \int d^3x \, (\bar{p}(x)\gamma^\mu n(x))(\bar{e}(x)\gamma_\mu \nu(x)) + h.c. \qquad (3.2)$$

The new fundamental constant G_F is called the Fermi constant. It was soon realized that a generalization of (3.2) is necessary to describe all observed β-decays [Gam36].

If we stay with a four-fermion interaction, the following question arises: How many Lorentz-invariant combinations of the two currents involved can be

33

Table 3.1. Possible operators and their transformation properties as well as their representation.

Operator	Transformation properties ($\Psi_f O \Psi_i$)	Representation with γ matrices
O_S (S)	scalar	$\mathbb{1}$
O_V (V)	vector	γ_μ
O_T (T)	tensor	$\gamma_\mu \gamma_\nu$
O_A (A)	axial vector	$i\gamma_\mu \gamma_5$
O_P (P)	pseudo-scalar	γ_5

built. The weak Hamiltonian H_β can be deduced from a Lagrangian $\mathcal{L}$ by

$$H_\beta = -\int d^3x \, \mathcal{L}(x). \tag{3.3}$$

The most general Lagrangian for β-decay, which transforms as a scalar under a Lorentz transformation, is given by

$$\mathcal{L}(x) = \sum_{j=1}^{5} [g_j \bar{p}(x) O_j n(x) \bar{e}(x) O'_j \nu(x) + g'_j \bar{p}(x) O_j n(x) \bar{e}(x) O'_j \gamma_5 \nu(x)] + h.c.$$

$$\tag{3.4}$$

with g_j, g'_j as arbitrary complex coupling constants and O_j, O'_j as operators. The possible invariants for the operators O are listed in table 3.1. The kind of coupling realized in nature was revealed by investigating allowed β-decay transitions (see chapter 6). From the absence of Fierz interference terms (for more details see [Sch66, Wu66] and chapter 6) it could be concluded that Fermi transitions are either of S or V type, while Gamow–Teller transitions could only be due to T- or A-type operators. P-type operators do not permit allowed transitions at all. After the discovery of parity violation, the measurements of electron–neutrino angular correlations in β-decay and the Goldhaber experiment (see chapter 1), it became clear that the combination $\gamma_\mu(1 - \gamma_5)$ represented all the data accurately. This is the (V–A) structure of weak interactions. After losing its leading role as a tool for probing weak interactions, current investigations of nuclear β-decay are used for searches S- and T-type contributions motivated by theories beyond the standard model and searches for a non-vanishing rest mass of the neutrino (see chapter 6). Models with charged Higgs particles, leptoquarks and supersymmetry (see chapter 5) might lead to such S,T contributions [Her95]. A compilation of current limits on S-type contributions is shown in figure 3.1. In summary, classical β-decay can be written in the form of two currents J (current–current coupling):

$$\mathcal{L}(x) = \frac{G_F}{\sqrt{2}} J_L \cdot J_H \tag{3.5}$$

where the leptonic current is given by (e, ν as spinor fields)

$$J_L = \bar{e}(x)\gamma_\mu(1 - \gamma_5)\nu(x) \tag{3.6}$$

as proposed by [Lan56, Sal57, Lee57] and the hadronic current by (using u, d quarks instead of proton and neutron)

$$J_H = \bar{u}(x)\gamma^\mu(1 - \gamma_5)d(x) \tag{3.7}$$

as first discussed by [Fey58, The58]. As we go from the quark level to nucleons, equation (3.7) must be rewritten due to renormalization effects in strong interactions as

$$J_H = \bar{p}(x)\gamma^\mu(g_V - g_A\gamma_5)n(x). \tag{3.8}$$

The coupling constants G_F, g_V and g_A have to be determined experimentally (see section 3.4.1). Measurements of G_F in muon decay are in good agreement with those in nuclear β-decay and lead to the concept of common current couplings (e–μ–τ universality, see figure 3.2), also justified in measurements of τ-decays. The total leptonic current is then given by

$$J_L = J_e + J_\mu + J_\tau \tag{3.9}$$

each of them having the form of (3.6). Analogous arguments hold for the quark currents which can be extended to three families as well. Furthermore, the existence of a universal Fermi constant leads to the hypothesis of conserved vector currents (CVC) [Ger56, Fey58] showing that there are no renormalization effects in the vector current. Also the observation that g_V and g_A are not too different (see section 3.4.2) shows that renormalization effects in the axial vector current are also small, leading to the concept of partially conserved axial currents (PCAC). For more details see [Gro90]. The formalism allows most of the observed weak interactions to be described. It contains maximal parity violation, lepton universality and describes charged current interactions (see chapter 4). How this picture is modified and embedded in the current understanding of gauge theories will be discussed next.

3.2 Gauge theories

All modern theories of elementary particles are gauge theories. We will, therefore, attempt to indicate the fundamental characteristics of such theories without going into the details of a complete presentation. Theoretical aspects such as renormalization, the derivation of Feynman graphs or the triangle anomalies will not be discussed here and we refer to standard textbooks such as [Qui83, Hal84, Ait89, Don92, Lea96]. However, it is important to realize that such topics do form part of the fundamentals of any such theory. One absolutely necessary requirement for such a theory is its *renormalizability*. Renormalization of the

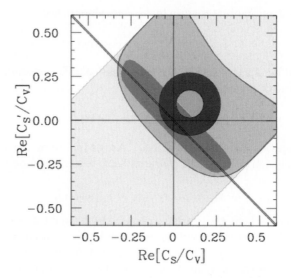

Figure 3.1. Comparison of constraints on scalar couplings in weak interactions. Limits included are from neutron decay alone (light shaded) and in combination with results from the polarization of electrons in ^{14}O and ^{10}C decay. The constraints obtained on Fierz terms coming from ^{22}Na and on the quantity a (see chapter 6) from ^{6}He the dark grey region results (see also chapter 6 for more explanations) are also added. The black circle corresponds to positron–neutrino correlation measurements in ^{32,33}Ar. The narrow area along the line at -45 degrees results from constraints on Fierz terms from $0^+ \rightarrow 0^+$ transitions (from [Adl99]).

fundamental parameters is necessary to produce a relation between calculable and experimentally measurable quantities. The fact that it can be shown that gauge theories are *always* renormalizable, as long as the gauge bosons are massless, is of fundamental importance [t'Ho72, Lee72]. Only after this proof, did gauge theories become serious candidates for modelling interactions. One well-known non-renormalizable theory is the general theory of relativity.

A further aspect of the theory is its *freedom from anomalies*. The meaning of anomaly in this context is that the classical invariance of the equations of motion or, equivalently, the Lagrangian no longer exists in quantum field theoretical perturbation theory. The reason for this arises from the fact that in such a case a consistent renormalization procedure cannot be found.

3.2.1 The gauge principle

The gauge principle can be explained by the example of classical electrodynamics. It is based on the Maxwell equations and the electric

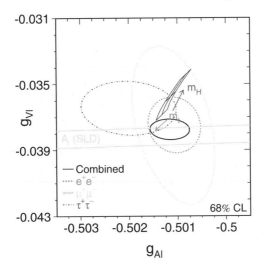

Figure 3.2. Lepton universality as probed in e^+e^- colliders at CERN and SLAC. The measured coupling constants g_V, g_A as obtained in the different flavours as well as the combined value are shown. The flat band shows the pull for varying the higgs and top mass. This precision measurement indicates a universal coupling of the charged leptons to the weak vector bosons, whose value favours a relatively light Higgs boson (with kind permission of the LEP EW working group).

and magnetic fields—measurable quantities which can be represented as the components of the field-strength tensor $F_{\mu\nu} = \partial_\mu A_\nu - \partial_\nu A_\mu$. Here the four-potential A_μ is given by $A_\mu = (\phi, A)$, and the field strengths are derived from it as $E = -\nabla\phi - \partial_t A$ and $B = \nabla \times A$. If $\rho(t, x)$ is a well-behaved, differentiable real function, it can be seen that under a transformation of the potential such as

$$\phi'(t, x) = \phi(t, x) + \partial_t \rho(t, x) \tag{3.10}$$

$$A'(t, x) = A(t, x) + \nabla\rho(t, x) \tag{3.11}$$

all observable quantities remain invariant. The fixing of ϕ and A to particular values in order to, for example, simplify the equations of motion, is called *fixing the gauge*.

In gauge theories, this gauge freedom for certain quantities is raised to a fundamental principle. The existence and structure of interactions is determined by the demand for such gauge-fixable but physically undetermined quantities. The inner structure of the gauge transformation is specified through a symmetry group.

As mentioned before, symmetries and behaviour under symmetry operations play a crucial role and will be considered next.

3.2.2 Global symmetries

Internal symmetries can be subdivided into discrete and continuous symmetries. We will concentrate on continuous symmetries. In quantum mechanics a physical state is described by a wavefunction $\psi(x, t)$. However, only the modulus squared appears as a measurable quantity. This means that as well as $\psi(x, t)$ the functions

$$\psi'(x, t) = e^{-i\alpha} \psi(x, t) \tag{3.12}$$

are also solutions of the Schrödinger equation, where α is a real (space and time independent) function. This is called a *global symmetry* and relates to the space and time independence of α. Consider the wavefunction of a charged particle such as the electron. The relativistic equation of motion is the Dirac equation:

$$i\gamma^\mu \partial_\mu \psi_e(x, t) - m\psi_e(x, t) = 0. \tag{3.13}$$

The invariance under the global transformation

$$\psi'_e(x, t) = e^{ie\alpha} \psi_e(x, t) \tag{3.14}$$

where e is a constant (for example, the electric charge), is clear:

$$e^{ie\alpha} i\gamma^\mu \partial_\mu \psi_e(x, t) = e^{ie\alpha} m\psi_e(x, t)$$
$$\Rightarrow i\gamma^\mu \partial_\mu e^{ie\alpha} \psi_e(x, t) = m e^{ie\alpha} \psi_e(x, t)$$
$$i\gamma^\mu \partial_\mu \psi'_e(x, t) = m\psi'_e(x, t). \tag{3.15}$$

Instead of discussing symmetries of the equations of motion, the Lagrangian $\mathcal{L}$ is often used. The equations of motion of a theory can be derived from the Lagrangian $\mathcal{L}(\phi, \partial_\mu\phi)$ with the help of the principle of least action (see e.g. [Gol80]). For example, consider a real scalar field $\phi(x)$. Its free Lagrangian is

$$\mathcal{L}(\phi, \partial_\mu\phi) = \tfrac{1}{2}(\partial_\mu\phi\partial^\mu\phi - m^2\phi^2). \tag{3.16}$$

From the requirement that the action integral S is stationary

$$\delta S[x] = 0 \quad \text{with } S[x] = \int \mathcal{L}(\phi, \partial_\mu\phi)\, dx \tag{3.17}$$

the equations of motion can be obtained:

$$\partial_\alpha \frac{\partial \mathcal{L}}{\partial(\partial_\alpha\phi)} - \frac{\partial \mathcal{L}}{\partial\phi} = 0. \tag{3.18}$$

The Lagrangian clearly displays certain symmetries of the theory. In general, it can be shown that the invariance of the field $\phi(x)$ under certain symmetry transformations results in the conservation of a four-current, given by

$$\partial_\alpha \left(\frac{\partial \mathcal{L}}{\partial(\partial_\alpha\phi)} \delta\phi \right) = 0. \tag{3.19}$$

This is generally known as *Noether's theorem* [Noe18]. Using this expression, time, translation and rotation invariance imply the conservation of energy, momentum and angular momentum respectively. We now proceed to consider the differences introduced by local symmetries, in which α in equation (3.12) is no longer a constant function but shows a space and time dependence.

3.2.3 Local (= gauge) symmetries

If the requirement for space and time independence of α is dropped, the symmetry becomes a local symmetry. It is obvious that under transformations such as

$$\psi_e'(x) = e^{ie\alpha(x)}\psi_e(x) \tag{3.20}$$

the Dirac equation (3.13) does not remain invariant:

$$(i\gamma^\mu \partial_\mu - m)\psi_e'(x) = e^{ie\alpha(x)}[(i\gamma^\mu \partial_\mu - m)\psi_e(x) + e(\partial_\mu\alpha(x))\gamma^\mu\psi_e(x)]$$
$$= e(\partial_\mu\alpha(x))\gamma^\mu\psi_e'(x) \neq 0. \tag{3.21}$$

The field $\psi_e'(x)$ is, therefore, not a solution of the free Dirac equation. If it were possible to compensate the additional term, the original invariance could be restored. This can be achieved by introducing a gauge field A_μ, which transforms itself in such a way that it compensates for the extra term. In order to achieve this, it is necessary to introduce a covariant derivative D_μ, where

$$D_\mu = \partial_\mu - ieA_\mu. \tag{3.22}$$

The invariance can be restored if all partial derivatives ∂_μ are replaced by the covariant derivative D_μ. The Dirac equation then becomes

$$i\gamma^\mu D_\mu \psi_e(x) = i\gamma^\mu(\partial_\mu - ieA_\mu)\psi_e(x) = m\psi_e(x). \tag{3.23}$$

If one now uses the transformed field $\psi_e'(x)$, it is easy to see that the original invariance of the Dirac equation can be restored if the gauge field transforms itself according to

$$A_\mu \to A_\mu + \partial_\mu\alpha(x). \tag{3.24}$$

The equations (3.20) and (3.24) describe the transformation of the wavefunction and the gauge field. They are, therefore, called *gauge transformations*. The whole of electrodynamics can be described in this way as a consequence of the invariance of the Lagrangian $\mathcal{L}$ or, equivalently, the equations of motion, under phase transformations $e^{ie\alpha(x)}$. The resulting conserved quantity is the electric charge, e. The corresponding theory is called quantum electrodynamics (QED) and, as a result of its enormous success, it has become a paradigm of a gauge theory. In the transition to classical physics, the gauge field A_μ becomes the classical vector potential of electrodynamics. The gauge field can be associated with the photon, which takes over the role of an exchange particle. It is found that generally in all gauge theories the gauge fields have to be massless. This is logical because a photon mass term would be proportional to $m_\gamma^2 A_\mu A^\mu$, which is obviously not invariant. Any required masses have to be built in subsequently. The case discussed here corresponds to the gauge theoretical treatment of electrodynamics. Group-theoretically the multiplication with a phase factor can be described by a unitary transformation, in this case the U(1) group. It has the unity operator as generator. The gauge principle can easily be generalized for Abelian gauge groups, i.e. groups whose generators commute with each other. It becomes somewhat more complex in the case of non-Abelian groups, as we will see in the next section.

3.2.4 Non-Abelian gauge theories (= Yang–Mills theories)

Non-Abelian means that the generators of the groups no longer commute, but are subject to certain commutator relations and the resulting non-Abelian gauge theories (Yang–Mills theories) [Yan54]. One example for commutator relations are the Pauli spin matrices σ_i,

$$[\sigma_i, \sigma_j] = i\hbar\sigma_k \tag{3.25}$$

which act as generators for the SU(2) group. Generally SU(N) groups possess $N^2 - 1$ generators. A representation of the SU(2) group is all unitary 2 × 2 matrices with determinant +1. Consider the electron and neutrino as an example. Apart from their electric charge and their mass these two particles behave identically with respect to the weak interaction, and one can imagine transformations such as

$$\left(\begin{array}{c} \psi_e(x) \\ \psi_\nu(x) \end{array} \right)' = U(x) \left(\begin{array}{c} \psi_e(x) \\ \psi_\nu(x) \end{array} \right) \tag{3.26}$$

where the transformation can be written as

$$U(a_1, a_2, a_3) = e^{i\frac{1}{2}(a_1\sigma_1 + a_2\sigma_2 + a_3\sigma_3)} = e^{i\frac{1}{2}a(x)\sigma}. \tag{3.27}$$

The particles are generally arranged in multiplets of the corresponding group (in (3.26) they are arranged as doublets). Considering the Dirac equation and substituting a covariant derivative for the normal derivative by introducing a gauge field $W_\mu(x)$ and a quantum number g in analogy to (3.22):

$$D_\mu = \partial_\mu + \frac{ig}{2} W_\mu(x) \cdot \sigma \tag{3.28}$$

does *not* lead to gauge invariance. Rather, because of the non-commutation of the generators, an additional term results, an effect which did not appear in the electromagnetic interaction. Only transformations of the gauge fields such as

$$W_\mu' = W_\mu + \frac{1}{g}\partial_\mu a(x) - W_\mu \times a(x) \tag{3.29}$$

Table 3.2. (a) Properties of the quarks ordered with increasing mass: I, isospin and its third component I_3, S, strangeness; C, charm; Q, charge; B, baryon number; B^*, bottom; T, top. (b) Properties of leptons. L_i flavour-related lepton number, $L = \sum_{i=e,\mu,\tau} L_i$.

(a) Flavour	Spin	B	I	I_3	S	C	B^*	T	$Q[e]$
u	1/2	1/3	1/2	1/2	0	0	0	0	2/3
d	1/2	1/3	1/2	−1/2	0	0	0	0	−1/3
s	1/2	1/3	0	0	−1	0	0	0	−1/3
c	1/2	1/3	0	0	0	1	0	0	2/3
b	1/2	1/3	0	0	0	0	−1	0	−1/3
t	1/2	1/3	0	0	0	0	0	1	2/3

(b) Lepton	$Q[e]$	L_e	L_μ	L_τ	L
e^-	−1	1	0	0	1
ν_e	0	1	0	0	1
μ^-	−1	0	1	0	1
ν_μ	0	0	1	0	1
τ^-	−1	0	0	1	1
ν_τ	0	0	0	1	1

supply the desired invariance. (Note the difference compared with (3.24).) The non-commutation of the generators causes the exchange particles to carry 'charge' themselves (contrary to the case of the photon, which does not carry electric charge) because of this additional term. Among other consequences, this results in a self-coupling of the exchange fields. We now proceed to discuss in more detail the non-Abelian gauge theories of the electroweak and strong interaction, which are unified in the *standard model of elementary particle physics*. The main interest of this book lies in neutrinos. Therefore, we concentrate on the electroweak part of the standard model.

3.3   The Glashow–Weinberg–Salam model

We now consider a treatment of electroweak interactions in the framework of gauge theories. The exposition will be restricted to an outline, for a more detailed discussion see the standard textbooks, for example [Hal84, Nac86, Ait89, Gre86a, Don92, Mar92, Lea96, Per00].

Theoretically, the standard model group corresponds to a direct product of three groups, SU(3)⊗SU(2)⊗U(1), where SU(3) belongs to the colour group of quantum chromodynamics (QCD), SU(2) to the weak isospin and U(1) belongs to the hypercharge. The particle content with its corresponding quantum numbers is given in table 3.2. The electroweak SU(2) ⊗ U(1) section, called the Glashow–Weinberg–Salam (GWS) model [Gla61, Wei67, Sal68] or quantum flavour dynamics (QFD) consists of the weak isospin SU(2) and the hypercharge group U(1). The concept of weak isospin is in analogy to isospin in strong interactions (see, e.g., [Gro90]). The elementary particles are arranged as doublets for chiral left-handed fields and singlets for right-handed fields in the form

$$\begin{pmatrix} u \\ d' \end{pmatrix}_L \quad \begin{pmatrix} c \\ s' \end{pmatrix}_L \quad \begin{pmatrix} t \\ b' \end{pmatrix}_L \quad \begin{pmatrix} e \\ v_e \end{pmatrix}_L \quad \begin{pmatrix} \mu \\ v_\mu \end{pmatrix}_L \quad \begin{pmatrix} \tau \\ v_\tau \end{pmatrix}_L$$
$$u_R \quad\;\; d_R \quad\;\; s_R \quad\;\; c_R \quad\;\; b_R \quad\;\; t_R \quad\;\; e_R \quad\;\; \mu_R \quad\;\; \tau_R. \quad (3.30)$$

We want to discuss the theory along the line taken in [Nac86] taking the first generation of the three known chiral lepton fields e_R, e_L and v_{eL} as an example. An extension to all three generations and quarks is straightforward. Neglecting any mass and switching off weak interactions and electromagnetism the Lagrangian for the free Dirac fields can be written as

$$\mathcal{L}(x) = (\bar{v}_{eL}(x), \bar{e}_L(x))(i\gamma^\mu \partial_\mu)\begin{pmatrix} v_{eL}(x) \\ e_L(x) \end{pmatrix} + \bar{e}_R(x)i\gamma^\mu \partial_\mu e_R(x). \quad (3.31)$$

This Lagrangian is invariant with respect to global SU(2) transformations on the fields v_{eL} and e_L. Going to a local SU(2) transformation, the Lagrangian clearly is not invariant but we can compensate for that by introducing a corresponding number of gauge vector fields. In the case of SU(2) we have three generators and, therefore, we need three vector fields called W_μ^1, W_μ^2, W_μ^3 (see section 3.2.4). The Lagrangian including the W-fields can then be written as

$$\mathcal{L}(x) = \tfrac{1}{2}\mathrm{Tr}(W_{\mu\rho}(x)W^{\mu\rho}(x)) + (\bar{v}_{eL}(x), \bar{e}_L(x))i\gamma^\mu(\partial_\mu + ig\,W_\mu)\begin{pmatrix} v_{eL} \\ e_L \end{pmatrix}$$
$$+ \bar{e}_R(x)i\gamma^\mu\partial_\mu e_R(x). \quad (3.32)$$

The introduced gauge group SU(2) is called the weak isospin. Introducing the fields $W_\mu^\pm$ as

$$W_\mu^\pm = \frac{1}{\sqrt{2}}(W_\mu^1 \mp iW_\mu^2) \quad (3.33)$$

from (3.32) the v–e–W coupling term can be obtained as

$$\mathcal{L} = -g(\bar{v}_{eL}, \bar{e}_L)\gamma^\mu W_\mu \frac{\sigma}{2}\begin{pmatrix} v_{eL} \\ e_L \end{pmatrix}$$
$$= -g(\bar{v}_{eL}, \bar{e}_L)\gamma^\mu \frac{1}{2}\begin{pmatrix} W_\mu^3 & \sqrt{2}W_\mu^+ \\ \sqrt{2}W_\mu^- & -W_\mu^3 \end{pmatrix}\begin{pmatrix} v_{eL} \\ e_L \end{pmatrix} \quad (3.34)$$
$$= -\frac{g}{2}\{W_\mu^3(\bar{v}_{eL}\gamma^\mu v_{eL} - \bar{e}_L\gamma^\mu e_L) + \sqrt{2}W_\mu^+ \bar{v}_{eL}\gamma^\mu e_L + \sqrt{2}W_\mu^- \bar{e}_L\gamma^\mu v_{eL}\}$$

with σ as the Pauli matrices. This looks quite promising because the last two terms already have the $\gamma^\mu(1 - \gamma_5)$ structure as discussed in section 3.1. Hence, by finding a method to make the W-boson very massive, at low energy the theory reduces to the Fermi four-point interaction (see chapter 6). Before discussing masses we want to add electromagnetism. The easiest assumption for associating the remaining field W_μ^3 with the photon field does not work, because W_μ^3 couples to neutrinos and not to e_R in contrast to the photon. Going back to (3.31) beside the SU(2) invariance one can recognize an additional invariance under two further U(1) transformations with quantum numbers y_L, y_R:

$$\begin{pmatrix} \nu_{eL}(x) \\ e_L(x) \end{pmatrix} \rightarrow e^{+iy_L\chi} \begin{pmatrix} \nu_{eL}(x) \\ e_L(x) \end{pmatrix} \tag{3.35}$$

$$e_R(x) \rightarrow e^{+iy_R\chi} e_R(x). \tag{3.36}$$

However, this would result in two 'photon-like' gauge bosons in contrast to nature from which we know there is only one. Therefore, we can restrict ourselves to one special combination of these phase transitions resulting in one U(1) transformation by choosing

$$y_L = -\tfrac{1}{2}. \tag{3.37}$$

y_R is fixed in (3.46). This U(1) group is called the weak hypercharge Y. We can make this U(1) into a gauge group as in QED, where the charge Q is replaced by the weak hypercharge Y. Between charge, hypercharge and the third component of the weak isospin, the following relation holds

$$Q = I_3 + \frac{Y}{2}. \tag{3.38}$$

The necessary real vector field is called B_μ and the corresponding gauge coupling constant g'. Now we are left with two massless neutral vector fields W_μ^3, B_μ and the question arises as to whether we can combine them in a way to account for weak neutral currents (see chapter 4) and electromagnetism. Let us define two orthogonal linear combinations resulting in normalized fields Z_μ and A_μ:

$$Z_\mu = \frac{1}{\sqrt{g^2 + g'^2}} (g W_\mu^3 - g' B_\mu) \tag{3.39}$$

$$A_\mu = \frac{1}{\sqrt{g^2 + g'^2}} (g' W_\mu^3 + g B_\mu). \tag{3.40}$$

By writing

$$\sin\theta_W = \frac{g'}{\sqrt{g^2 + g'^2}} \tag{3.41}$$

$$\cos\theta_W = \frac{g}{\sqrt{g^2 + g'^2}} \tag{3.42}$$

we can simplify the expressions to

$$Z_\mu = \cos\theta_W W_\mu^3 - \sin\theta_W B_\mu \tag{3.43}$$

$$A_\mu = \sin\theta_W W_\mu^3 + \cos\theta_W B_\mu. \tag{3.44}$$

The angle $\sin \theta_W$ is called the Weinberg angle and is one of the fundamental parameters of the standard model. Replacing the fields W_μ^3, B_μ in (3.34) by Z_μ, A_μ results in

$$
\begin{aligned}
\mathcal{L} = &- \frac{g}{\sqrt{2}}(W_\mu^+ \bar{\nu}_{eL} \gamma^\mu e_L + W_\mu^- \bar{e}_L \gamma^\mu \nu_{eL}) \\
&- \sqrt{g^2 + g'^2} Z_\mu \{ \tfrac{1}{2} \bar{\nu}_{eL} \gamma^\mu \nu_{eL} - \tfrac{1}{2} \bar{e}_L \gamma^\mu e_L \\
&- \sin^2 \theta_W (-\bar{e}_L \gamma^\mu e_L + y_R \bar{e}_R \gamma^\mu e_R) \} \\
&- \frac{gg'}{\sqrt{g^2 + g'^2}} A_\mu (-\bar{e}_L \gamma^\mu e_L + y_R \bar{e}_R \gamma^\mu e_R).
\end{aligned}
\tag{3.45}
$$

One can note that the Z_μ coupling results in neutral currents. However, A_μ no longer couples neutrinos and is, therefore, a good candidate to be associated with the photon field. To reproduce electromagnetism we have to choose the following

$$
y_R = -1 \qquad \frac{gg'}{\sqrt{g^2 + g'^2}} = e
\tag{3.46}
$$

which immediately yields another important relation by using (3.41)

$$
\sin \theta_W = \frac{e}{g}.
\tag{3.47}
$$

This finally allows us to write the Lagrangian using electromagnetic, charged and neutral currents:

$$
\begin{aligned}
\mathcal{L} = &- e \Big\{ A_\mu J_{em} + \frac{1}{\sqrt{2} \sin \theta_W} (W_\mu^+ \bar{\nu}_{eL} \gamma^\mu e_L + W_\mu^- \bar{e}_L \gamma^\mu \nu_{eL}) \\
&+ \frac{1}{\sin \theta_W \cos \theta_W} Z_\mu J_{NC}^\mu \Big\}
\end{aligned}
\tag{3.48}
$$

with the currents

$$
J_{em}^\mu = - \bar{e}_L \gamma^\mu e_L - \bar{e}_R \gamma^\mu e_R = -\bar{e} \gamma^\mu e
\tag{3.49}
$$

$$
J_{NC}^\mu = \tfrac{1}{2} \bar{\nu}_{eL} \gamma^\mu \nu_{eL} - \tfrac{1}{2} \bar{e}_L \gamma^\mu e_L - \sin^2 \theta_W J_{em}^\mu.
\tag{3.50}
$$

3.3.1 Spontaneous symmetry breaking and the Higgs mechanism

In the formulation of the theory all particles have to be massless to guarantee gauge invariance. The concept of spontaneous symmetry breaking is then used for particles to receive mass through the so-called *Higgs mechanism* [Hig64, Kib67]. Spontaneous symmetry breaking results in the ground state of a system having no longer the full symmetry corresponding to the underlying Lagrangian. Consider the following classical Lagrangian

$$
\mathcal{L} = (\partial_\mu \Phi)^\dagger (\partial^\mu \Phi) - \mu^2 \Phi^\dagger \Phi - \lambda (\Phi^\dagger \Phi)^2
\tag{3.51}
$$

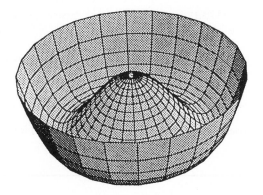

Figure 3.3. Schematic view of the Higgs potential ('Mexican hat') and its minimum for $\mu^2 < 0$.

where $\Phi(x)$ is a complex scalar field. $\mathcal{L}$ is invariant under the group U(1) of *global* transformations equivalent to (3.14). The kinetic energy term is positive and can vanish only if Φ = constant. The ground state of the system will be obtained when the value of the constant corresponds to the minimum of the potential

$$V(\Phi) = \mu^2 \Phi^\dagger \Phi + \lambda (\Phi^\dagger \Phi)^2. \qquad (3.52)$$

If $\mu^2 > 0$ and $\lambda > 0$, a mimimum configuration occurs at the origin and we have a symmetric ground-state configuration. If, however, $\mu^2 < 0$, the minimum is at

$$\rho = \Phi \Phi^\dagger = -\mu^2/2\lambda \qquad (3.53)$$

which means that there is a whole ring of radius

$$|\Phi| \equiv \frac{v}{\sqrt{2}} = \sqrt{-\mu^2/2\lambda} \qquad (3.54)$$

in the complex plane (see figure 3.3). There are infinitely many ground states, degenerate with each other but none shows the original symmetry of the Lagrangian any longer. The symmetry is broken spontaneously. Generally, it can be shown that spontaneous symmetry breaking is connected with the degeneracy of the ground state. Now we impose invariance under a *local* gauge transformation, as it is implemented in the standard model. In the electroweak model the simplest way of spontaneous symmetry breaking is achieved by introducing a doublet of complex scalar fields, one charged, one neutral:

$$\phi = \begin{pmatrix} \phi^\dagger \\ \phi^0 \end{pmatrix} \qquad (3.55)$$

where the complex fields are given by

$$\phi^\dagger = \frac{\phi_1 + i\phi_2}{\sqrt{2}}$$

$$\phi^0 = \frac{\phi_3 + i\phi_4}{\sqrt{2}}. \tag{3.56}$$

Adding a kinetic term to the potential (3.52) leads to the following expression for the Lagrangian:

$$\mathcal{L}_{\text{Higgs}} = (\partial_\mu \phi)^\dagger (\partial^\mu \phi) - \mu^2 \phi^\dagger \phi - \lambda (\phi^\dagger \phi)^2. \tag{3.57}$$

Proceeding as before, the potential $V(\phi)$ has a minimimum for $\mu^2 < 0$ at

$$\phi^\dagger \phi = \frac{-\mu^2}{2\lambda} = \frac{v^2}{2}. \tag{3.58}$$

Here again the minima, corresponding to the vacuum expectation values of ϕ lie on a circle with $\langle \phi \rangle \equiv v/\sqrt{2} = \sqrt{-\mu^2/2\lambda}$. This ground state is degenerate and its orientation in two-dimensional isospin space is not defined. It can choose any value between $[0, 2\pi]$. From this infinite number of possible orientations we choose a particular field configuration which is defined as the vacuum state as

$$\phi_0 = \frac{1}{\sqrt{2}} \begin{pmatrix} 0 \\ v \end{pmatrix} \tag{3.59}$$

which is no longer invariant under SU(2) transformations. The upper component is motivated by the fact that a vacuum is electrically neutral. The field $\phi(x)$ can now be expanded around the vacuum

$$\phi = \frac{1}{\sqrt{2}} \begin{pmatrix} 0 \\ v + H(x) \end{pmatrix} \tag{3.60}$$

where a perturbation theory for $H(x)$ can be formulated as usual. Now consider the coupling of this field to fermions first. Fermions get their masses through coupling to the vacuum expectation value (vev) of the Higgs field. To conserve isospin invariance of the coupling, the Higgs doublet has to be combined with a fermion doublet and singlet. The resulting coupling is called Yukawa coupling and has the typical form (given here for the case of electrons)

$$\begin{aligned}
\mathcal{L}_{\text{Yuk}} &= -c_e \bar{e}_R \phi^\dagger \begin{pmatrix} \nu_{eL} \\ e_L \end{pmatrix} + h.c. \\
&= -c_e \left[\bar{e}_R \phi_0^\dagger \begin{pmatrix} \nu_{eL} \\ e_L \end{pmatrix} + (\bar{\nu}_e, \bar{e}_L) \phi_0 e_R \right] \\
&= -c_e \left[\bar{e}_R \frac{1}{\sqrt{2}} \nu_{eL} + \bar{e}_L \frac{1}{\sqrt{2}} \nu_{eR} \right] \\
&= -c_e v \frac{1}{\sqrt{2}} (\bar{e}_R e_L + \bar{e}_L e_R) \\
&= -c_e \frac{v}{\sqrt{2}} \bar{e} e.
\end{aligned} \tag{3.61}$$

Here c_e is an arbitrary coupling constant. This corresponds exactly to a mass term for the electron with an electron mass of

$$m_e = c_e \frac{v}{\sqrt{2}}. \tag{3.62}$$

The same strategy holds for the other charged leptons and quarks with their corresponding coupling constant c_i. In this way fermions obtain their masses within the GWS model.

Neutrinos remain massless because with the currently accepted particle content there are no right-handed ν_R singlet states and one cannot write down couplings like (3.61). With the evidence for massive neutrinos described later, one is forced to generate the masses in another way such as using Higgs triplets or adding right-handed neutrino singlets (see chapter 5).

Substituting the covariant derivative for the normal derivative in $\mathcal{L}$ as in (3.22) leads directly to the coupling of the Higgs field with the gauge fields. For details see [Nac86, Gun90]. The gauge bosons then acquire masses of

$$m_W^2 = \frac{g^2 v^2}{4} = \frac{e^2 v^2}{4 \sin^2 \theta_W} \tag{3.63}$$

$$m_Z^2 = \frac{(g^2 + g'^2) v^2}{4} = \frac{e^2 v^2}{4 \sin^2 \theta_W \cos^2 \theta_W} \tag{3.64}$$

resulting in

$$\frac{m_W}{m_Z} = \cos \theta_W. \tag{3.65}$$

An interesting quantity deduced from this relation is the ρ-parameter defined as

$$\rho = \frac{m_W}{m_Z \cos \theta_W}. \tag{3.66}$$

Any experimental signature for a deviation from $\rho = 1$ would be a hint for new physics. An estimate for v can be given by (3.63) resulting in

$$v = (\sqrt{2} G_F)^{-1/2} \approx 246 \text{ GeV}. \tag{3.67}$$

The inclusion of spontaneous symmetry breaking with the help of a complex scalar field doublet has another consequence, namely the existence of a new scalar particle called the Higgs boson, with a mass of m_H, such that

$$m_H^2 = 2\lambda v^2. \tag{3.68}$$

This is the only unobserved particle of the standard model and many efforts are made to prove its existence (see section 3.4.5). To obtain invariance under hypercharge transformations, we have to assign a hypercharge of $y_H = 1/2$ to the Higgs.

3.3.2 The CKM mass matrix

It has been experimentally proved that the mass eigenstates for quarks are not identical to flavour eigenstates. This is shown by the fact that transitions between the various generations are observed. Thus, the mass eigenstates of the d and s quark are not identical to the flavour eigenstates d' and s', which take part in the weak interaction. They are connected via

$$\begin{pmatrix} d' \\ s' \end{pmatrix} = \begin{pmatrix} \cos\theta_C & \sin\theta_C \\ -\sin\theta_C & \cos\theta_C \end{pmatrix} \begin{pmatrix} d \\ s \end{pmatrix}. \tag{3.69}$$

The *Cabibbo angle* θ_C is about $13°$ ($\sin\theta_C = 0.222 \pm 0.003$). The extension to three generations leads to the so-called *Cabibbo–Kobayashi–Maskawa matrix* (CKM) [Kob73]

$$\begin{pmatrix} d' \\ s' \\ b' \end{pmatrix} = \begin{pmatrix} V_{ud} & V_{us} & V_{ub} \\ V_{cd} & V_{cs} & V_{cb} \\ V_{td} & V_{ts} & V_{tb} \end{pmatrix} \times \begin{pmatrix} d \\ s \\ b \end{pmatrix} = U \times \begin{pmatrix} d \\ s \\ b \end{pmatrix} \tag{3.70}$$

which can be parametrized with three mixing angles and a single phase:

$$U = \begin{pmatrix} c_{12}c_{13} & s_{12}c_{13} & s_{13}e^{-i\delta} \\ -s_{12}c_{23} - c_{12}s_{23}s_{13}e^{i\delta} & c_{12}c_{23} - s_{12}s_{23}s_{13}e^{i\delta} & s_{23}c_{13} \\ s_{12}s_{23} - c_{12}s_{23}s_{13}e^{i\delta} & -c_{12}s_{23} - s_{12}c_{23}s_{13}e^{i\delta} & c_{23}c_{13} \end{pmatrix} \tag{3.71}$$

where $s_{ij} = \sin\theta_{ij}, c_{ij} = \cos\theta_{ij}$ ($i, j = 1, 2, 3$). The individual matrix elements describe transitions between the different quark flavours and have to be determined experimentally. The present experimental results in combination with the constraint of unitarity of U give the values (90% CL) [PDG00]:

$$|U| = \begin{pmatrix} 0.9745\ldots0.9757 & 0.219\ldots0.224 & 0.002\ldots0.005 \\ 0.218\ldots0.224 & 0.9736\ldots0.9750 & 0.036\ldots0.046 \\ 0.004\ldots0.014 & 0.034\ldots0.046 & 0.9989\ldots0.9993 \end{pmatrix}. \tag{3.72}$$

However, some deviation might be seen using neutron decay [Abe02]. The Wolfenstein parametrization of U [Wol83], an expansion with respect to $\lambda = \sin\theta_{12}$ accurate up to third order in λ

$$U = \begin{pmatrix} 1 - \frac{1}{2}\lambda^2 & \lambda & A\lambda^3(\rho - i\eta) \\ -\lambda & 1 - \frac{1}{2}\lambda^2 & A\lambda^2 \\ A\lambda^3(1 - \rho - i\eta) & -A\lambda^2 & 1 \end{pmatrix} \tag{3.73}$$

is useful. Such a parametrization might not be useful in the leptonic sector, because it assumes hierarchical matrix elements, with the diagonal ones being the strongest. This case is probably not realized in the leptonic sector as we will see later. A useful concept are geometrical presentations in the complex (η, ρ) plane

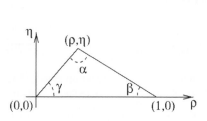
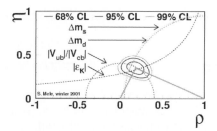

Figure 3.4. Left: Schematic picture of the unitarity triangle in the complex plane using the Wolfenstein parameters η, ρ. Right: Existing experimental limits constraining the apex of the triangle (from [Mel01]).

called unitarity triangles (figure 3.4). The unitarity of the CKM matrix leads to various relations among the matrix elements where, in particular,

$$V_{ud}V_{ub}^* + V_{cd}V_{cb}^* + V_{td}V_{tb}^* = 0 \tag{3.74}$$

is usually quoted as 'the unitarity triangle'. The relations form triangles in the complex plane, with the feature that all triangles have the same area. A rescaled triangle is obtained by making $V_{cd}V_{cb}^*$ real (one side is then aligned to the real axis) and dividing the lengths of all sides by $V_{cd}V_{cb}^*$ (given the side along the real axis length 1). Two vertices are then fixed at $(0,0)$ and $(1,0)$. The third vertex is then given by the Wolfenstein parameters (ρ, η). With all the available data, one finds [Nir01] that

$$A = 0.827 \pm 0.058 \qquad \lambda = 0.2221 \pm 0.0021 \tag{3.75}$$
$$\rho = 0.23 \pm 0.11 \qquad \eta = 0.37 \pm 0.08 \tag{3.76}$$
$$\sin 2\beta = 0.77 \pm 0.08 \qquad \sin 2\alpha = -0.21 \pm 0.56 \qquad 0.43 < \sin^2 \gamma < 0.91 \tag{3.77}$$

3.3.3 *CP violation*

The phase $e^{i\delta}$ in (3.71) can be linked to CP violation. The necessary condition for CP invariance of the Lagrangian is that the CKM matrix and its complex conjugate are identical, i.e. its elements are real. While this is always true for two families, for three families it is only true in the previous parametrization if $\delta = 0$ or $\delta = \pi$. This means that if δ does not equal one of those values, then the CKM matrix is a source of CP violation (see, e.g., [Nac86]). The first observation of CP violation has been observed in the kaon system [Chr64]. The experimentally observed particles K_S and K_L are only approximately identical to the CP eigenstates K_1 and K_2, so that it is necessary to define the observed states

K_L ($\simeq K_2$) and K_S ($\simeq K_1$) as (see, e.g., [Com83]):

$$|K_S\rangle = (1 + |\epsilon|^2)^{-1/2}(|K_1\rangle - \epsilon|K_2\rangle) \qquad (3.78)$$

$$|K_L\rangle = (1 + |\epsilon|^2)^{-1/2}(|K_2\rangle + \epsilon|K_1\rangle). \qquad (3.79)$$

CP violation caused by this mixing can be characterized by the parameter ϵ. The ratio of the amplitudes for the decay into charged pions may be used as a measure of CP violation [Per00, PDG00]:

$$|\eta_{+-}| = \frac{A(K_L \to \pi^+\pi^-)}{A(K_S \to \pi^+\pi^-)} = (2.285 \pm 0.019) \times 10^{-3}. \qquad (3.80)$$

A similar relation is obtained for the decay into two neutral pions, characterized in analogy as η_{00}. The ϵ appearing in equations (3.78) and (3.79), together with a further parameter ϵ' can be connected with η via the relation

$$\eta_{+-} = \epsilon + \epsilon' \qquad (3.81)$$

$$\eta_{00} = \epsilon - 2\epsilon' \qquad (3.82)$$

from which it can be deduced (see e.g. [Com83]) that

$$\left|\frac{\eta_{00}}{\eta_{+-}}\right| \approx 1 - 3\,\mathrm{Re}\left(\frac{\epsilon'}{\epsilon}\right). \qquad (3.83)$$

Evidence for a non-zero ϵ' would show that CP is violated *directly* in the decay, i.e. in processes with $\Delta S = 1$, and does not only depend on the existence of mixing [Com83]. The experimental status is shown in table 3.3, establishing that ϵ' is indeed different from zero.

Other important decays that will shed some light on CP violation are the decays $K^+ \to \pi^+\nu\bar{\nu}$ and $K_L \to \pi^0\nu\bar{\nu}$ which have small theoretical uncertainties. Two events of the first reaction have been observed by the E787 experiment at BNL [Adl00, Che02].

CP violation in combination with CPT invariance requires also T violation. T violation was directly observed for the first time in the kaon system by the CPLEAR experiment at CERN [Ang98].

CP violation might also show up in B-meson decays. The gold-plated channel for investigation is $B_d \to J/\Psi + K_S$ because of the combination of the experimentally clean signature and exceedingly small theoretical uncertainties. It allows a measurement of $\sin 2\beta$. The current experimental status is shown in table 3.3. The B factories at SLAC at Stanford (BaBar experiment [Bab95]) and at KEK in Japan (Belle experiment [Bel95b]) have already observed CP violation in the B system and provide important results [Aub01, Aba01].

In the leptonic sector the issue could be similar: massive neutrinos will lead to a CKM-like matrix in the leptonic sector often called the Maki–Nakagawa–Sakata (MNS) matrix [Mak62] and, therefore, to CP violation. Furthermore,

Table 3.3. Current (spring 2003) status of CP violation in kaon decay (expressed as the ratio ϵ'/ϵ) and in B-meson decays (expressed as the angle $\sin 2\beta$). Here the first error is statistical and the second systematic.

Experiment	ϵ'/ϵ
NA 31	$(23 \pm 7) \times 10^{-4}$
E 731	$(7.4 \pm 8.1) \times 10^{-4}$
KTeV	$(20.7 \pm 2.8) \times 10^{-4}$
NA 48	$(14.7 \pm 2.2) \times 10^{-4}$

Experiment	$\sin 2\beta$
BaBar	$0.741 \pm 0.067 \pm 0.034$
Belle	$0.719 \pm 0.074 \pm 0.035$
CDF	$0.79^{+0.41}_{-0.44}$
OPAL	$3.2^{+1.8}_{-2.0} \pm 0.5$
Aleph	$0.93^{+0.64+0.36}_{-0.88-0.24}$

if neutrinos are Majorana particles, there would already be the possibility of CP violation with two families and in three flavours three phases will show up [Wol81] (see section 5.5). A chance to probe one phase of CP violation in the leptonic sector exists with the planned neutrino factories (see chapter 4). The Majorana phases have direct impact on the observables in neutrinoless double β-decay (see chapter 7).

3.4 Experimental determination of fundamental parameters

Although it has been extraordinarily successful, not everything can be predicted by the standard model. In fact it has 18 free parameters as input all of which have to be measured (see chapter 5). A few selected measurements are discussed now in a little more detail.

3.4.1 Measurement of the Fermi constant G_F

The Fermi constant G_F has been of fundamental importance in the history of weak interaction. Within the context of the current GWS model, it can be expressed as

$$\frac{G_F}{\sqrt{2}} = \frac{g^2}{8m_W^2}. \tag{3.84}$$

In the past the agreement of measurements of G_F in β-decay (now called G_β) and in μ-decay (now called G_μ) lead to the hypothesis of conserved vector currents

(CVC hypothesis, see section 3.1); nowadays, the measurements can be used to test the universality of weak interactions. A small deviation between the two is expected anyway because of the Cabibbo-mixing, which results in

$$\frac{G_\beta}{G_\mu} \simeq \cos\theta_C \approx 0.98. \tag{3.85}$$

In general, precision measurements of the fundamental constants including the Fermi constant, allow us to restrict the physics beyond the standard model [Her95, Mar99].

The best way to determine G_F which can also be seen as a definition of G_F ($G_F := G_\mu$) is the measurement of the muon lifetime τ:

$$\tau^{-1} = \Gamma(\mu \to e\nu_\mu\nu_e) = \frac{G_F^2 m_\mu^5}{192\pi^3}(1 + \Delta\rho) \tag{3.86}$$

where $\Delta\rho$ describe radiative corrections. Equation (3.86) can be expressed as [Rit00]

$$\tau^{-1} = \Gamma(\mu \to e\nu_\mu\nu_e) = \frac{G_F^2 m_\mu^5}{192\pi^3} F\left(\frac{m_e^2}{m_\mu^2}\right)\left(1 + \frac{3}{5}\frac{m_\mu^2}{m_W^2}\right)$$
$$\times \left(1 + \frac{\alpha(m_\mu)}{2\pi}\left(\frac{25}{4} - \pi^2\right)\right) \tag{3.87}$$

with $(x = m_e^2/m_\mu^2)$

$$F(x) = 1 - 8x - 12x^2\ln x + 8x^3 - x^4 \tag{3.88}$$

and

$$\alpha(m_\mu)^{-1} = \alpha^{-1} - \frac{2}{3\pi}\left(\ln\frac{m_e}{m_\mu}\right) + \frac{1}{6\pi} \approx 136. \tag{3.89}$$

The second term in (3.87) is an effect of the W propagator and the last term is the leading contribution of the radiative corrections. Unfortunately, the experimental value of [Bar84, Gio84]

$$G_F = 1.16637(1) \times 10^{-5} \text{ GeV}^{-2} \tag{3.90}$$

still has an error of 18 ppm. Therefore, at PSI a new experiment has been approved to improve the value of the muon lifetime by a factor of about 20 [Car99, Kir99a]. This will finally result in a total experimental uncertainty of 0.5 ppm on G_F and will have much more sensitivity on new physics effects.

3.4.2   Neutrino–electron scattering and the coupling constants g_V and g_A

A fundamental electroweak process to study is νe scattering, which can be of the form

$$\nu_\mu e \to \nu_\mu e \qquad \bar\nu_\mu e \to \bar\nu_\mu e \tag{3.91}$$
$$\nu_e e \to \nu_e e \qquad \bar\nu_e e \to \bar\nu_e e. \tag{3.92}$$

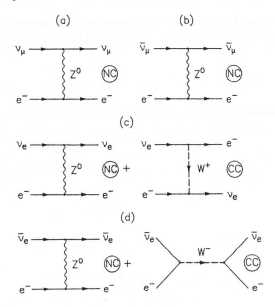

Figure 3.5. Feynman diagrams for neutrino–electron NC and CC reactions: $\nu_\mu e$ NC (*a*), $\bar{\nu}_\mu e$ NC (*b*), $\nu_e e$ NC + CC (*c*) and $\bar{\nu}_e e$ NC + CC scattering (*d*).

While the first reaction can only happen via neutral current (NC) interactions, for the second both neutral current and charged current (CC) are possible (figure 3.5), see also [Pan95].

3.4.2.1 Theoretical considerations

The Lagrangian for the first reaction (3.91) is

$$\mathcal{L} = -\frac{G_F}{\sqrt{2}}[\bar{\nu}_\mu \gamma^\alpha (1 - \gamma_5)\nu_\mu][\bar{e}\gamma_\alpha(g_V - g_A\gamma_5)e] \tag{3.93}$$

with the prediction from the GWS model of

$$g_V = -\tfrac{1}{2} + 2\sin^2\theta_W \qquad g_A = -\tfrac{1}{2}. \tag{3.94}$$

A similar term can be written for the second type of interaction. In addition, the CC contribution can be written as

$$\mathcal{L} = -\frac{G_F}{\sqrt{2}}[\bar{e}\gamma^\alpha (1 - \gamma_5)\nu_e][\bar{\nu}_e\gamma_\alpha(1 - \gamma_5)e] \tag{3.95}$$

$$= -\frac{G_F}{\sqrt{2}}[\bar{\nu}_e\gamma^\alpha (1 - \gamma_5)\nu_e][\bar{e}\gamma_\alpha(1 - \gamma_5)e] \tag{3.96}$$

where in the second step a Fierz transformation was applied (see [Bil94] for technical information on this). The predictions of the GWS model for the chiral couplings g_L and g_R are:

$$g_L = \tfrac{1}{2}(g_V + g_A) = -\tfrac{1}{2} + \sin^2 \theta_W \qquad g_R = \tfrac{1}{2}(g_V - g_A) = \sin^2 \theta_W. \quad (3.97)$$

A detailed calculation [Sch97] leads to the expected cross sections which are given by (see also chapter 4)

$$\frac{\mathrm{d}\sigma}{\mathrm{d}y}(\overset{(-)}{\nu_\mu} e) = \frac{G_F^2 m_e}{2\pi} E_\nu \left[(g_V \pm g_A)^2 + (g_V \mp g_A)^2 (1 - y)^2 + \frac{m_e}{E_\nu}(g_A^2 - g_V^2)y \right] \quad (3.98)$$

and

$$\frac{\mathrm{d}\sigma}{\mathrm{d}y}(\overset{(-)}{\nu_e} e) = \frac{G_F^2 m_e}{2\pi} E_\nu \left[(G_V \pm G_A)^2 + (G_V \mp G_A)^2 (1 - y)^2 \right.$$
$$\left. + \frac{m_e}{E_\nu}(G_A^2 - G_V^2)y \right] \quad (3.99)$$

with $G_V = g_V + 1$ and $G_A = g_A + 1$. The upper(lower) sign corresponds to $\nu e(\bar{\nu} e)$ scattering. The quantity y is called the inelasticity or the Bjorken y and is given by

$$y = \frac{T_e}{E_\nu} \approx \frac{E_e}{E_\nu} \quad (3.100)$$

where T_e is the kinetic energy of the electron. Therefore, the value of y is restricted to $0 \le y \le 1$. The cross sections are proportional to E_ν. An integration with respect to y leads to total cross sections of

$$\sigma(\overset{(-)}{\nu_\mu} e) = \sigma_0(g_V^2 + g_A^2 \pm g_V g_A) \quad (3.101)$$

$$\sigma(\overset{(-)}{\nu_e} e) = \sigma_0(G_V^2 + G_A^2 \pm G_V G_A) \quad (3.102)$$

with

$$\sigma_0 = \frac{2G_F^2 m_e}{3\pi}(\hbar c)^2 E_\nu = 5.744 \times 10^{-42} m_M \text{ cm}^2 \frac{E_\nu}{\text{GeV}}. \quad (3.103)$$

(3.101) can be reformulated into

$$g_V^2 + g_A^2 = [\sigma(\nu_\mu e) + \sigma(\bar{\nu}_\mu e)]/2\sigma_0 \quad (3.104)$$

$$g_V g_A = [\sigma(\nu_\mu e) - \sigma(\bar{\nu}_\mu e)]/2\sigma_0. \quad (3.105)$$

By measuring the four cross sections (3.101) the constants g_V and g_A and additionally using (3.94), $\sin^2 \theta_W$ can also be determined. For each fixed measured value of $\sigma(\nu e)/\sigma_0$ one obtains an ellipsoid in the g_V, g_A plane with the

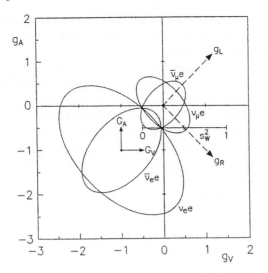

Figure 3.6. Schematic drawing of the four ellipses for fixed $\sigma(\nu e)/\sigma_0$ values in the (g_V, g_A) plane for the various νe scattering processes. The directions of the g_L and g_R axis under $45°$ are shown as dashed lines and the GWS prediction $-\frac{1}{2} < g_V < \frac{3}{2}, g_A = -\frac{1}{2}$ for $0 < \sin^2\theta_W < 1$ are also shown (after [Shu97]).

main axis orientated in the direction of 45 degrees, i.e. along the g_R, g_L directions (figure 3.6).

In $\nu_e e$ scattering there is interference because of the presence of both amplitudes (NC and CC) in the interactions. The cross sections are given by

$$\sigma(\nu_e e) = (g_V^2 + g_A^2 + g_V g_A)\sigma_0 + 3\sigma_0 + 3(g_V + g_A)\sigma_0 \qquad (3.106)$$

$$\sigma(\bar{\nu}_e e) = (g_V^2 + g_A^2 - g_V g_A)\sigma_0 + \sigma_0 + (g_V + g_A)\sigma_0 \qquad (3.107)$$

where the interference term is given by

$$I(\nu_e e) = 3I(\bar{\nu}_e e) = 3(g_V + g_A)\sigma_0 = 3(2\sin^2\theta_W - 1)\sigma_0. \qquad (3.108)$$

The small cross section requires experiments with a large mass and a high intensity neutrino beam. The signature of this type of event is a single electron in the final state. At high energies the electron is boosted in the forward direction and, besides a good energy resolution, a good angular resolution is required for efficient background discrimination (see [Pan95] for details).

3.4.2.2 $\nu_\mu e$ scattering

The same experimental difficulties also occur in measuring $\nu_\mu e$ scattering cross sections. Accelerators provide neutrino beams with energies in the MeV–several GeV range (see chapter 4). Experiments done in the 1980s consisted of calorimeters of more than 100 t mass (CHARM, E734 and CHARM-II), of which CHARM-II has, by far, the largest dataset [Gei93, Vil94]. With good spatial and energy resolution, good background discrimation was possible. The dominant background stems basically from ν_e CC reactions due to beam contamination with ν_e and NC π^0 production, with $\pi^0 \rightarrow \gamma\gamma$ which could mimic electrons. The latter can be discriminated either by having a different shower profile in the calorimeter or by having a wider angular distribution. The results are shown in figure 3.7. However, there is still an ambiguity which fortunately can be solved by using data from forward–backward asymmetry measurements in elastic e^+e^- scattering (γZ interference) measured at LEP and SLC. The resulting solution is then [Sch97] (figure 3.7)

$$g_V = -0.035 \pm 0.017 \qquad g_A = -0.503 \pm 0.017. \qquad (3.109)$$

This is in good agreement with GWS predictions (3.94) assuming $\sin^2 \theta_W = 0.23$.

3.4.2.3 $\nu_e e$ and $\bar{\nu}_e e$ scattering

Results on $\bar{\nu}_e e$ scattering rely on much smaller datasets. Using nuclear power plants as strong $\bar{\nu}_e$ sources cross sections of [Rei76]

$$\sigma(\bar{\nu}_e e) = (0.87 \pm 0.25) \times \sigma_0 \qquad 1.5 < E_e < 3.0 \text{ MeV} \qquad (3.110)$$
$$\sigma(\bar{\nu}_e e) = (1.70 \pm 0.44) \times \sigma_0 \qquad 3.0 < E_e < 4.5 \text{ MeV} \qquad (3.111)$$

were obtained, where σ_0 is the predicted integrated V–A cross section (3.103) folded with the corresponding antineutrino flux.

Elastic $\nu_e e(\bar{\nu}_e e)$ scattering was investigated by E225 at LAMPF [All93]. Using muon-decay at rest, resulting in an average neutrino energy of $\langle E_\nu \rangle = 31.7$ MeV, 236 events were observed giving a cross section of

$$\sigma(\nu_e e) = (3.18 \pm 0.56) \times 10^{-43} \text{ cm}^2. \qquad (3.112)$$

By using $\langle E_\nu \rangle = 31.7$ MeV and the GWS prediction

$$\sigma(\nu_e e) = \sigma_0(\tfrac{3}{4} + 3\sin^2\theta_W + 4\sin^4\theta_W) = 9.49 \times 10^{-42} \text{ cm}^2 \frac{E_\nu}{\text{GeV}} \qquad (3.113)$$

and $g_V, g_A = 0$ in (3.106) these are in good agreement. The interference term was determined to be

$$I(\nu_e e) = (-2.91 \pm 0.57) \times 10^{-43} \text{cm}^2 = (-1.60 \pm 0.32)\sigma_0. \qquad (3.114)$$

A new measurement was performed by LSND (see chapter 8) resulting in [Aue01]

$$\sigma(\nu_e e) = [10.1 \pm 1.1(\text{stat.}) \pm 1.0(\text{sys.})] \times 10^{-45} \frac{E_\nu}{\text{MeV}} \qquad (3.115)$$

also in good agreement with E225 and the GWS prediction.

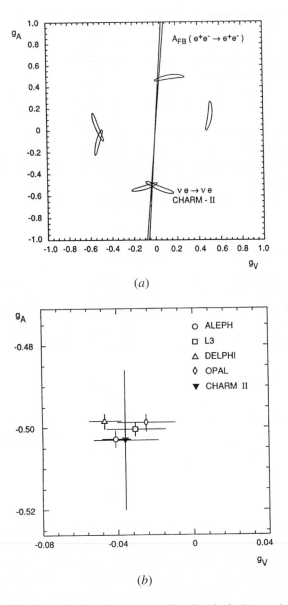

Figure 3.7. (*a*) Allowed regions (90% CL) of combinations in the (g_V, g_A) plane obtained with the CHARM-II data. Only statistical errors are considered. The small straight areas are the regions allowed by forward–backward asymmetry measurements in elastic e^+e^- scattering. Together they select a single solution consistent with $g_A = -\frac{1}{2}$. (*b*) Solution of the ambiguities. Together with the four LEP experiments a unique solution can be found. They are shown together with the CHARM-II result (from [Vil94]).

3.4.2.4   Neutrino tridents

A chance to observe interference for the second generation is given by neutrino trident production (using ν_μ beams), the generation of a lepton pair in the Coulomb field of a nucleus

$$\nu_\mu N \rightarrow \nu_\mu \ell^+ \ell^- N. \tag{3.116}$$

A reduction in the cross section of about 40% is predicted in the case of interference with respect to pure (V–A) interactions. Searches are done with high-energy neutrino beams (see chapter 4) for events with low hadronic energy E_{had} and small invariant masses of the $\ell^+\ell^-$ pair. Trident events were observed in several experiments [Gei90, Mis91]. Here also an interference effect could be observed.

3.4.3   Measurement of the Weinberg angle

One fundamental parameter of the GWS model is the Weinberg angle $\sin^2 \theta_W$. In the language of higher-order terms, the definition for $\sin^2 \theta_W$ has to be done very carefully [PDG00] because radiative corrections modify the mass and charge on different energy scales (see chapter 5). The most popular ones are the on-shell and $\overline{\text{MS}}$ definitions (see [PDG00]). The on-shell definition relies on the tree level formula

$$\sin^2 \theta_W = 1 - \frac{m_W^2}{m_Z^2} \tag{3.117}$$

obtained by dividing (3.63) and (3.64) so that it is also valid for the renormalized $\sin^2 \theta_W$ in all orders of perturbation theory. The modified minimal subtraction $\overline{\text{MS}}$ scheme (see [Lea96] for details) uses (see (3.41))

$$\sin^2 \theta_W(\mu) = \frac{g'^2(\mu)}{g^2(\mu) + g'^2(\mu)} \tag{3.118}$$

where the coupling constants are defined by modified minimal subtraction and the scale chosen, μm_Z, is convenient for electroweak processes.

The Weinberg angle can be measured in various ways. The determinations of the coupling constants g_V and g_A mentioned in (3.94) provide a way of determining $\sin^2 \theta_W$. Another possibility is νN scattering (for more details, see chapter 4). Here are measured the NC *versus* CC ratios (see (4.121) and (4.122)), given by

$$R_\nu = \frac{\sigma_{NC}(\nu N)}{\sigma_{CC}(\nu N)} = \frac{1}{2} - \sin^2 \theta_W + \frac{20}{27} \sin^4 \theta_W \tag{3.119}$$

$$R_{\bar{\nu}} = \frac{\sigma_{NC}(\bar{\nu} N)}{\sigma_{CC}(\bar{\nu} N)} = \frac{1}{2} - \sin^2 \theta_W + \frac{20}{9} \sin^4 \theta_W. \tag{3.120}$$

A low-energy measurement is the observation of atomic parity violation in heavy atoms [Mas95, Blu95]. Their measuring quantity is the weak charge given by

$$Q_W \approx Z(1 - 4\sin^2\theta_W) - N \qquad (3.121)$$

with Z being the number of protons and N the number of nucleons in the atom. However, the most precise measurements come from observables using the Z-pole, especially asymmetry measurements. These include the left–right asymmetry

$$A_{LR} = \frac{\sigma_L - \sigma_R}{\sigma_L + \sigma_R} \qquad (3.122)$$

with $\sigma_L (\sigma_R)$ being the cross section for left(right)-handed incident electrons. This has been measured precisely by SLD at SLAC. The left–right forward–backward asymmetry is defined as

$$A_{LR}^{FB}(f) = \frac{\sigma_{LF}^f - \sigma_{LB}^f - \sigma_{RF}^f + \sigma_{RB}^f}{\sigma_{LF}^f + \sigma_{LB}^f + \sigma_{RF}^f + \sigma_{RB}^f} = \frac{3}{4}A_f \qquad (3.123)$$

where, e.g., σ_{LF}^f is the cross section for a left-handed incident electron to produce a fermion f in the forward hemisphere. The Weinberg angle enters because A_f depends only on the couplings g_V and g_A:

$$A_f = \frac{2g_V g_A}{g_V^2 g_A^2}. \qquad (3.124)$$

A compilation of $\sin^2\theta_W$ measurements is shown in table 3.4.

3.4.4   Measurement of the gauge boson masses m_W and m_Z

The accurate determination of the mass of the Z-boson was one of the major goals of LEP and SLC. The Z^0 shows up as a resonance in the cross section in e^+e^- scattering (figure 3.8). With an accumulation of several million Z^0-bosons, the current world average is given by [PDG00]

$$m_Z = 91.1874 \pm 0.0021 \text{ GeV}. \qquad (3.125)$$

Until 1996 the determination of the W-boson mass was the domain of $p\bar{p}$ machines like the SppS at CERN ($\sqrt{s} = 630$ GeV) and the Tevatron at Fermilab ($\sqrt{s} = 1.8$ TeV). The combined limit of the results is given by

$$m_W = 80.452 \pm 0.091 \text{ GeV}. \qquad (3.126)$$

With the start of LEP2, independent measurements at e^+e^- colliders became possible by W-pair production. Two effects could be used for an m_W measurement: the cross sections near the threshold of W-pair production

Table 3.4. Compilation of measurements of the Weinberg angle $\sin^2\theta_W$ (on-shell and in the $\overline{\text{MS}}$ scheme from various observables assuming global best-fit values (for $m_H = m_Z$) $m_t = 173 \pm 4$ GeV and $\alpha_S = 0.1214 \pm 0.0031$ (after [PDG00]).

Reaction	$\sin^2\theta_W$ (on-shell)	$\sin^2\theta_W$ ($\overline{MS}$)
m_Z	0.2231 ± 0.0005	0.2313 ± 0.0002
m_W	0.2228 ± 0.0006	0.2310 ± 0.0005
A_{FB}	0.2225 ± 0.0007	0.2307 ± 0.0006
LEP asymmetries	0.2235 ± 0.0004	0.2317 ± 0.0003
A_{LR}	0.2220 ± 0.0005	0.2302 ± 0.0004
DIS (isoscalar)	0.226 ± 0.004	0.234 ± 0.004
$\nu_\mu(\bar{\nu}_\mu)\text{p} \to \nu_\mu(\bar{\nu}_\mu)\text{p}$	0.203 ± 0.032	0.211 ± 0.032
$\nu_\mu(\bar{\nu}_\mu)\text{e} \to \nu_\mu(\bar{\nu}_\mu)\text{e}$	0.221 ± 0.008	0.229 ± 0.008
Atomic parity violation	0.220 ± 0.003	0.228 ± 0.003
SLAC eD	0.213 ± 0.019	0.222 ± 0.018
All data	0.2230 ± 0.0004	$0.231\,24 \pm 0.000\,17$

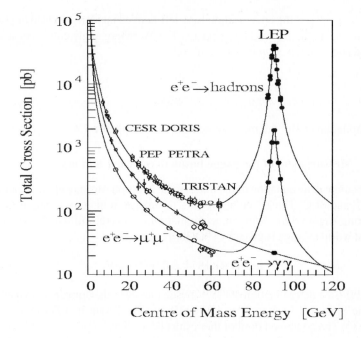

Figure 3.8. Cross sections ($\text{e}^+\text{e}^- \to$ hadrons), ($\text{e}^+\text{e}^- \to \mu^+\mu^-$)($\text{e}^+\text{e}^- \to \gamma\gamma$) as a function of the centre-of-mass energy. The sharp spike at the Z^0 resonance is clearly visible.

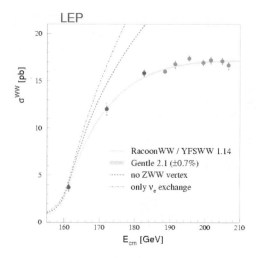

Figure 3.9. Measurements of the cross section ($e^+e^- \rightarrow W^+W^-$) as a function of the centre-of-mass energy obtained while LEP2 was running. The threshold behaviour can be used to determine the W-mass and the behaviour shows the effect of self-coupling of the gauge bosons. Scenarios with no ZWW vertex and pure ν_e exchange are clearly excluded. Predictions of Monte Carlo simulations are shown as lines (with kind permission of LEP EW working group).

(figure 3.9) and the shape of the invariant mass distribution of the W-pair. The combined LEP value is [Gle00]:

$$m_W = 80.350 \pm 0.056 \, \text{GeV} \tag{3.127}$$

resulting in a world average of

$$m_W = 80.398 \pm 0.041 \, \text{GeV}. \tag{3.128}$$

For a detailed discussion see [Gle00].

3.4.5 Search for the Higgs boson

The only particle of the standard model not yet discovered is the Higgs boson. However, information on the Higgs mass can be obtained by electroweak precision measurements due to its contribution to radiative corrections (figure 3.10). A best-fit value of 90^{+55}_{-47} GeV could be determined as shown in figure 3.11 or an upper limit of 200 GeV with 95% CL. In the late phase of LEP2 a limit of $m_H > 114.4$ GeV could be obtained. Here, the dominant production mechanism at LEP is 'Higgs-strahlung':

$$e^+ + e^- \rightarrow Z^* \rightarrow Z + H^0 \tag{3.129}$$

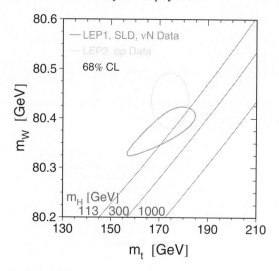

Figure 3.10. Combination of experimental data in the m_W and m_t plane from precision measurements. The Higgs boson mass enters through radiative corrections, whose contributions are shown for three representative masses as straight lines. As can be seen the experiments prefer a rather light Higgs (with kind permission of LEP EW working group).

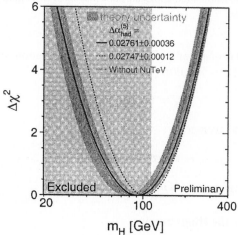

Figure 3.11. χ^2 distribution of global fits on electroweak data as a function of the Higgs mass. A best-fit value of $m_H = 90$ GeV results, already in contradiction with the direct experimental lower limit (with kind permission of LEP EW working group).

resulting in a Higgs and a Z-boson [Gun90]. The dominant signatures are two b-jets coming from the Higgs decay and two jets from the Z (60% of total decays) or missing energy because of a Z $\rightarrow$ $\nu\bar{\nu}$ decay (18% of total decays). Further investigations will be made at the Tevatron RunII started recently at

Fermilab and at the LHC at CERN starting in 2007. Again for light Higgses (m_H < 160 GeV) the H $\rightarrow$ b$\bar{\text{b}}$ decay will be dominant, which for heavier Higgses changes into gauge boson decays H $\rightarrow$ WW, ZZ. From experimental considerations, at LHC the light Higgs search might be conducted in the decay channel H $\rightarrow$ $\gamma\gamma$ because of the large background in other channels. The Higgs sector gets more complicated as more Higgs doublets are involved as in the minimal supersymmetric standard model discussed in chapter 5.

Chapter 4

Neutrinos as a probe of nuclear structure

Before exploring the intrinsic properties of neutrinos, we want to discuss how neutrinos can be used for measuring other important physical quantities. They allow a precise determination of various electroweak parameters and can be used to probe the structure of the nucleon via neutrino–nucleon scattering, as they are a special case of lepton–nucleon scattering. To perform systematic studies with enough statistics, artificial neutrino beams have to be created. Such sources are basically high-energy particle accelerators. Further information on this subject can be found in [Com83, Bil94, Lea96, Sch97, Con98, Per00].

4.1 Neutrino beams

Because of the small cross section of neutrino interactions, to gain a reasonable event rate R (events per second), the target mass of the detector (expressed in numbers of nucleons in the target N_T) has to be quite large and the intensity I (ν per $cm^{-2}s^{-1}$) of the beam should be as high as possible. An estimate of the expected event rate is then given by

$$R = N_T \sigma I \qquad (4.1)$$

with σ being the appropriate cross section (cm^2). Let us focus on the beams first.

4.1.1 Conventional beams

Neutrino beams have to be produced as secondary beams, because no direct, strongly focused, high-energy neutrino source is available. A schematic layout of a typical neutrino beam-line is shown in figure 4.1. A proton synchrotron delivers bunches of high-energy protons (of the order 10^{13} protons per bunch) on a fixed target (therefore the commonly used luminosity unit is protons on target—pot), resulting in a high yield of secondary mesons, predominantly pions and kaons. By using beam optical devices (dipole or quadrupole magnets or magnetic horns) secondaries of a certain charge sign are focused into a long decay tunnel. There,

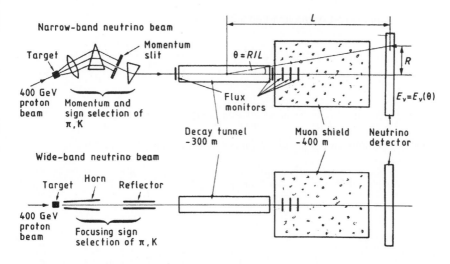

Figure 4.1. Schematic arrangements of neutrino beams: top, narrow-band beams; bottom, wide-band beams (from [Eis86]).

the secondaries decay mostly via the reaction (assuming focusing of positive secondaries)

$$M^+ \rightarrow \mu^+ + \nu_\mu \qquad (M \equiv \pi, K) \qquad (4.2)$$

with a branching ratio of 100% for pions and 63.5% for kaons. As can be seen, a beam dominantly of ν_μ is produced (or, accordingly, a $\bar{\nu}_\mu$ beam if the oppositely signed charged mesons are focused). Only a fraction of the produced mesons decay in the tunnel with length L_D. The probability P for decay is given as

$$P = 1 - \exp(-L_D/L_0) \qquad (4.3)$$

with

$$L_0 = \beta c \times \gamma \tau_M = \frac{p_M}{m_M} \times c\tau_M = \begin{cases} 55.9 \, \text{m} & \times & p_\pi/\text{GeV} \\ 7.51 \, \text{m} & \times & p_K/\text{GeV}. \end{cases} \qquad (4.4)$$

For $p_M = 200$ GeV and $L_D = 300$ m this implies: $L_0 = 11.2$ km, $P = 0.026$ (pions) and $L_0 = 1.50$ km, $P = 0.181$ (kaons). These probabilities have to be multiplied with the muonic branching ratios given earlier to get the number of neutrinos. To get a certain fraction of meson decays, L_D must increase proportional to momentum (energy) because of relativistic time dilation. At the end of the decay tunnel there is a long muon shield, to absorb the remaining πs and Ks via nuclear reactions and stop the muons by ionization and radiation losses. The experiments are located after this shielding. The neutrino spectrum can be determined from the kinematics of the two-body decay of the mesons. Energy (E_ν) and angle ($\cos\theta_\nu$) in the laboratory frame are related to the same quantities in the rest frame (marked with $*$) by

$$E_\nu = \bar\gamma E_\nu^*(1 + \bar\beta \cos\theta_\nu^*) \qquad \cos\theta_\nu = \frac{\cos\theta_\nu^* + \bar\beta}{1 + \bar\beta \cos\theta_\nu^*} \qquad (4.5)$$

with

$$\bar\beta = \frac{p_M}{E_M} \qquad \bar\gamma = \frac{E_M}{m_M} \qquad \text{and} \qquad E_\nu^* = \frac{m_M^2 - m_\mu^2}{2m_M}. \qquad (4.6)$$

The two extreme values are given for $\cos\theta_\nu^* = \pm 1$ and result in

$$E_\nu^{\min} = \frac{m_M^2 - m_\mu^2}{2m_M^2}(E_M - p_M) \approx \frac{m_M^2 - m_\mu^2}{4E_M} \approx 0 \qquad (4.7)$$

and

$$E_\nu^{\max} = \frac{m_M^2 - m_\mu^2}{2m_M^2}(E_M + p_M) \approx \left(1 - \frac{m_\mu^2}{m_M^2}\right) \times E_M = \begin{cases} 0.427 & \times & E_\pi \\ 0.954 & \times & E_K \end{cases}$$
$$(4.8)$$

using $E_M \gg m_M$. With a meson energy spectrum $\phi_M(E_M)$ between $E_M^{\min}$ and $E_M^{\max}$ the resulting neutrino spectrum and flux is given by

$$\phi_\nu(E_\nu) \propto \int_{E_M^{\min}}^{E_M^{\max}} dE_M\, \phi_M(E_M)\frac{1}{p_M}\left(\frac{m_M^2 - m_\mu^2}{m_M^2}E_M - E_\nu\right). \qquad (4.9)$$

Using (4.5) the following relation in the laboratory frame holds:

$$E_\nu(\theta_\nu) = \frac{m_M^2 - m_\mu^2}{2(E_M - p_M\cos\theta_\nu)} \approx E_M\frac{m_M^2 - m_\mu^2}{m_M^2 + E_M^2\theta_\nu^2} \approx E_\nu^{\max}\frac{1}{1 + \bar\gamma_M^2\theta_\nu^2}. \qquad (4.10)$$

As can be seen for typical configurations (the radius R of the detector much smaller than distance L to the source, meaning $\theta_\nu < R/L$) only the high-energy part of the neutrino spectrum hits the detector ($E_\nu(0) = E_{\nu\max}$). Two types of beams can be produced—different physics goals require the corresponding beam optical system. One is a narrow-band beam (NBB) using momentum-selected secondaries, the other one is a wide-band beam (WBB) having a much higher intensity.

A realization of a different type of beam with lower neutrino energies and, therefore, lower proton beam energies is based on meson decay at rest within the proton target leading to isotropic neutrino emission. This will be discussed in chapter 8.

4.1.1.1 Narrow-band beams (NBB)

An NBB collects the secondaries of interest coming from the target via quadrupole magnets. By using additional dipoles, it selects and focuses particles

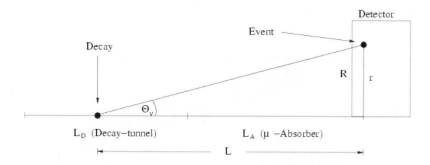

Figure 4.2. Geometric relation in a NBB between the position of meson decay (distance from the detector), decay angle θ_ν and radial position of the event in the detector.

of a certain charge and momentum range (typically $\Delta p_M/p_M \approx 5\%$) that are leaving this area into the decay tunnel as a parallel secondary beam. Because of these two features (parallel and momentum selected), there is a unique relation between the radial distance with respect to the beam axis of a neutrino event in a detector and the neutrino energy for a given decay length (figure 4.2). There is only an ambiguity because two mesons (π,K) are present in the beam. Furthermore, the decay length is distributed along the decay tunnel, which results in a smearing into two bands. This is shown in figure 4.3 for data obtained with the CDHSW experiment [Ber87]. For this reason NBBs are sometimes called dichromatic beams.

The main advantages of such a beam are a rather flat neutrino flux spectrum, the possibility of estimating E_ν from the radial position in the detector and a rather small contamination from other neutrino species. However, the intensity is orders of magnitude smaller than the one obtained in wide-band beams. A schematic energy spectrum from a NBB is shown in figure 4.4.

4.1.1.2 Wide-band beams (WBB)

In a WBB the dipoles and quadrupoles are replaced by a system of so called magnetic horns. They consist of two horn-like conductors which are pulsed with high currents synchronously with the accelerator pulse. This generates a magnetic field in the form of concentric circles around the beam axis, which focuses particles with the appropriate charge towards the beam axis. To increase this effect, a second horn, called the reflector, is often installed behind. Here, the prediction of the absolute neutrino energy spectrum and composition is a difficult task. Detailed Monte Carlo simulations are required to simulate the whole chain from meson production at the target towards the neutrino flux at a detector. Instrumentation along the beam-line helps to determine accurate input parameters for the simulation. Particularly in the case of West Area Neutrino

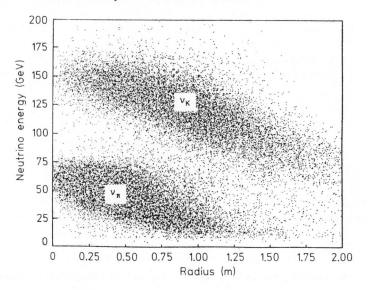

Figure 4.3. Scatter plot of E_ν with respect to radial event position for CC events as obtained with the CDHS detector at the CERN SPS. The dichromatic structure of the narrow-band beam (NBB) with $E_M = 160$ GeV is clearly visible and shows the neutrino events coming from pion and kaon decay (from [Ber87]).

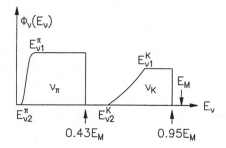

Figure 4.4. Schematic energy spectrum of neutrinos in a NBB hitting a detector. The contributions from pions and kaons are clearly separated.

Facility (WANF) at CERN, the SPY experiment was performed to measure the secondary particle yield [Amb99], due to insufficient data from previous experiments [Ath80]. While in the NBB, because of the correlation of radial distance and neutrino energy, a reasonable estimation of E_ν can be performed, in a WBB this is more difficult. In addition to beam-line simulations the observed event rates and distributions can be used to extract the neutrino flux by using known cross sections ('empirical parametrization', see [Con98]). Furthermore, the beam can be polluted by other neutrino flavours like ν_e coming from decays, e.g. the K_{e3}-decay

$$K^{\pm} \to \pi^0 e^{\pm} \overset{(-)}{\nu_e} \tag{4.11}$$

with a branching ratio of 4.8%, muon decays and decays from mesons produced in the absorber.

4.1.2 ν_τ beams

A completely different beam was necessary for the DONUT (E872) experiment at Fermilab (FNAL) [Kod01]. Their goal was to detect ν_τ CC reactions

$$\nu_\tau + N \to \tau^- + X \tag{4.12}$$

and, therefore, they needed a ν_τ beam. This was achieved by placing the detector only 36 m behind the 1 m long tungsten target, hit by a 800 GeV proton beam. The ν_τ beam results from the decay of the produced D_S-mesons via

$$D_S \to \tau \bar{\nu}_\tau (BR = 6.4 \pm 1.5\%) \quad \text{and} \quad \tau \to \nu_\tau + X. \tag{4.13}$$

They observed five event candidates for the process (4.13). New beam concepts might be realized in the future.

4.1.3 Neutrino beams from muon decay

Currently three new neutrino beams are considered for future accelerators (see also chapter 8). Among them is a high intensity beam as just described but with a lower energy of only about 500 MeV ('Superbeam') significantly reducing the ν_e component. In addition a pure beam of ν_e is proposed by accelerating β-unstable isotopes ('beta beam') to a few hundred MeV [Zuc01]. A third concept considers muon decay as a source for well-defined neutrino beams in the form of a muon storage ring ('neutrino factory') [Gee98]. Instead of using the neutrinos from the decay of secondaries, the idea is to collect the associated muons and put them, after some acceleration, into a storage ring. The decay

$$\mu^+ \to e^+ \nu_e \bar{\nu}_\mu \tag{4.14}$$

is theoretically and experimentally well understood and, therefore, the energy spectrum as well as the composition of the beam is accurately known. The neutrino spectrum from μ^+ decay is given in the muon rest frame by

$$\frac{d^2N}{dx\, d\Omega} = \frac{1}{4\pi}(2x^2(3 - 2x) - P_\mu 2x^2(1 - 2x)\cos\theta) \quad \text{(for } \bar{\nu}_\mu \text{ and e)} \tag{4.15}$$

$$\frac{d^2N}{dx\, d\Omega} = \frac{1}{4\pi}(12x^2(1 - x) - P_\mu 12x^2(1 - x)\cos\theta) \quad \text{(for } \nu_e) \tag{4.16}$$

with $x = 2E_\nu/m_\mu$, P_μ the average muon polarization along the muon beam direction and θ the angle between the neutrino momentum vector and the muon spin direction. The spectrum for unpolarized muons is shown in figure 4.27. For a detector at a large distance, the spectrum looks identical but the energy scale is multiplied by a Lorentz boost factor $2E_\mu/m_\mu$. The ν_e plays a special role because it is always emitted in the opposite direction to the muon polarization. Therefore 100% polarized muons with the right sign could produce a beam free of ν_e. Opposite-flavour beams are produced if the μ^- decay is used for the beam, resulting in a change of sign in (4.15) and (4.16).

4.2   Neutrino detectors

A second important component is the detector. The small cross sections involved in neutrino physics require detectors of large size and mass to get a reasonable event rate. Several requirements should be fulfilled by such a detector:

- identification of a charged lepton to distinguish CC and NC events,
- measurement of energy and the scattering angle of the charged lepton to determine the kinematic variables of the event,
- measurement of the total hadronic energy, e.g. to reconstruct E_ν,
- identification of single secondary hadrons and their momenta to investigate in detail the hadronic final state,
- detection of short living particles and
- use of different target materials.

Some of these requirements are exclusive of each other and there is no single detector to fulfil all of them. The actual design depends on the physics questions under study. In the following three examples of the most common detector concepts are discussed. Information about other types of experiments can be found for the bubble chamber BEBC in [Bar83], and the fine-grained calorimeters CHARM and CHARM-II in [Jon82, Gei93, Pan95].

4.2.1   CDHS

The CERN–Dortmund–Heidelberg–Saclay (CDHS) experiment (figure 4.5) was a heavy (1150 t) and about 22 m long sampling calorimeter, serving as detector and target [Hol78]. It consisted of 21 modules, made of iron plates (3.75 m diameter) and planes of plastic scintillators (3.6 m by 3.6 m). The iron served as the target as well as initiating a hadronic shower (the typical size of a shower is about 1 m in length and 25 cm in radius). Part of the shower energy is converted into light within the scintillators which is then read out by photomultipliers. This allows the hadronic energy E_{had} to be reconstructed. In between were hexagonal drift chambers for measuring muon tracks. The iron plates were toroidally magnetized by a field of 1.6 T, which allowed the muon momenta to be measured via their radius of curvature. Having measured the muon and hadronic energy, the visible or neutrino energy could be determined by

$$E_\nu \approx E_{\text{Vis}} = E_\mu + E_{\text{had}}. \tag{4.17}$$

Figure 4.5. Photograph of the CDHS detector at CERN (with kind permission of CERN).

The CCFR [Sak90] and, later on, the NuTeV [Bol90] experiment worked in a similar fashion, with the exception that the complete muon spectrometer followed after the calorimeter. The result was a smaller acceptance for muons but a better angular resolution. The MINOS experiment will work in a similar way (see chapter 8).

4.2.2 NOMAD

NOMAD (Neutrino Oscillation MAgnetic Detector) [Alt98] at the WANF at CERN used drift chambers as the target and tracking medium, with the chamber walls as interaction targets and the chambers for precise particle tracking. They were optimized to fulfil the two contradictory requirements of being as heavy as possible to obtain a large number of neutrino interactions and being as light as possible to reduce multiple scattering. In total there were 44 chambers with a fiducial mass of 2.7 t and an active area of 2.6 m × 2.6 m. They were followed by a transition radiation detector (TRD) for e/π separation (a π-rejection of more than 10^3 for 90% electron efficiency was achieved). Further electron identification was done with a preshower detector and an electromagnetic calorimeter consisting of 875 lead-glass Cerenkov counters. Behind that, a hadronic calorimeter in the form of an iron-scintillator sampling calorimeter and a set of 10 drift chambers for muon identification followed. The detector was located within a magnetic field of 0.4 T, perpendicular to the beam axis (figure 4.6) for momentum determination. In front of the drift chambers another iron-scintillator calorimeter of about 20 t target mass was installed, working as the detectors described in section 4.2.1.

The idea of having a very light target follows the detection principle for taking data as in bubble chamber experiments like Gargamelle (see chapter 1) and BEBC, namely to measure all tracks precisely. The planned ICARUS experiment is going to work in the same spirit (see chapter 8).

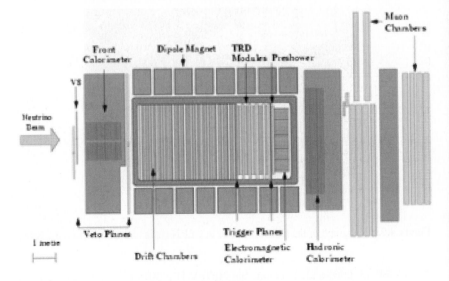

Figure 4.6. Schematic view of the NOMAD detector at CERN (from [Alt98]).

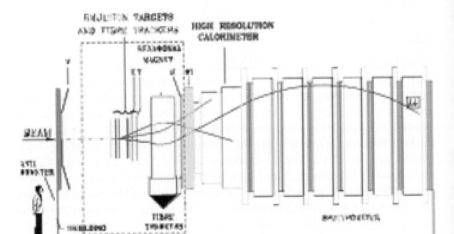

Figure 4.7. Schematic view of the CHORUS detector at CERN.

4.2.3 CHORUS

A second detector just in front of NOMAD was CHORUS (CERN Hybrid Oscillation Search Apparatus) (figure 4.7) [Esk97]. Here, the main active target consisted of four blocks of nuclear emulsions (in total a mass of 770 kg), with a surface area of 1.42 m × 1.44 m and a thickness of 2.8 cm each. The thickness of a single emulsion sheet was 350 μm. The main advantage of emulsions is the excellent spatial resolution of a few μm (necesssary to fulfil the idea of detecting τ-leptons created via reaction (4.12)), but they also make very short tracks visible and show possible decays as kinks in the track. For timing purposes and for extrapolating the tracks back into the emulsions, a scintillating fibre tracker was interleaved, consisting of 500 μm diameter fibres 2.3 m in length. Behind the target complex followed a hexagonal spectrometer magnet (0.12 T) for momentum measurement, a high-resolution spaghetti calorimeter for measuring hadronic showers and a muon spectrometer in the form of toroidal modules made of magnetized iron which are interleaved with drift chambers, limited streamer tubes and scintillators. After the run period the emulsions are scanned with high-speed CCD microscopes. Emulsions are also used in the DONUT and OPERA experiments.

Having discussed neutrino beams and detectors, we now proceed to experimental results.

4.3 Total cross section for neutrino–nucleon scattering

The total neutrino and antineutrino cross sections for νN scattering have been measured in a large number of experiments [Mac84, Ber87, All88]. They can proceed (assuming ν_μ beams) via charged currents (CC) involving W-exchange and neutral current (NC) processes with Z-exchange

$$\nu_\mu N \to \mu^- X \qquad \bar{\nu}_\mu N \to \mu^+ X \quad \text{(CC)} \tag{4.18}$$

$$\nu_\mu N \to \nu_\mu X \qquad \bar{\nu}_\mu N \to \bar{\nu}_\mu X \quad \text{(NC)} \tag{4.19}$$

with N $\equiv$ p, n or an isoscalar target (average of neutrons and protons) and X as the hadronic final state. The total CC neutrino–nucleon cross section on isoscalar targets[1] as a function of E_ν was determined dominantly by CCFR and CDHSW. Both were using NBB and an iron target. Except for small deviations at low energies ($E_\nu < 30$ GeV) a linear rise in the cross section with E_ν was observed (figure 4.8). If we include the data from the CHARM experiment, the current world averages are given as [Con98]

$$\sigma(\nu N) = (0.677 \pm 0.014) \times 10^{-38} \text{ cm}^2 \times E_\nu/(\text{GeV}) \tag{4.20}$$

$$\sigma(\bar{\nu} N) = (0.334 \pm 0.008) \times 10^{-38} \text{ cm}^2 \times E_\nu/(\text{GeV}). \tag{4.21}$$

The study of CC events at the HERA collider allowed a measurement equivalent to a fixed target beam energy of 50 TeV, where even the propagator effect becomes visible (figure 4.9). The linear rise of the cross section with E_ν as observed in

---

[1] A correction factor has to be applied for heavy nuclei because of a neutron excess there. For Fe it was determined to be -2.5% for $\sigma(\nu N)$ and $+2.3\%$ for $\sigma(\bar{\nu} N)$.

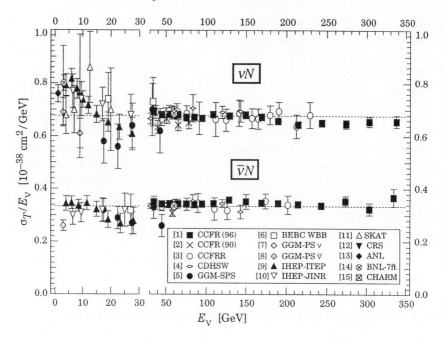

Figure 4.8. Compilation of σ/E_ν in νN and $\bar\nu$N scattering as a function of E_ν obtained by several experiments (from [PDG02]).

hard νN scattering is direct evidence for scattering on point-like objects within the nucleon. This assumption is the basis of the quark–parton–model (QPM, see section 4.8), which predicts that deep-inelastic νN scattering can be seen as an incoherent superposition of quasi-elastic neutrino–(anti)quark scattering. At low energies ($E_\nu < 30$ GeV), the ratio $R = \sigma(\nu N)/\sigma(\bar\nu N) \approx 3$ agrees with the simple QPM prediction without sea-quark contributions. That R is about 2 at higher energies and is a direct hint for their contribution (see section 4.9 for more details). The total cross section for CC reactions on protons and neutrons was measured, for example, with bubble chambers like BEBC, filled with liquid hydrogen (WA21) and deuterium (WA25). The results are [All84, Ade86]:

$$\sigma(\nu p) = (0.474 \pm 0.030)^{-38} \text{ cm}^2 \times E_\nu/(\text{GeV}) \qquad (4.22)$$

$$\sigma(\bar\nu p) = (0.500 \pm 0.032) \times 10^{-38} \text{ cm}^2 \times E_\nu/(\text{GeV}) \qquad (4.23)$$

$$\sigma(\nu n) = (0.84 \pm 0.07) \times 10^{-38} \text{ cm}^2 \times E_\nu/(\text{GeV}) \qquad (4.24)$$

$$\sigma(\bar\nu n) = (0.22 \pm 0.02) \times 10^{-38} \text{ cm}^2 \times E_\nu/(\text{GeV}) \qquad (4.25)$$

Averaging the protons and neutrons results in good agreement with (4.20) and (4.21). To obtain more information about the structure of the nucleon, we have to look at deep inelastic scattering(DIS).

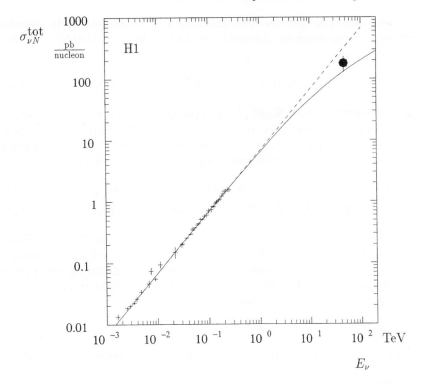

Figure 4.9. Compilation of $\sigma(E_\nu)$ from νN scattering (crosses) and from the H1 experiment at DESY. The dotted curve corresponds to a prediction without a W-propagator ($m_W = \infty$), the full line is a prediction with W-propagator ($m_W = 80$ GeV) (from [Ahm94].

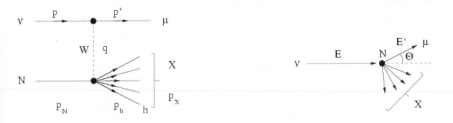

Figure 4.10. Kinematics of the CC reaction $\nu_\mu N \rightarrow \mu X$ via W-exchange: left, the underlying Feynman graph; right, variables in the laboratory system.

4.4 Kinematics of deep inelastic scattering

In deep inelastic lepton–nucleon scattering, leptons are used as point-like probes of nucleon structure. Reactions, especially those focusing on weak interaction properties, are done with neutrinos according to (4.18) and (4.19). In a similar fashion the electromagnetic structure is explored via deep inelastic scattering with charged leptons

$$e^{\pm} + N \rightarrow e^{\pm} + X \qquad \mu^{\pm} + N \rightarrow \mu^{\pm} + X. \tag{4.26}$$

Let us discuss the kinematics of CC interactions (4.18) on fixed targets as shown in figure 4.10. The 4-momenta, p, p', $q = p - p'$, p_N, p_X and p_h of the incoming ν, the outgoing μ, the exchanged W, the incoming nucleon N, outgoing hadronic final state X and of an outgoing hadron h are given in the laboratory frame as

$$p = (E_\nu, p_\nu) \qquad p' = (E_\mu, p_\mu) \qquad q = (\nu, q) \tag{4.27}$$

$$p_N = (M, 0) \qquad p_X = (E_X, p_X) \qquad p_h = (E_h, p_h) \tag{4.28}$$

with M being the nucleon mass. Measured observables in the laboratory frame are typically the energy $E' = E_\mu$ and the scattering angle $\theta = \theta_\mu$ of the outgoing muon (in analogy with the outgoing lepton in eN/μN scattering) for a given neutrino energy $E = E_\nu$. These two quantities can be used to measure several important kinematic event variables.

- The total centre-of-mass energy $\sqrt{s}$:

$$s = (p + p_N)^2 = 2ME + M^2 \approx 2ME. \tag{4.29}$$

- The (negative) 4-momentum transfer:

$$\begin{aligned} Q^2 &= -q^2 = -(p - p')^2 = -(E - E')^2 + (p - p')^2 \\ &= 4EE' \sin^2 \tfrac{1}{2}\theta > 0 \end{aligned} \tag{4.30}$$

- The energy transfer in the laboratory frame:

$$\nu = \frac{q \times p_N}{M} = E - E' = E_X - M. \tag{4.31}$$

- The Bjorken scaling variable x:

$$x = \frac{-q^2}{2q \times p_N} = \frac{Q^2}{2M\nu} \qquad \text{with } 0 \le x \le 1. \tag{4.32}$$

- The relative energy transfer (inelasticity) y (often called the Bjorken y)

$$y = \frac{q \times p_N}{p \times p_N} = \frac{\nu}{E} = 1 - \frac{E'}{E} = \frac{Q^2}{2MEx}. \tag{4.33}$$

- The total energy of the outgoing hadrons in their centre-of-mass frame

$$W^2 = E_X^2 - p_X^2 = (E - E' + M)^2 - (p - p')^2 = -Q^2 + 2M\nu + M^2. \tag{4.34}$$

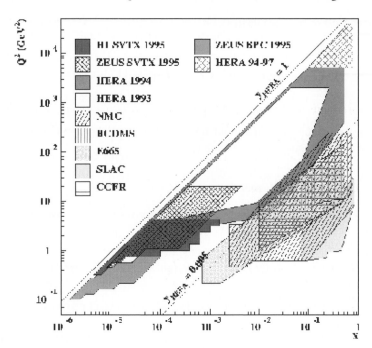

Figure 4.11. Allowed kinematic regions in the (x, Q^2) plane that can be explored by various experiments.

Equations (4.29) and (4.33) can be combined to give the useful relation

$$xy = \frac{Q^2}{2ME} = \frac{Q^2}{s - M^2}. \tag{4.35}$$

At a fixed energy E, inelastic reactions can, therefore, be characterized by two variables such as (E', θ), (Q^2, ν), (x, Q^2) or (x, y). For quasi-elastic reactions $(x = 1)$, one variable $(E', \theta, Q^2$ or $\nu)$ is sufficient. Figure 4.11 shows the parameter space covered by current experiments. As can be seen, the ep collider HERA at DESY ($\sqrt{s} \approx 320$ GeV) is able to probe a unique region in parameter space, because its centre-of-mass energy would correspond to 50 TeV beam energy in fixed-target experiments.

4.5 Quasi-elastic neutrino–nucleon scattering

Quasi-elastic (QEL) reactions are characterized by the fact that the nucleon does not break up and, therefore, $x \approx 1$. Reactions of the form $\nu + n \rightarrow \ell^- + p$ are quasi-elastic or, being more specific, in QEL $\nu_\mu N$ scattering the following reactions have to be considered:

$$\nu_\mu + n \rightarrow \mu^- + p \qquad E_{\text{Thr}} = 110 \text{ MeV} \qquad (4.36)$$

$$\bar{\nu}_\mu + p \rightarrow \mu^+ + n \qquad E_{\text{Thr}} = 113 \text{ MeV} \qquad (4.37)$$

$$\overset{(-)}{\nu_\mu} + p \rightarrow \overset{(-)}{\nu_\mu} + p. \qquad (4.38)$$

Corresponding reactions also hold for ν_e. The quasi-elastic NC scattering on neutrons is, in practice, not measurable.

4.5.1 Quasi-elastic CC reactions

The most general matrix element in V–A theory for (4.36) is given by [Lle72, Com83, Str03]

$$ME = \frac{G_F}{\sqrt{2}} \times \bar{u}_\mu(p')\gamma_\alpha(1 - \gamma_5)u_\nu(p) \times \langle p(P')|J_\alpha^{CC}|n(P)\rangle \qquad (4.39)$$

with u_μ, u_ν as the leptonic spinors and the hadronic current given as

$$\langle p(P')|J_\alpha^{CC}|n(P)\rangle = \cos\theta_C \bar{u}_p(P')\Gamma_\alpha^{CC}(Q^2)u_n(P). \qquad (4.40)$$

p, p', P and P' are the 4-momenta of ν, μ, n, p and the term Γ_α^{CC} contains six *a priori* unknown complex form factors $F_S(Q^2)$, $F_P(Q^2)$, $F_V(Q^2)$, $F_A(Q^2)$, $F_T(Q^2)$, $F_M(Q^2)$ for the different couplings:

$$\Gamma_\alpha^{CC} = \gamma_\alpha F_V + \frac{i\sigma_{\alpha\beta}q_\beta}{2M}F_M + \frac{q_\alpha}{M}F_S + \left[\gamma_\alpha F_A + \frac{i\sigma_{\alpha\beta}q_\beta}{2M}F_T + \frac{q_\alpha}{M}F_P\right]\gamma_5$$

$$q = P' - P = p - p' \qquad Q^2 = -q^2 \qquad \sigma_{\alpha\beta} = \frac{1}{2i}(\gamma_\alpha\gamma_\beta - \gamma_\beta\gamma_\alpha). \ (4.41)$$

The terms associated with F_T and F_S are called second class currents and F_M corresponds to weak magnetism. Assuming T-invariance and charge symmetry, the scalar and tensor form factors F_T and F_S have to vanish. Furthermore, terms in cross sections containing pseudo-scalar interactions are always multiplied by m_μ^2 [Lle72] and can be neglected for high energies ($E_\nu \gg m_\mu$). Under these assumptions, (4.41) is shortened to

$$\Gamma_\alpha^{CC} = \gamma_\alpha(F_V - F_A\gamma_5) + \frac{i\sigma_{\alpha\beta}q_\beta}{2M}F_M \qquad (4.42)$$

containing vector and axial vector contributions as well as weak magnetism. Using the CVC hypothesis (see section 3.1), F_V and F_M can be related to the electromagnetic form factors (G_E, G_M) of the nucleons, appearing in the Rosenbluth formula for the differential cross section of elastic eN → eN (N = p, n) scattering via [Lea96]

$$F_V = \frac{G_E^V + \tau G_M^V}{1 + \tau} \tag{4.43}$$

$$F_M = \frac{G_M^V - \tau G_E^V}{1 + \tau} \tag{4.44}$$

with $\tau + Q^2/4M^2$. They have an experimentally determined dipole form given as

$$G_{E,M}(Q^2) = \frac{G_{E,M}(0)}{(1 + Q^2/M_V^2)^2} \qquad \text{with } M_V = 0.84 \text{ GeV} \tag{4.45}$$

with the normalization at $Q^2 = 0$:

$$G_E^p(0) = 1 \qquad G_E^n(0) = 1 \tag{4.46}$$

$$G_E^V(0) = 1 \qquad G_M^V(0) = \mu_p - \mu_n = 4.706 \tag{4.47}$$

with μ_p, μ_n as magnetic moments in units of the nuclear magneton. Assuming the same dipole structure for F_A and taking $F_A(0) = g_A/g_V = -1.2670 \pm 0.0030$ from neutron decay [PDG02], the only free parameter is M_A. It is measured in quasi-elastic νN scattering and has the average value of $M_A = (1.026 \pm 0.020)$ GeV [Ber02a] (figure 4.12). Recently new data from ep and eD scattering showed that (4.45) is only accurate to 10–20% and more sophisticated functions have to be used [Bos95, Bra02, Bud03]. An accurate understanding of the quasi-elastic regime is essential for newly planned neutrino superbeams (see section 8.10.4).

Taking it all together, the quasi-elastic cross sections are given by [Sch97]

$$\frac{d\sigma_{QE}}{dQ^2}\begin{pmatrix} \nu_\mu n \to \mu^- p \\ \bar{\nu}_\mu p \to \mu^+ n \end{pmatrix} = \frac{M^2 G_F^2 \cos^2 \theta_c}{8\pi E_\nu^2}\left(A_1(Q^2) \pm A_2(Q^2)\frac{s-u}{M^2} \right.$$
$$\left. + A_3(Q^2)\frac{(s-u)^2}{M^4} \right) \tag{4.48}$$

where $s - u = 4ME_\nu - Q^2$ and M is the mass of the nucleon. The functions A_1, A_2 and A_3 depend on the form factors F_A, F_V, F_M and Q^2. The most generalized expressions are given in [Mar69]. Equation (4.48) is analogous to the Rosenbluth formula describing elastic eN scattering.

4.5.2 (Quasi-)elastic NC reactions

The matrix element for the NC nucleon current is analogous to (4.39) neglecting again S, P and T terms. For $d\sigma/dQ^2$ (4.48) holds but the form factors have to be replaced by the corresponding NC form factors (figure 4.13). Several experiments have measured the cross section for this process (see [Man95]). $\sin^2\theta_W$ and M_A

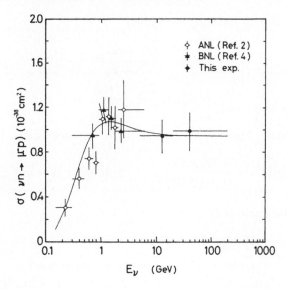

Figure 4.12. Compilation of results for $\sigma(\nu_\mu n \to \mu p)$ of various experiments. The curve shows the prediction of V–A theory with $M_A = 1.05$ GeV (from [Kit83]).

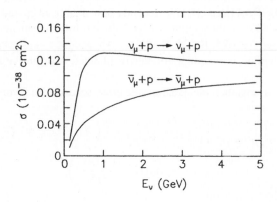

Figure 4.13. GWS prediction of the cross sections $\sigma(\nu_\mu p \to \nu_\mu p)$ and $\sigma(\bar{\nu}_\mu p \to \bar{\nu}_\mu p)$ as a function of E_ν with the parameters $m_A = 1.00$ GeV and $\sin^2 \theta_W = 0.232$ (from [Hor82]).

serve as fit parameters. Values obtained with the BNL experiment E734 result in [Ahr87]

$$M_A = (1.06 \pm 0.05) \text{ GeV} \qquad \sin^2 \theta_W = 0.218^{+0.039}_{-0.047}. \qquad (4.49)$$

4.6 Coherent, resonant and diffractive production

Beside quasi-elastic and deep inelastic scattering, there are other mechanisms which can contribute to the neutrino cross section. Among them are diffractive, resonance and coherent particle production. Typical resonance reactions, in which intermediate resonance states like $\Delta(1232)$ are produced, are

$$\nu_\mu p \rightarrow \mu^- p \pi^+ \tag{4.50}$$

$$\nu_\mu n \rightarrow \mu^- n \pi^+ \tag{4.51}$$

$$\nu_\mu n \rightarrow \mu^- p \pi^0 \tag{4.52}$$

or NC reactions

$$\nu_\mu p \rightarrow \nu_\mu p \pi^0 \qquad \nu_\mu p \rightarrow \nu_\mu n \pi^+ \tag{4.53}$$

$$\nu_\mu n \rightarrow \nu_\mu n \pi^0 \qquad \nu_\mu n \rightarrow \nu_\mu p \pi^- \tag{4.54}$$

will not be discussed in more detail here (see [Pas00]). As an example we briefly mention coherent π^0 production which directly probes the Lorentz structure of NC interactions. Helicity conserving V, A interactions will result in a different angular distribution of the produced π^0 than the ones from helicity changing S, P, T interactions. For more extensive details see [Win00]. Coherent π^0 production

$$\nu + (A, Z) \rightarrow \nu + \pi^0 + (A, Z) \tag{4.55}$$

leaves the nucleus intact. Because of helicity conservation in NC events, the π^0 is emitted at small angles in contrast to incoherent and resonant production. Several experiments have measured this process [Ama87, Cos88] and the results are compiled in figure 4.14. The ratio of ν and $\bar{\nu}$ induced production is deduced to be

$$\frac{\sigma(\nu(A, Z) \rightarrow \nu \pi^0 (A, Z))}{\sigma(\bar{\nu}(A, Z) \rightarrow \bar{\nu} \pi^0 (A, Z))} = 1.22 \pm 0.33 \tag{4.56}$$

still with a rather large error but they are in agreement with theoretical expectations which predict a ratio of one [Rei81]. Improved measurements will be done by the K2K experiment (see chapter 9). This process is the main background to experiments studying elastic $\nu_\mu e$ scattering (see section 3.4.2.2). However, it serves as an important tool for measuring total NC rates in atmospheric neutrino experiments (see chapter 9).

Diffractive processes are characterized by leaving the nucleus intact implying low momentum transfer. This can be described by a new kinematic variable t, being the square of the 4-momentum transferred to the target

$$t = (p - p')^2. \tag{4.57}$$

At low Q^2 and large ν, a virtual hadronic fluctuation of the gauge bosons, in the case of neutrinos the weak bosons W and Z, may interact with matter before

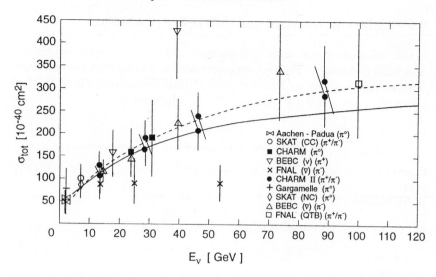

Figure 4.14. Compilation of results on coherent single-π production cross sections in CC ν_μ and $\bar{\nu}_\mu$ interactions. The curve shows the prediction of the Rein–Sehgal model [Rei83] for $m_A = 1.3$ GeV/c^2 (full line) and the Bel'kov–Kopeliovich approach (dashed line). The results of the experiments are scaled according to both models to allow comparison (from [Win00]).

being reabsorbed. Diffractive production of mesons on a target might produce real mesons in the final state, e.g.

$$\nu_\mu N \rightarrow \mu^- \rho^+ N. \tag{4.58}$$

In an analogous way the NC diffractive production of neutral vector mesons (V^0) such as $\rho^0, \omega, \Phi, J/\Psi \ldots$ can also be considered (figure 4.15). The elementary nature of the interaction is still unknown. It can be described by the exchange of a colour singlet system, called Pomeron. In νN scattering diffractive production of $\pi, \overline{\rho^\pm, a_1}$ and D_S^* mesons have been observed, while in lepto- and photoproduction also ρ^0, ω, ϕ and J/Ψ have been seen due to the higher statistics. A revival of interest in diffractive phenomena took place with the observation of 'rapidity gap' events at the ep collider HERA.

After discusssing quasi-elastic and a short review of resonance and diffractive production, which dominate the cross section at low energies, we now want to focus on deep inelastic scattering which leads to the concept of structure functions.

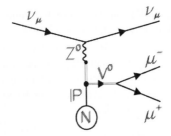

Figure 4.15. Feynman graph for diffractive vector meson production via the exchange of a pomeron $\mathcal{P}$.

4.7   Structure function of nucleons

The double differential cross section for CC reactions is given (using protons as nucleons) by

$$
\begin{aligned}
\frac{d\sigma^{\nu,\bar{\nu}}}{dQ^2\,d\nu} &= \frac{G_F^2}{2\pi}\frac{E'}{E}\left(2W_1^{\nu,\bar{\nu}}(Q^2,\nu)\times\sin^2\frac{\theta}{2}+W_2^{\nu,\bar{\nu}}(Q^2,\nu)\times\cos^2\frac{\theta}{2}\right.\\
&\qquad\left.\pm W_3^{\nu,\bar{\nu}}(Q^2,\nu)\frac{E+E'}{M}\sin^2\frac{\theta}{2}\right)\\
&= \frac{G_F^2}{2\pi}(xy^2\frac{M}{\nu}W_1(x,y)+\left(1-y-\frac{Mxy}{2E}\right)W_2(x,y)\\
&\qquad\pm xy\left(1-\frac{y}{2}W_3(x,y)\right).
\end{aligned}
\tag{4.59}
$$

Equation (4.59) can be deduced from more general arguments (see [Clo79, Lea96]). With the formulae given for the kinematic variables (4.29)–(4.33), this can be translated into other quantities as well:

$$
\frac{d\sigma}{dx\,dy}=2ME\nu\times\frac{d\sigma}{dQ^2\,d\nu}=\frac{M\nu}{E'}\times\frac{d\sigma}{dE'\,d\cos\theta}=2MEx\times\frac{d\sigma}{dx\,dQ^2}.
\tag{4.60}
$$

The three structure functions W_i describe the internal structure of the proton as seen in neutrino–proton scattering. At very high energies, the W-propagator term can no longer be neglected and in (4.59) the replacement

$$
G_F^2\rightarrow G_F^2\bigg/\left(1+\frac{Q^2}{m_W^2}\right)^2
\tag{4.61}
$$

has to be made. The description for ep/μp scattering is similar with the exception that there are only two structure functions. The term containing W_3 is missing because it is parity violating. By investigating inelastic ep scattering at SLAC in the late 1960s [Bre69], it was found that at values of Q^2 and ν not too small

$(Q^2 > 2 \text{ GeV}^2, \nu > 2 \text{ GeV})$, the structure functions did not depend on two variables independently but only on the dimensionless combination in form of the Bjorken scaling variable $x = Q^2/2M\nu$ (4.32). This behaviour was predicted by Bjorken [Bjo67] for deep inelastic scattering and is called scaling invariance (or Bjorken scaling). A physical interpretation was given by Feynman as discussed in the next section. The same scaling behaviour is observed in high-energy neutrino scattering, leading to the replacements

$$MW_1(Q^2, \nu) = F_1(x) \tag{4.62}$$

$$\nu W_2(Q^2, \nu) = F_2(x) \tag{4.63}$$

$$\nu W_3(Q^2, \nu) = F_3(x). \tag{4.64}$$

4.8   The quark–parton model, parton distribution functions

The basic idea behind the parton model is the following [Fey69, Lea96, Sch97]: in elastic electromagnetic scattering of a point-like particle on an extended target, the spatial extension can be described by a form factor $F(Q^2)$. This form factor can be seen as the Fourier transform of the spatial charge or magnetic moment distribution of the target. Form factors independent of Q^2 imply hard elastic scattering on point-like target objects. The SLAC results can then be interpreted, since the scaling invariance implies that deep inelastic ep scattering can be seen as an incoherent superposition of hard elastic electron-parton scattering. The parton is kicked out of the proton, while the remaining partons (the proton remnant) act as spectators and are not involved in the interaction (figure 4.16). After that the processes of fragmentation and hadronization follow, producing the particles observable in high-energy experiments. In this model, the variable x can be given an intuitive interpretation: assuming a proton with 4-momentum $p_p = (E_p, P_p)$, then the parton has the 4-momentum $xp_p = (xE_p, xP_p)$ before its interaction. This means the variable $x(0 < x < 1)$ describes the fraction of the proton momentum and energy of the interacting parton (figure 4.17). After several experiments on deep inelastic lepton–nucleon scattering, the result was that the partons are identical to the quarks proposed by Gell-Mann and Zweig in their SU(3) classification of hadrons [Gel64, Zwe64]. In addition to the valence quarks (a proton can be seen as a combination of uud-quarks, a neutron as of udd-quarks), the gluons also contribute, because, according to the Heisenberg uncertainty principle, they can fluctuate into quark–antiquark pairs for short times. These are known as the sea-(anti)quarks.

The picture described, called the quark–parton model (QPM), is today the basis for the description of deep inelastic lepton–nucleon scattering. For high Q^2 and the scattering on spin-$\frac{1}{2}$ particles, the Callan–Gross relation [Cal69]

$$2x F_1(x) = F_2(x) \tag{4.65}$$

holds between the first two structure functions. For a derivation see [Lea96].

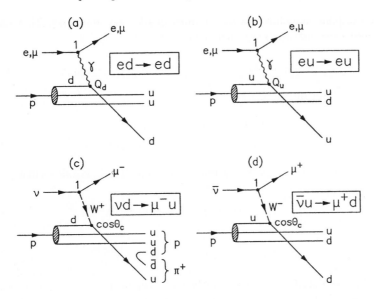

Figure 4.16. Graphs for the dominant processes in DIS in ep/μp scattering (a, b), νp scattering (c) and $\bar{\nu}$p scattering (d).

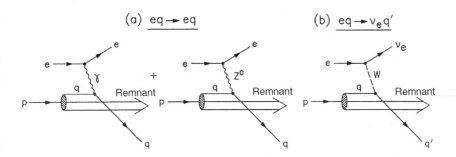

Figure 4.17. Deep inelastic ep scattering as described in the quark–parton model via photon and Z^0 exchange (neutral currents) and W exchange (charged currents).

4.8.1 Deep inelastic neutrino proton scattering

First, let us define the parton distribution functions (PDF) within a proton. As an example, take $u(x)$:

$$u(x)\,\mathrm{d}x = \text{Number of u-quarks in the proton with momentum}$$
$$\text{fraction between } x \text{ and } x + \mathrm{d}x \qquad (4.66)$$

and corresponding definitions for the other quarks and antiquarks. They can be split into a valence- and a sea-quark contribution

$$u(x) = u_V(x) + u_S(x) \qquad d(x) = d_V(x) + d_S(x).$$ (4.67)

Symmetry of the $q\bar{q}$ sea requires

$$u_S(x) = \bar{u}(x) \qquad s(x) = \bar{s}(x)$$
$$d_S(x) = \bar{d}(x) \qquad c(x) = \bar{c}(x).$$ (4.68)

Because of the valence quark structure of the proton (uud), it follows that

$$\int_0^1 u_V(x)\,dx = \int_0^1 [u(x) - \bar{u}(x)]\,dx = 2$$ (4.69)

$$\int_0^1 d_V(x)\,dx = \int_0^1 [d(x) - \bar{d}(x)]\,dx = 1.$$ (4.70)

The QPM predicts deep inelastic scattering as an incoherent sum of (quasi)-elastic lq or l$\bar{q}$ scattering on partons. The double differential cross section can be written as

$$\frac{d\sigma}{dx\,dy}(lp \to l'X) = \sum_{q,q'} q(x)\frac{d\sigma}{dy}(lq \to l'q') + \sum_{\bar{q},\bar{q}'} \bar{q}(x)\frac{d\sigma}{dy}(l\bar{q} \to l'\bar{q}').$$ (4.71)

Using fundamental Feynman rules, one gets the following relations:

$$\frac{d\sigma}{dy}(eq \to eq) = \frac{d\sigma}{dy}(e\bar{q} \to e\bar{q}) = \frac{8\pi\alpha^2}{Q^4}m_q E q_q^2 \left(1 - y + \frac{y^2}{2}\right)$$ (4.72)

$$\frac{d\sigma}{dy}(\nu q \to \mu^- q') = \frac{d\sigma}{dy}(\bar{\nu}\bar{q} \to \mu^+\bar{q}') = \frac{2G_F^2}{\pi}m_q E$$ (4.73)

$$\frac{d\sigma}{dy}(\nu\bar{q} \to \mu^-\bar{q}') = \frac{d\sigma}{dy}(\bar{\nu}q \to \mu^+q') = \frac{2G_F^2}{\pi}m_q E(1 - y)^2$$ (4.74)

where $y = 1 - E'/E = 1/2(1 - \cos\theta^*)$ and q_q is the charge of the quark. Equation (4.72) describes electromagnetic interactions via photon exchange, while (4.73) and (4.74) follow from V–A theory ignoring the W-propagator. The additional term $(1 - y)^2$ follows from angular momentum conservation because scattering with $\theta^* = 180°$ ($y = 1$) is not allowed. The corresponding cross sections can then be written using the QPM formulae as

$$\frac{d\sigma}{dx\,dy}(\nu p) = \sigma_0 \times 2x[[d(x) + s(x)] + [\bar{u}(x) + \bar{c}(x)](1 - y)^2]$$ (4.75)

$$\frac{d\sigma}{dx\,dy}(\bar{\nu}p) = \sigma_0 \times 2x[[u(x) + c(x)](1 - y)^2 + [\bar{d}(x) + \bar{s}(x)]]$$ (4.76)

with (using (4.29))

$$\sigma_0 = \frac{G_F^2 M E}{\pi} = \frac{G_F^2 s}{2\pi} = 1.583 \times 10^{-38} \text{ cm}^2 \times E/\text{GeV}. \qquad (4.77)$$

Equation (4.75) together with scaling invariance and the Callan–Cross relation (4.65) allows the derivation of the following relations:

$$F_2^{vp}(x) = 2x[d(x) + \bar{u}(x) + s(x) + \bar{c}(x)]$$
$$x F_3^{vp}(x) = 2x[d(x) - \bar{u}(x) + s(x) - \bar{c}(x)]$$
$$F_2^{\bar{v}p}(x) = 2x[u(x) + c(x) + \bar{d}(x) + \bar{s}(x)]$$
$$x F_3^{\bar{v}p}(x) = 2x[u(x) + c(x) - \bar{d}(x) - \bar{s}(x)]. \qquad (4.78)$$

In a similar way, neutron structure functions can be written in terms of the proton PDFs by invoking isospin invariance:

$$u_n(x) = d_p(x) = d(x)$$
$$d_n(x) = u_p(x) = u(x)$$
$$s_n(x) = s_p(x) = s(x)$$
$$c_n(x) = c_p(x) = c(x). \qquad (4.79)$$

The corresponding structure functions are then

$$F_2^{vn}(x) = 2x[u(x) + \bar{d}(x) + s(x) + \bar{c}(x)]$$
$$x F_3^{vn}(x) = 2x[u(x) - \bar{d}(x) + s(x) - \bar{c}(x)]. \qquad (4.80)$$

Finally the cross section for lepton scattering on an isoscalar target N is obtained by averaging

$$\frac{d\sigma}{dx\,dy}(\text{lN}) = \frac{1}{2}\left(\frac{d\sigma}{dx\,dy}(\text{lp}) + \frac{d\sigma}{dx\,dy}(\text{ln})\right) \qquad F_i^{lN} = \frac{1}{2}(F_i^{lp} + F_i^{ln}). \quad (4.81)$$

Combining (4.78), (4.79) and (4.81) and assuming $s = \bar{s}, c = \bar{c}$ results in

$$F_2^{e(\mu)N} = \tfrac{5}{18}x(u + d + \bar{u} + \bar{d}) + \tfrac{1}{9}x(s + \bar{s}) + \tfrac{4}{9}x(c + \bar{c})$$
$$F_2^{vN} = F_2^{\bar{v}N} = x[u + d + s + c + \bar{u} + \bar{d} + \bar{s} + \bar{c}] = x[q + \bar{q}]$$
$$x F_3^{vN} = x[u + d + 2s - \bar{u} - \bar{d} - 2\bar{c}] = x[q - \bar{q} + 2(s - c)]$$
$$x F_3^{\bar{v}N} = x[u + d + 2c - \bar{u} - \bar{d} - 2\bar{s}] = x[q - \bar{q} - 2(s - c)] \qquad (4.82)$$
$$\text{with } q = u + d + s + c, \ \bar{q} = \bar{u} + \bar{d} + \bar{s} + \bar{c}.$$

As can be seen, the structure function F_2^{vN} measures the density distribution of all quarks and antiquarks within the proton, while the $v/\bar{v}$ averaged structure function F_3^{vN} measures the valence-quark distribution. Reordering (4.81) shows that F_2

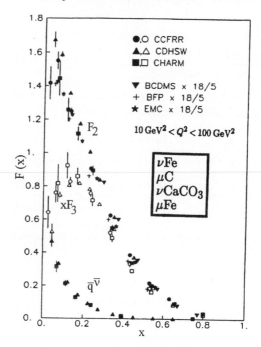

Figure 4.18. Compilation of the structure functions $F_2^{\nu N}$ and $x F_3^{\nu N}$ from $\nu/\bar{\nu}$ scattering as well as of $5/18 F_2^{\mu N}$ from μ scattering on isoscalar targets and the distribution function $\bar{q}^{\bar{\nu}} = x(\bar{q} - \bar{s} - \bar{c})$ (from [PDG02]).

and F_3 can be basically determined by the sum and difference of the differential cross sections.

Experimentally the procedure is as follows (for details see [Die91, Con98]). Using the equations given earlier the structure functions are determined from the differential cross sections. From these, the single-quark distribution functions as well as the gluon structure function $xg(x)$ can be extracted. Figure 4.18 shows a compilation of such an analysis. As can be seen, the sea quarks are concentrated at low x ($x < 0.4$) values, while the valence quarks extend to higher values. It should be noted that the numbers are given for a fixed Q^2. Extensive measurements over a wide range of x and Q^2, increasing the explored parameter space by two orders of magnitude, are performed at HERA. Recently CCFR, published a new low x, low Q^2 analysis based on neutrino scattering data [Fle01].

4.8.1.1 QCD effects

As already mentioned, measurements of structure functions over a wide range of Q^2 show a deviation from scaling invariance for fixed x:

$$F_i(x) \rightarrow F_i(x, Q^2). \tag{4.83}$$

For higher Q^2, $F_i(x, Q^2)$ rises at small x and gets smaller at high x (figure 4.19). This can be understood by QCD. Higher Q^2 implies a better time and spatial resolution. Therefore, more and more partons from the sea with smaller and smaller momentum fractions can be observed, leading to a rise at small x. Quantitatively, this Q^2 evolution of the structure functions can be described by the DGLAP (named after Dokshitzer, Gribov, Lipatov, Altarelli and Parisi) equations [Alt77, Dok77, Gri72]. They are given by

$$\frac{dq_i(x, Q^2)}{d \ln Q^2} = \frac{\alpha_S(Q^2)}{2\pi} \int_x^1 \frac{dy}{y} \left[q_i(y, Q^2) \times P_{qq}\left(\frac{x}{y}\right) \right.$$
$$\left. + g(y, Q^2) \times P_{qg}\left(\frac{x}{y}\right) \right] \tag{4.84}$$

$$\frac{dg_i(x, Q^2)}{d \ln Q^2} = \frac{\alpha_S(Q^2)}{2\pi} \int_x^1 \frac{dy}{y} \left[\sum_{j=1}^{N_f} [q_j(y, Q^2) + \bar{q}_j(y, Q^2)] \times P_{gq}\left(\frac{x}{y}\right) \right.$$
$$\left. + g(y, Q^2) \times P_{gg}\left(\frac{x}{y}\right) \right]. \tag{4.85}$$

The splitting functions $P_{ij}(x/y)$ (with $i, j = q, g$) give the probability that parton j with momentum y will be resolved as parton i with momentum $x < y$. They can be calculated within QCD. Therefore, from measuring the structure function at a fixed reference value Q_0^2, their behaviour with Q^2 can be predicted with the DGLAP equations. *H* compilation of structure functions is shown in figure 4.20.

Non-perturbative QCD processes that contribute to the structure function measurements are collectively termed higher-twist effects. These effects occur at small Q^2 where the impulse approximation (treating the interacting parton as a free particle) of scattering from massless non-interacting quarks is no longer valid. Examples include target mass effects, diquark scattering and other multiparton effects. Because neutrino experiments use heavy targets in order to obtain high interaction rates, nuclear effects (like Fermi motion) must also be considered. For more detailed treatments see [Con98].

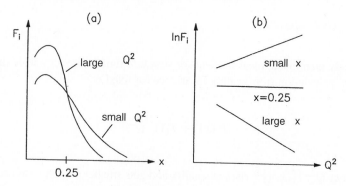

Figure 4.19. Schematic drawing of the Q^2 dependence of structure functions as predicted by QCD: left, $F(x, Q^2)$ as a function of x for small and large Q^2; right, $\ln F(x, Q^2)$ as a function of Q^2 for fixed x (from [PDG02]).

4.9   y distributions and quark content from total cross sections

Corresponding to (4.69) the fraction of the proton momentum carried by u-quarks is defined by

$$U = \int_0^1 xu(x)\,\mathrm{d}x \qquad (4.86)$$

and in a similar way for the other quarks. Using this notation, the y distributions are then given by

$$\frac{\mathrm{d}\sigma}{\mathrm{d}y}(\nu N) = \sigma_0 \times [[Q + S] + [\bar{Q} - S](1 - y)^2] \approx \sigma_0 \times [Q + \bar{Q}(1 - y)^2]$$

$$(4.87)$$

$$\frac{\mathrm{d}\sigma}{\mathrm{d}y}(\bar{\nu} N) = \sigma_0 \times [[Q - S](1 - y)^2 + [\bar{Q} + S]] \approx \sigma_0 \times [Q(1 - y)^2 + \bar{Q}]$$

$$(4.88)$$

Neglecting the s and c contributions, the ratio of both y distributions is approximately about one for $y = 0$. Figure 4.21 shows the measured y distributions from the CDHS experiment resulting in (taking into account radiative corrections) [Gro79]

$$\frac{\bar{Q}}{Q + \bar{Q}} = 0.15 \pm 0.03 \qquad \frac{S}{Q + \bar{Q}} = 0.00 \pm 0.03 \qquad \frac{\bar{Q} + S}{Q + \bar{Q}} = 0.16 \pm 0.01.$$

$$(4.89)$$

A further integration with respect to y results in the following values for the total cross sections:

$$\sigma(\nu N) = \frac{\sigma_0}{3} \times [3Q + \bar{Q} + 2S] \approx \frac{\sigma_0}{3} \times [3Q + \bar{Q}] \qquad (4.90)$$

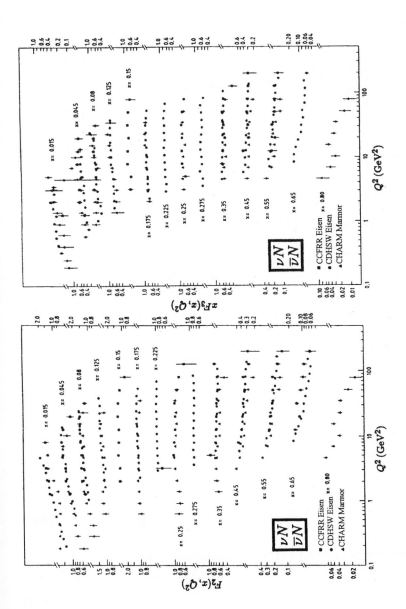

Figure 4.20. Compilation of the structure functions $F_2(x, Q^2)$ and $xF_3(x, Q^2)$ as obtained in νN and $\bar{\nu}N$ scattering with the CDHSW, CCFRR and CHARM experiments. The Q^2 dependence is plotted for fixed x (from [PDG02]).

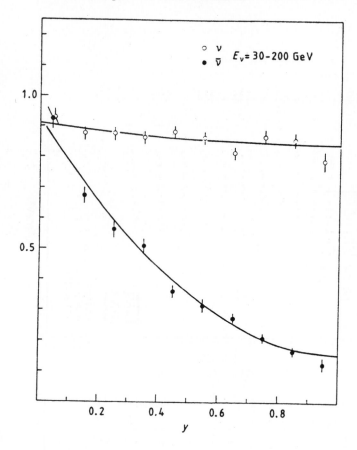

Figure 4.21. The differential cross sections versus y as obtained by CDHS for νN and $\bar{\nu}$N CC scattering. The dominant flat distribution for neutrinos and $(1 - y)^2$ behaviour for antineutrinos show that left- and right-handed couplings are different. The distributions are explained by dominant scattering from valence quarks with left-handed couplings (from [Eis86]).

$$\sigma(\bar{\nu}N) = \frac{\sigma_0}{3} \times [Q + 3\bar{Q} + 2S] \approx \frac{\sigma_0}{3} \times [Q + 3\bar{Q}]. \qquad (4.91)$$

Using the ratio $R = \sigma(\nu N)/\sigma(\bar{\nu}N)$, this can be written as

$$\frac{\bar{Q}}{Q} = \frac{3 - R}{3R - 1}. \qquad (4.92)$$

A measurement of $R < 3$ is a direct hint of the momentum contribution $\bar{Q}$ of the sea quarks (see section 4.11). Using the measured values (4.89) resulting in

$R = 2.02$, it follows that $Q \approx 0.41$ and $\bar{Q} \approx 0.08$. Therefore,

$$\int_0^1 F_2^{\nu N}(x)\, dx = Q + \bar{Q} \approx 0.49$$

$$Q_V = Q - \bar{Q} \approx 0.33 \qquad Q_S = \bar{Q}_S = \bar{Q} \approx 0.08$$

$$\frac{\bar{Q}}{Q + \bar{Q}} \approx 0.16 \qquad \frac{\bar{Q}}{Q} \approx 0.19. \tag{4.93}$$

This shows that quarks and antiquarks carry about 49% of the proton momentum, whereas valence quarks contribute about 33% and sea quarks about 16%. Half of the proton spin has to be carried by the gluons. For more extensive reviews on nucleon structure see [Con98, Lam00].

The QPM formulae allow predictions to be made about the different structure functions, which can serve as important tests for the model. As an example, the electromagnetic and weak structure functions for an isoscalar nucleon are related by

$$F_2^{\mu N, eN} = \tfrac{5}{18} F_2^{\nu N} - \tfrac{1}{6} x[s + \bar{s} - c - \bar{c}] \approx \tfrac{5}{18} F_2^{\nu N} - \tfrac{1}{6} x[s + \bar{s}] \approx \tfrac{5}{18} F_2^{\nu N} \tag{4.94}$$

neglecting $c(x)$ and $s(x)$, which is small at large x. This means

$$\frac{F_2^{\mu N, eN}}{F_2^{\nu N}} = \frac{5}{18}\left(1 - \frac{3}{5} \times \frac{s + \bar{s} - c - \bar{c}}{q + \bar{q}}\right) \approx \frac{5}{18}. \tag{4.95}$$

This is an important test for QPM especially for the fractional charge of quarks, because the factor 5/18 is the average of the squared quark charges (1/9 and 4/9).

4.9.1 Sum rules

Using the QPM relations important sum rules (integrations of structure functions with respect to x) are obtained, which can be tested experimentally. The total number of quarks and antiquarks in a nucleon are given by

$$\frac{1}{2}\int_0^1 \frac{1}{x}(F_2^\nu(x) + F_2^{\bar{\nu}}(x))\, dx = \int_0^1 [q(x) + \bar{q}(x)]\, dx. \tag{4.96}$$

The Gross–Llewellyn Smith (GLS) [Gro69] sum rule gives the QCD expectation for the integral of the valence quark densities. To leading order in perturbative QCD, the integral $\int \frac{dx}{x}(x F_3)$ is the number of valence quarks in the proton and should equal three. QCD corrections to this integral result in a dependence on α_s

$$S_{\text{GLS}} = \frac{1}{2}\int_0^1 (F_3^\nu(x) + F_3^{\bar{\nu}}(x))\, dx = \int_0^1 \bar{F}_3(x)\, dx = \int_0^1 [q(x) - \bar{q}(x)]\, dx$$

$$= 3\left[1 - \frac{\alpha_s}{\pi} - a(n_f)\left(\frac{\alpha_s}{\pi}\right)^2 - b(n_f)\left(\frac{\alpha_s}{\pi}\right)^3\right]. \tag{4.97}$$

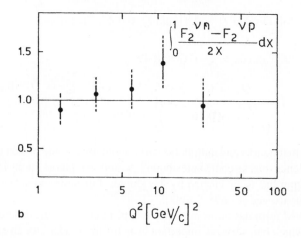

Figure 4.22. Test of Adler sum rule. The estimated uncertainties (dashed lines) are shown separately (from [All85a]).

In this equation, a and b are known functions of the number of quark flavours n_f which contribute to scattering at a given x and Q^2. This is one of the few QCD predictions that are available to order α_s^3. The world average is [Con98]

$$\int_0^1 F_3(x)\,dx = 2.64 \pm 0.06 \tag{4.98}$$

which is consistent with the next-to-next-to-leading order evaluation of (4.97) with the QCD parameter $\Lambda_{QCD} = 250 \pm 50$ MeV.

A further important sum rule is the Adler sum rule [Adl66]. This predicts the difference between the quark densities of the neutron and the proton, integrated over x (figure 4.22). It is given at high energies (in all orders of QCD) by

$$S_A = \frac{1}{2}\int_0^1 \frac{1}{x}(F_2^{\nu n}(x) - F_2^{\nu p}(x))\,dx = \int_0^1 [u_V(x) - d_V(x)]\,dx = 1. \tag{4.99}$$

Common to the determination of sum rules is the experimental difficulty of measuring them at very small x, the part dominating the integral.

For completeness, two more sum rules should be mentioned. The analogue to the Adler sum rule for charged-lepton scattering is the Gottfried sum rule [Got67]:

$$S_G = \int_0^1 \frac{1}{x}(F_2^{\mu p}(x) - F_2^{\mu n}(x))\,dx = \frac{1}{3}\int_0^1 [u(x) + \bar{u}(x) - d(x) - \bar{d}(x)]\,dx$$

$$= \frac{1}{3}\left(1 + 2\int_0^1 [\bar{u}(x) - \bar{d}(x)]\,dx\right) = \frac{1}{3} \tag{4.100}$$

The experimental value is $S_G = 0.235 \pm 0.026$ [Arn94]. This is significantly different from expectation and might be explained by an isospin asymmetry of the

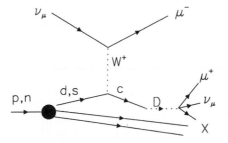

Figure 4.23. Feynman Graph for dimuon production due to charm production in charged current ν_μN interactions.

sea, i.e. $\bar{u}(x) \neq \bar{d}(x)$, strongly supported by recent measurements [Ack98a]. Note that this assumption $\bar{u} = \bar{d}$ was not required in the Adler sum rule. Furthermore, there is the Bjorken sum rule [Bjo67]

$$S_B = \int_0^1 [(F_1^{\bar{\nu}p}(x) - F_1^{\nu p}(x)] \, dx = 1 - \frac{2\alpha_S(Q^2)}{3\pi}. \qquad (4.101)$$

We now continue to discuss a few more topics investigated in neutrino nucleon scattering. Because of the richness of possible observable quantities we restrict ourselves to a few examples. For more details see [Sch97, Con98].

4.10   Charm physics

An interesting topic to investigate is charm production which allows us to measure the mass of the charm quark. In the case of neutrino scattering, the underlying process is a neutrino interacting with an s or d quark, producing a charm quark that fragments into a charmed hadron. The charmed hadrons decay semi-leptonically ($BR \approx 10\%$) and produce a second muon of opposite sign (the so called OSDM events) (figure 4.23)

$$\nu_\mu + N \longrightarrow \mu^- + c + X \qquad (4.102)$$
$$\hookrightarrow s + \mu^+ + \nu_\mu.$$

However, the large mass m_c of the charm quark gives rise to a threshold behaviour in the dimuon production rate at low energies. This is effectively described by the slow rescaling model [Bar76, Geo76] in which x is replaced by the slow rescaling variable ξ given by

$$\xi = x \left(1 + \frac{m_c^2}{Q^2}\right). \qquad (4.103)$$

Table 4.1. Compilation of the mass of the charm quark and the strange sea parameter κ obtained by leading order fits in various experiments. The experiments are ordered with respect to increasing average neutrino energy.

Experiment	m_c (GeV)	κ
CDHS	—	$0.47 \pm 0.08 \pm 0.05$
NOMAD	$1.3 \pm 0.3 \pm 0.3$	$0.48^{+0.09+0.17}_{-0.07-0.12}$
CHARMII	$1.8 \pm 0.3 \pm 0.3$	$0.39^{+0.07+0.07}_{-0.06-0.07}$
CCFR	$1.3 \pm 0.2 \pm 0.1$	$0.44^{+0.09+0.07}_{-0.07-0.02}$
FMMF	—	$0.41^{+0.08+0.103}_{-0.08-0.069}$

The differential cross section for dimuon production is then expressed generally as

$$\frac{d^3\sigma(\nu_\mu N \to \mu^- \mu^+ X)}{d\xi \, dy \, dz} = \frac{d^2\sigma(\nu_\mu N \to cX)}{d\xi \, dy} D(z) B_c(c \to \mu^+ X) \qquad (4.104)$$

where the function $D(z)$ describes the hadronization of charmed quarks and B_c is the weighted average of the semi-leptonic branching ratios of the charmed hadrons produced in neutrino interactions. As mentioned before, in leading order charm is produced by direct scattering of the strange and down quarks in the nucleon. The leading order differential cross section for an isoscalar target, neglecting target mass effects, is

$$\frac{d^3\sigma(\nu_\mu N \to cX)}{d\xi \, dy \, dz} = \frac{G_F^2 M E_\nu \xi}{\pi} [u(\xi, Q^2) + d(\xi, Q^2)]|V_{cd}|^2$$

$$+ 2s(\xi, Q^2)|V_{cs}|^2 \left(1 - y + \frac{xy}{\xi}\right) D(z) B_c. \qquad (4.105)$$

Therefore, by measuring the ratio of dimuon production *versus* single muon production as a function of neutrino energy, m_c can be determined from the threshold behaviour (figure 4.24). The production of opposite-sign dimuons is also governed by the proportion of strange to non-strange quarks in the nucleon sea, $\kappa = 2\bar{s}/(\bar{u} + \bar{d})$, the CKM matrix elements V_{cd} and V_{cs} and B_c. Table 4.1 shows a compilation of such measurements.

The study of open charm production in the form of D-meson production is another important topic, especially to get some insight into the fragmentation process. Recently CHORUS performed a search for D^0 production [Kay02]. In total 283 candidates are observed, with an expected background of 9.2 events coming from K- and Λ-decay. The ratio $\sigma(D^0)/\sigma(\nu_\mu CC)$ is found to be $(1.99 \pm 0.13(\text{stat.}) \pm 0.17(\text{syst.})) \times 10^{-2}$ at 27 GeV average ν_μ energy (figure 4.24). NOMAD performed a search for D^{*+}-production using the decay

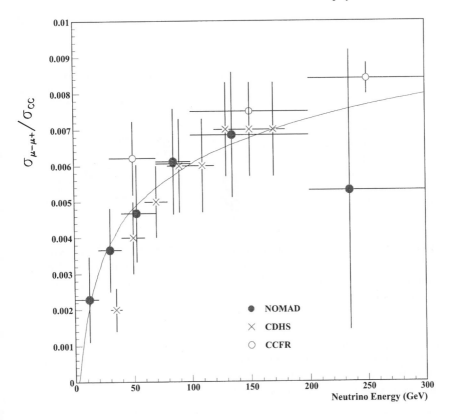

Figure 4.24. Compilation of observed dimuon *versus* single muon rates as a function of the visible energy E_{vis} obtained by CDHS, CCFR and NOMAD. The threshold behaviour due to the charm quark mass is clearly visible (from [Ast00]).

chain $D^{*+} \rightarrow D^0 + \pi^+$ followed by $D^0 \rightarrow K^- + \pi^+$. In total 35 ± 7.2 events could be observed resulting in a D^{*+} yield in ν_μ CC interactions of $(0.79 \pm 0.17(\text{stat.}) \pm 0.10(\text{syst.}))\%$ [Ast02]. Another measurement related to charm is the production of bound charm–anticharm states like the J/Ψ. Due to the small cross section, the expected number of events in current experiments is rather small. It can be produced via NC reations by boson–gluon fusion as shown in figure 4.25. They were investigated by three experiments (CDHS [Abr82], CHORUS [Esk00] and NuTeV [Ada00]) with rather inconclusive results. Their production in νN scattering can shed some light on the theoretical description of heavy quarkonium systems, which is not available in other processes [Pet99, Kni02].

The charm quark can be produced from strange quarks in the sea. This allows $s(x)$ to be measured by investigating dimuon production. It is not only possible to measure the strange sea of the nucleon but also to get information

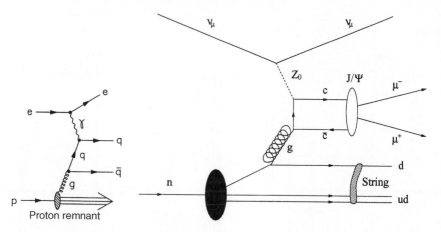

Figure 4.25. Feynman graph of boson gluon fusion. Left: Photon gluon fusion as obtained in e, μN scattering producing J/Ψ mesons. This is a direct way to measure the gluon structure function $xg(x)$. Right: Z^0 gluon fusion responsible for neutral current J/Ψ production in νN scattering.

about its polarization. This is done by measurements of the Λ-polarization. The polarization is measured by the asymmetry in the angular distributions of the protons in the parity-violating decay process $\Lambda \rightarrow p\pi^-$. In the Λ rest frame the decay protons are distributed as follows

$$\frac{1}{N}\frac{dN}{d\Omega} = \frac{1}{4\pi}(1 + \alpha_\Lambda Pk)$$ (4.106)

where P is the Λ polarization vector, $\alpha_\Lambda = 0.642 \pm 0.013$ is the decay asymmetry parameter and k is the unit vector along the proton decay direction. Since NOMAD is unable to distinguish protons from pions in the range relevant for this search, any search for neutral strange particles (V^0) should rely on the kinematics of the V^0-decay. The definition of the kinematic variables and the so called Armenteros plot are shown in figure 4.26. Their recent results on Λ and $\bar\Lambda$ polarization can be found in [Ast00, Ast01].

4.11   Neutral current reactions

Inelastic NC reactions νN $\rightarrow \nu$N are described by the QPM as elastic NC events such as

$$\nu q \rightarrow \nu q \qquad \nu\bar q \rightarrow \nu\bar q$$ (4.107)

$$\bar\nu q \rightarrow \bar\nu q \qquad \bar\nu\bar q \rightarrow \bar\nu\bar q.$$ (4.108)

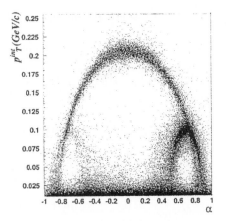

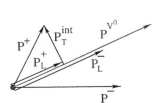

Figure 4.26. Left: Definition of kinematic variables. Right: Armenteros plot for neutral decaying particles V^0 as observed by the NOMAD experiment, showing clearly the distribution of kaons (big parabola), $\bar{\Lambda}$s (small parabola left-hand corner) and Λs (small parabola right-hand corner) (from [Ast00]).

The differential cross sections are given by

$$\frac{d\sigma}{dy}(\nu q) = \frac{d\sigma}{dy}(\bar{\nu}\bar{q}) = \frac{G_F^2 m_q}{2\pi} E_\nu \left[(g_V + g_A)^2 + (g_V - g_A)^2 (1-y)^2 \right.$$
$$\left. + \frac{m_q}{E_\nu} (g_A^2 - g_V^2) y \right]$$
$$= \frac{2G_F^2 m_q}{\pi} E_\nu \left[g_L^2 + g_R^2 (1-y)^2 - \frac{m_q}{E_\nu} g_L g_R y \right] \qquad (4.109)$$

$$\frac{d\sigma}{dy}(\bar{\nu}q) = \frac{d\sigma}{dy}(\nu\bar{q}) = \frac{G_F^2 m_q}{2\pi} E_\nu \left[(g_V - g_A)^2 + (g_V + g_A)^2 (1-y)^2 \right.$$
$$\left. + \frac{m_q}{E_\nu} (g_A^2 - g_V^2) y \right]$$
$$= \frac{2G_F^2 m_q}{\pi} E_\nu \left[g_R^2 + g_L^2 (1-y)^2 - \frac{m_q}{E_\nu} g_L g_R y \right]. \qquad (4.110)$$

For the following, the last term will be neglected because of $E_\nu \gg m_q$. The GWS predictions for the coupling constants are:

$$g_V = \tfrac{1}{2} - \tfrac{4}{3} \sin^2 \theta_W \qquad g_A = \tfrac{1}{2} \qquad \text{for } q \equiv u, c$$
$$g'_V = -\tfrac{1}{2} + \tfrac{2}{3} \sin^2 \theta_W \qquad g'_A = -\tfrac{1}{2} \qquad \text{for } q \equiv d, s \qquad (4.111)$$

and

$$g_L = \tfrac{1}{2} - \tfrac{2}{3} \sin^2 \theta_W \qquad g_R = -\tfrac{2}{3} \sin^2 \theta_W \qquad \text{for } q \equiv u, c$$

$$g'_L = -\tfrac{1}{2} + \tfrac{1}{3}\sin^2\theta_W \qquad g'_R = \tfrac{1}{3}\sin^2\theta_W \qquad \text{for } q \equiv d, s. \quad (4.112)$$

According to the QPM a similar relation holds as in CC events (4.71)

$$\frac{d\sigma}{dx\,dy}(\overset{(-)}{\nu}\, p \to \overset{(-)}{\nu}\, X) = \sum_q q(x)\frac{d\sigma}{dy}(\overset{(-)}{\nu}\, q) + \sum_{\bar{q}} \bar{q}(x)\frac{d\sigma}{dy}(\overset{(-)}{\nu}\, \bar{q}). \quad (4.113)$$

The corresponding proton structure functions are then obtained:

$$F_2^{\nu p,\bar{\nu}p} = 2x[(g_L^2 + g_R^2)[u + c + \bar{u} + \bar{c}] + (g_L'^2 + g_R'^2)[d + s + \bar{d} + \bar{s}]]$$
$$= x[(g_A^2 + g_V^2)[u + c + \bar{u} + \bar{c}] + (g_A'^2 + g_V'^2)[d + s + \bar{d} + \bar{s}]] \quad (4.114)$$
$$xF_3^{\nu p,\bar{\nu}p} = 2x[(g_L^2 - g_R^2)[u + c - \bar{u} - \bar{c}] + (g_L'^2 - g_R'^2)[d + s - \bar{d} - \bar{s}]]$$
$$= 2x[g_V g_A[u + c - \bar{u} - \bar{c}] + g_V' g_A'[d + s - \bar{d} - \bar{s}]]. \quad (4.115)$$

The neutron structure functions are obtained with the replacements given in (4.79) which leads to the structure functions for an isoscalar target:

$$F_2^{\nu N,\bar{\nu}N} = x[(g_L^2 + g_R^2)[u + d + 2c + \bar{u} + \bar{d} + 2\bar{c}]$$
$$+ (g_L'^2 + g_R'^2)[u + d + 2s + \bar{u} + \bar{d} + 2\bar{s}]]$$
$$xF_3^{\nu N,\bar{\nu}N} = x(g_L^2 - g_R^2)[u + d + 2c - \bar{u} - \bar{d} - 2\bar{c}]]$$
$$+ (g_L'^2 - g_R'^2)[u + d + 2s - \bar{u} - \bar{d} - 2\bar{s}]]. \quad (4.116)$$

Neglecting the s and c sea quarks the corresponding cross sections can be written as

$$\frac{d\sigma}{dx\,dy}(\nu N) = \sigma_0 \times x[(g_L^2 + g_L'^2)[q + \bar{q}(1 - y)^2] + (g_R^2 + g_R'^2)[\bar{q} + q(1 - y)^2]]$$

$$(4.117)$$

$$\frac{d\sigma}{dx\,dy}(\bar{\nu} N) = \sigma_0 \times x[(g_R^2 + g_R'^2)[q + \bar{q}(1 - y)^2] + (g_L^2 + g_L'^2)[\bar{q} + q(1 - y)^2]]$$

$$(4.118)$$

with $q = u + d$ and $\bar{q} = \bar{u} + \bar{d}$ and σ_0 given by (4.77). Comparing these cross sections with the CC ones, integrating with respect to x and y and using the measureable ratios

$$R_\nu^N = \frac{\sigma_{NC}(\nu N)}{\sigma_{CC}(\nu N)} \qquad R_{\bar{\nu}}^N = \frac{\sigma_{NC}(\bar{\nu}N)}{\sigma_{CC}(\bar{\nu}N)} \qquad r = \frac{\sigma_{CC}(\bar{\nu}N)}{\sigma_{CC}(\nu N)} \quad (4.119)$$

leads to the following interesting relations for the couplings

$$g_L^2 + g_L'^2 = \frac{R_\nu^N - r^2 R_{\bar{\nu}}^N}{1 - r^2} \qquad g_R^2 + g_R'^2 = \frac{r(R_{\bar{\nu}}^N - R_\nu^N)}{1 - r^2}. \quad (4.120)$$

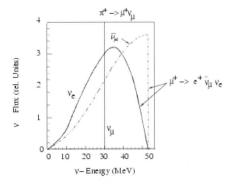

Figure 4.27. Energy spectrum of neutrinos coming from π^+ decay at rest. Beside a monoenergetic line of ν_μ at 29.8 MeV coming from pion decay there are the continous spectra of ν_e and $\bar{\nu}_\mu$ with equal intensity and energies up to 52.8 MeV from muon decay.

Using the GWS predictions for the couplings the precise measurements of R_ν^N or $R_{\bar{\nu}}^N$ allows a measurement of the Weinberg angle ($r = 0.5$ and using (4.112))

$$R_\nu^N = (g_L^2 + g_L'^2) + r(g_R^2 + g_R'^2) = \tfrac{1}{2} - \sin^2\theta_W + (1+r)\tfrac{5}{9}\sin^4\theta_W \quad (4.121)$$

$$R_{\bar{\nu}}^N = (g_L^2 + g_L'^2) + \frac{1}{r}(g_R^2 + g_R'^2) = \frac{1}{2} - \sin^2\theta_W + \left(1+\frac{1}{r}\right)\frac{5}{9}\sin^4\theta_W \quad (4.122)$$

These ratios were measured by several experiments, the most accurate ones being CHARM, CDHSW and CCFR [All87, Hai88, Blo90, Arr94]. The values obtained by CDHSW are:

$$R_\nu^N = 0.3072 \pm 0.0033 \qquad R_{\bar{\nu}}^N = 0.382 \pm 0.016. \quad (4.123)$$

For a precision measurement of $\sin^2\theta_W$ several correction factors have to be taken into account. The analyses for the three experiments result in values for $\sin^2\theta_W$ of 0.236 ± 0.006 ($m_c = 1.5$ GeV), 0.228 ± 0.006 ($m_c = 1.5$ GeV) and 0.2218 ± 0.0059 ($m_c = 1.3$ GeV). A recent measurement of NuTeV came up with a value 3σ away from the standard model expectation [Zel02] and awaits future confirmation. For a compilation of measurements of the Weinberg angle see section 3.4.3.

As a general summary of all the observed results it can be concluded that the GWS predictions are in good agreement with the experimental results.

4.12   Neutrino cross section on nuclei

After extensively discussing neutrino–nucleon scattering, it is worthwhile taking a short look at neutrino reactions with nuclei. This is quite important not only for low-energy tests of electroweak physics but also for neutrino astrophysics,

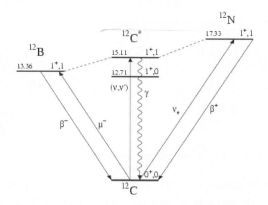

Figure 4.28. The $A = 12$ isobaric analogue triplet together with various possible transitions involving the ^{12}C ground state.

either in the astrophysical process itself or in the detection of such neutrinos. This kind of neutrino spectroscopy has to be done with lower energy (a few MeV) neutrinos, typically coming from pion decay at rest (DAR) and subsequent muon decay (see figure 4.27), giving rise to equal numbers of ν_e, ν_μ and $\bar{\nu}_\mu$. The study of such reactions allows important low-energy tests of NC and CC couplings and measurements of nuclear form factors. Consider, as an example, transitions between the ground state of ^{12}C and the isobaric analogue triplet states of the $A = 12$ system, i.e. ^{12}B, ^{12}C*, ^{12}N shown in figure 4.28. It has well-defined quantum numbers and contains simultaneous spin and isospin flips $\Delta I = 1$, $\Delta S = 1$. Such neutrino reactions on carbon might be important for all experiments based on organic scintillators. The most stringent signature is the inverse β-reaction ^{12}C(ν_e, e^-) ^{12}N$_{gs}$, where ^{12}N$_{gs}$ refers to the ground state of ^{12}N. A coincidence signal can be formed by the prompt electron together with the positron from the ^{12}N$_{gs}$ β^+-decay with a lifetime of 15.9 ms. With appropriate spatial and time cuts, KARMEN (see chapter 8) observed 536 such ν_e-induced CC events. The cross section is dominated by the form factor F_A (see (4.41)), which is given using a dipole parametrization, the CVC hypothesis and scaling between F_M and F_A (see [Fuk88] for more details) by

$$\frac{F_A(Q^2)}{F_A(0)} = \frac{1}{(1 - \frac{1}{12}R_A^2 Q^2)^2}.$$ (4.124)

The radius of the weak axial charge distribution R_A has been determined by a fit as [Bod94]

$$R_A = (3.8^{+1.4}_{-1.8}) \text{ fm}$$ (4.125)

and the form factor at zero momentum transfer as

$$F_A(0) = 0.73 \pm 0.11$$ (4.126)

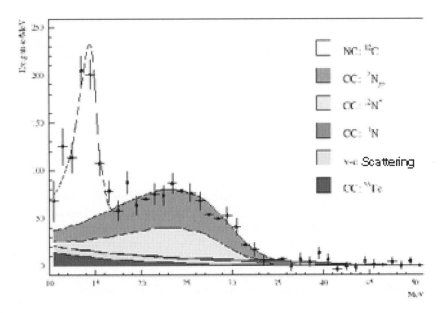

Figure 4.29. Energy spectrum of single prong events within the μ-decay time window (0.5–3.5 μs as obtained by KARMEN. The peak corresponds to the reaction ^{12}C(ν, ν') ^{12}C*$(1^+, 1; 15.1$ MeV). The bump for energies larger than 16 MeV comes from a variety of ν_e-induced CC reactions on carbon and iron. The largest contribution is the CC contributions into excited states of ^{12}N.

in good agreement with values obtained from the ft values (see chapter 6) of ^{12}B and ^{12}N β-decay. For comparison, muon capture on ^{12}C is only able to measure the form factor at a fixed or zero momentum transfer.

Another reaction of interest is the NC inelastic scattering process ^{12}C(ν, ν') ^{12}C*$(1^+, 1; 15.1$ MeV). The signal is a 15.1 MeV gamma ray. This peak is clearly visible in the data of KARMEN (figure 4.29). CC and NC reactions differ only by a Clebsch–Gordan coefficient of 1/2 and the fact that the ν_e and ν_μ spectra are almost identical allows the μ–e universality of the ν–Z^0 coupling at low energies to be tested. This can be done by looking at the ratio $R = \langle\sigma_{NC}(\nu_e + \bar{\nu}_\mu)\rangle / \langle\sigma_{CC}(\nu_e)\rangle$ which should be close to one. The measured value of KARMEN is

$$R = 1.17 \pm 0.11 \pm 0.12. \qquad (4.127)$$

Using the NC inelastic scattering process, a test on the Lorentz structure of the weak interactions could also be performed. In the same way, the electron energy spectrum from muon decay is governed by the Michel parameter ρ, the ν_e energy spectrum depends on an analogous quantity ω_L. KARMEN measured

$$\omega_L = 2.7^{+3.8}_{-3.2} \pm 3.1 \times 10^{-2} \qquad (4.128)$$

Table 4.2. Compilation of various nuclear cross sections obtained by KARMEN and LSND in the $A = 12$ system averaged over the corresponding neutrino energies.

Reaction	σ (cm^2) KARMEN	σ (cm^2) LSND
$\langle\sigma({}^{12}\text{C}(\nu_e, e^-){}^{12}\text{N}_{gs})\rangle$	$9.3 \pm 0.4 \pm 0.8 \times 10^{-42}$	$9.1 \pm 0.4 \pm 0.9 \times 10^{-42}$
$\langle\sigma({}^{12}\text{C}(\nu, \nu'){}^{12}\text{C}^*)\rangle(\nu = \nu_e, \bar{\nu}_\mu)$	$10.9 \pm 0.7 \pm 0.8 \times 10^{-42}$	—
$\langle\sigma({}^{12}\text{C}(\nu_e, e^-){}^{12}\text{N}^*)\rangle$	$5.1 \pm 0.6 \pm 0.5 \times 10^{-42}$	$5.7 \pm 0.6 \pm 0.6 \times 10^{-42}$
$\langle\sigma({}^{12}\text{C}(\nu_\mu, \mu^-){}^{12}\text{N}_{gs})\rangle$	—	$6.6 \pm 1.0 \pm 1.0 \times 10^{-41}$

in good agreement with the GWS prediction of $\omega_L = 0$ [Arm98]. A compilation of results from KARMEN and LSND (for both see chapter 8) is shown in table 4.2. Other examples will be discussed in the corresponding context.

After discussing neutrinos as probes of nuclear structure we now want to proceed to investigate neutrino properties especially in the case of non-vanishing neutrino masses. For that reason we start with a look at the physics beyond the standard model and the possibility of implementing neutrino masses.

Chapter 5

Neutrino masses and physics beyond the standard model

In spite of its enormous success in describing the available experimental data with high accuracy, the standard model discussed in chapter 3 is generally not believed to be the last step in unification. In particular, there are several parameters which are not predicted as you would expect from theory. For example the standard model contains 18 free parameters which have to be determined experimentally:

- the coupling constants e, α_S, $\sin^2 \theta_W$,
- the boson masses m_W, m_H,
- the lepton masses m_e, m_μ, m_τ,
- the quark masses m_u, m_d, m_s, m_c, m_b, m_t and
- the CKM matrix parameters: three angles and a phase δ.

Including massive neutrinos would add further parameters. In addition, the mass hierarchy remains unexplained, left-handed and right-handed particles are treated very differently and the quantization of the electric charge and the equality of the absolute values of proton and electron charge to a level better than 10^{-21} is not predicted.

However, what has undoubtedly succeeded has been the unification of two of the fundamental forces at higher energies, namely weak interactions and electromagnetism. The question arises as to whether there is another more fundamental theory which will explain all these quantities and whether a further unification of forces at still higher energies can be achieved. The aim is now to derive *all* interactions from the gauge transformations of *one simple* group G and, therefore, one coupling constant α (we will refrain here from discussing other, more specific solutions). Such theories are known as grand unified theories (GUTs). The grand unified group must contain the $SU(3) \otimes SU(2) \otimes U(1)$ group as a subgroup, i.e.

$$G \supset SU(3) \otimes SU(2) \otimes U(1). \tag{5.1}$$

The gauge transformations of a simple group, which act on the particle multiplets characteristic for this group, result in an interaction between the elements within a multiplet which is mediated by a similarly characteristic number of gauge bosons. The three well-known and completely different coupling constants can be derived in the end from a single one only if the symmetry associated with the group G is broken in nature. The hope of achieving this goal is given by the experimental fact that it is known that the coupling constants are not really constants. For more extensive reviews on GUTs see [Fuk03, Lan81, Moh86, 92, Ros84].

5.1   Running coupling constants

In quantum field theories like QED and QCD, dimensionless physical quantities $\mathcal{P}$ are expressed by a perturbation series in powers of the coupling constant α. Assume the dependence of $\mathcal{P}$ on a single coupling constant and energy scale Q. Renormalization introduces another scale μ where the subtraction of the UV divergences is actually performed and, therefore, both $\mathcal{P}$ and α become functions of μ. Since $\mathcal{P}$ is dimensionless, it only depends on the ratio Q^2/μ^2 and on the renormalized coupling constant $\alpha(\mu^2)$. Because the choice of μ is arbitrary, any explicit dependence of $\mathcal{P}$ on μ must be cancelled by an appropriate μ-dependence of α. It is natural to identify the renormalization scale with the physical energy scale of the process, $\mu^2 = Q^2$. In this case, α transforms into a running coupling constant $\alpha(Q^2)$ and the energy dependence of $\mathcal{P}$ enters only through the energy dependence of $\alpha(Q^2)$.

In general, there are equations in gauge theories which describe the behaviour of coupling constants α_i as a function of Q^2. These so-called 'renormalization group equations' have the general form

$$\frac{\partial \alpha_i(Q^2)}{\partial \ln Q^2} = \beta(\alpha_i(Q^2)). \tag{5.2}$$

The perturbative expansion of the beta function β depends on the group and the particle content of the theory. In lowest order the coupling constants are given by

$$\alpha_i(Q^2) = \frac{\alpha_i(\mu^2)}{1 + \alpha_i(\mu^2)\beta_0 \ln(Q^2/\mu^2)}. \tag{5.3}$$

As an example in QCD, the lowest term is given by

$$\beta_0 = \frac{33 - 2N_f}{12\pi} \tag{5.4}$$

with N_f as the number of active quark flavours. Alternatively, quite often another parametrization is used in form of

$$\alpha_i(Q^2) = \frac{1}{\beta_0 \ln(Q^2/\Lambda^2)} \tag{5.5}$$

which is equivaltent to (5.3) if

$$\Lambda^2 = \frac{\mu^2}{\exp(1/\beta_0 \alpha_i(\mu^2))}.$$ (5.6)

In the standard model (see chapter 3) strong and weak interactions are described by non-Abelian groups and, as a consequence, there is a decrease in the coupling constant with increasing energy, the so called asymptotic freedom (figure 5.1). This is due to the fact that the force-exchanging bosons like gluons and W, Z are carriers of the corresponding charge of the group itself, in contrast to QED, where photons have no electric charge. The starting points for the extrapolation are the values obtained at the Z^0 resonance given by

$$\alpha_S(m_Z^2) = 0.1184 \pm 0.0031$$ (5.7)
$$\alpha_{em}^{-1}(m_Z^2) = 127.9 \pm 0.1$$ (5.8)

and $\sin^2 \theta_W$ as given in table 3.4. These values are taken from [PDG00, Bet00]. After the extrapolation is carried out, all three coupling constants should meet at a point roughly on a scale of 10^{16} GeV (see, however, section 5.4.3) and from that point on an unbroken symmetry with a single coupling constant should exist. As previously mentioned the particle contents also influence the details of the extrapolation and any new particles introduced as, e.g., in supersymmetry would modify the Q^2 dependence of the coupling constants.

The simplest group with which to realize unification is SU(5). We will, therefore, first discuss the minimal SU(5) model (Georgi–Glashow model) [Geo74], even if it is no longer experimentally preferred.

5.2 The minimal SU(5) model

For massless fermions the gauge transformations fall into two independent classes for left- and right-handed fields, respectively. Let us assume the left-handed fields are the elementary fields (the right-handed transformations are equivalent and act on the corresponding charge conjugated fields). We simplify matters by considering only the first family, consisting of u, d, e and ν_e, giving 15 elementary fields, with c indicating antiparticles:

$$\begin{matrix} u_r, u_g, u_b, \nu_e \\ u_r^c, u_g^c, u_b^c, d_r^c, d_g^c, d_b^c \quad e^+ \\ d_r, d_g, d_b, e^- \end{matrix}$$ (5.9)

with r, g, b as the colour index of QCD. The obvious step would be to arrange the particles in three five-dimensional representations, which is the fundamental SU(5) representation. However, only particles within a multiplet can be transformed into each other and it is known that six of them, $u_r, u_g, u_b, d_r, d_g, d_b$,

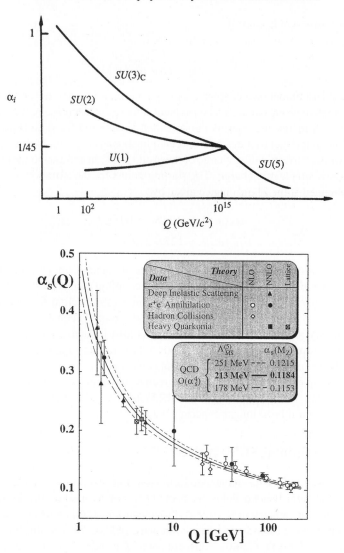

Figure 5.1. Top: Qualitative evolution with Q^2 of the three running coupling constants within the grand unification scale SU(5). Bottom: The clearest effect of running coupling with achievable energies is observed in the strong coupling α_S. Various experimental quantities can be used for its determination (from [Bet00]).

are transformed into each other via SU(2) and SU(3) transformations. Therefore, the fields have to be arranged in higher representations as a 10- and a $\bar{5}$- dimensional representation (the representation complementary to the fundamental representation 5, although this is not significant for our current purposes). The actual arrangement of fields into the multiplets results from the just mentioned quark transformations and the condition that the sum of the charges in every multiplet has to be zero:

$$
\bar{5} = \begin{pmatrix} d_g^c \\ d_r^c \\ d_b^c \\ e^- \\ -\nu_e \end{pmatrix}
\qquad
10 = \frac{1}{\sqrt{2}} \begin{pmatrix}
0 & -u_b^c & +u_r^c & +u_g & +d_g \\
+u_b^c & 0 & -u_g^c & +u_r & +d_r \\
-u_r^c & +u_g^c & 0 & +u_b & +d_b \\
-u_g & -u_r & -u_b & 0 & +e^+ \\
-d_g & -d_r & -d_b & -e^+ & 0
\end{pmatrix}. \quad (5.10)
$$

The minus signs in these representations are conventional. SU(5) has 24 generators T_j (SU(N) groups have $N^2 - 1$ generators), with a corresponding 24 gauge fields B_j, which can be written in matrix form as

$$
\begin{pmatrix}
G_{11} - \frac{2B}{\sqrt{30}} & G_{12} & G_{13} & X_1^c & Y_1^c \\
G_{21} & G_{22} - \frac{2B}{\sqrt{30}} & G_{23} & X_2^c & Y_2^c \\
G_{31} & G_{32} & G_{33} - \frac{2B}{\sqrt{30}} & X_3^c & Y_3^c \\
X_1 & X_2 & X_3 & \frac{W^3}{\sqrt{2}} + \frac{3B}{\sqrt{30}} & W^+ \\
Y_1 & Y_2 & Y_3 & W^- & -\frac{W^3}{\sqrt{2}} + \frac{3B}{\sqrt{30}}
\end{pmatrix}.
$$

$$(5.11)$$

Here the 3×3 submatrix G characterizes the gluon fields of QCD and the 2×2 submatrix W, B contains the gauge fields of the electroweak theory. In addition to the gauge bosons known to us, there are, however, a further 12 gauge bosons X, Y, which mediate transitions between baryons and leptons. The SU(5) symmetry has, however, to be broken in order to result in the standard model. Here also the break occurs through the coupling to the Higgs fields which also has to be an SU(5) multiplet. SU(5) can be broken through a 24-dimensional Higgs multiplet with a vacuum expectation value (vev) of about 10^{15}–10^{16} GeV. This means that all particles receiving mass via this breaking (e.g. the X, Y bosons) have a mass which is of the order of magnitude of the unification scale. By suitable SU(5) transformations we can ensure that only the X and Y bosons couple to the vacuum expectation value of the Higgs, while the other gauge bosons remain massless. An SU(5)-invariant mass term of the 24-dimensional Higgs field with the $\bar{5}$ and 10 representations of the fermions is not possible, so that the latter also remain massless. To break $SU(2) \otimes U(1)$ at about 100 GeV a further, independent five-dimensional Higgs field is necessary, which gives the W, Z bosons and the fermions their mass.

We now leave this simplest unifying theory and consider its predictions. For a more detailed description see, e.g., [Lan81]. A few predictions can be drawn from (5.10):

(i) Since the sum of charges has to vanish in a multiplet, the quarks have to have 1/3 multiples of the electric charge. For the first time the appearance of non-integer charges is required.

(ii) From this immediately follows the equality of the absolute value of the electron and proton charge also.
(iii) The relation between the couplings of the $\mathcal{B}$-field to a SU(2) doublet (see equation (3.30)) and that of the W^3-field is, according to equation (5.11), given by $(3/\sqrt{15}) : 1$. This gives a prediction for the value of the Weinberg angle $\sin^2 \theta_W$ [Lan81]:

$$\sin^2 \theta_W = \frac{g'^2}{g'^2 + g^2} = \frac{3}{8}. \tag{5.12}$$

This value is only valid for energies above the symmetry breaking. If renormalization effects are taken into consideration, at lower energies a slightly smaller value of

$$\sin^2 \theta_W = (0.218 \pm 0.006) \ln \left(\frac{100 \text{ MeV}}{\Lambda_{\text{QCD}}} \right) \tag{5.13}$$

results. This value is in agreement with the experimentally determined value (see section 3.4.3).
(iv) Probably the most dramatic prediction is the transformation of baryons into leptons due to X, Y exchange. This would, among other things, permit the decay of the proton and with it ultimately the instability of all matter.

Because of the importance of the last process it will be discussed in a little more detail.

5.2.1   Proton decay

As baryons and leptons are in the same multiplet, it is possible that protons and bound neutrons can decay. The main decay channels in accordance with the SU(5) model are [Lan81]:

$$p \rightarrow e^+ + \pi^0 \tag{5.14}$$

and

$$n \rightarrow \nu + \omega. \tag{5.15}$$

Here the baryon number is violated by one unit. We specifically consider proton decay. The process $p \rightarrow e^+ + \pi^0$ should amount to about 30–50% of all decays. The proton decay can be calculated analogously to the muon decay, resulting in a lifetime [Lan81]

$$\tau_p \approx \frac{M_X^4}{\alpha_5^2 m_p^5} \tag{5.16}$$

with $\alpha_5 = g_5^2/4\pi$ as the SU(5) coupling constant. Using the renormalization group equations (5.2) with standard model particle contents, the two quantities

M_X and α_5 can be estimated as [Lan81]

$$M_X \approx 1.3 \times 10^{14} \, \text{GeV} \frac{\Lambda_{\text{QCD}}}{100 \, \text{MeV}} \pm (50\%) \tag{5.17}$$

$$\alpha_5(M_X^2) = 0.0244 \pm 0.0002.$$

The minimal SU(5) model thus leads to the following prediction for the dominant decay channel [Lan86]:

$$\tau_p(\text{p} \rightarrow \text{e}^+\pi^0) = 6.6 \times 10^{28 \pm 0.7} \left[\frac{M_X}{1.3 \times 10^{14} \, \text{GeV}} \right]^4 \text{yr}$$

or

$$\tau_p(\text{p} \rightarrow \text{e}^+\pi^0) = 6.6 \times 10^{28 \pm 1.4} \left[\frac{\Lambda_{\text{QCD}}}{100 \, \text{MeV}} \right]^4 \text{yr}. \tag{5.18}$$

With $\Lambda_{\text{QCD}} = 200$ MeV the lifetime becomes $\tau_p = 1.0 \times 10^{30 \pm 1.4}$ yr. For reasonable assumptions on the value of Λ_{QCD}, the lifetime should, therefore, be smaller than 10^{32} yr. Besides the uncertainty in Λ_{QCD}, additional sources of error in the form of the quark wavefunctions in the proton must be considered. These are contained in the error on the exponent and a conservative upper limit of $\tau_p = 1.0 \times 10^{32}$ yr can be assumed.

The experimental search for this decay channel is dominated by Super-Kamiokande, a giant water Cerenkov detector installed in the Kamioka mine in Japan (see chapter 8). The decay should show the signature schematically shown in figure 5.2. By not observing this decay a lower limit of $\tau_p/BR(\text{p} \rightarrow \text{e}^+\pi^0) > 5.4 \times 10^{33}$ yr (90% CL) for the decay $\text{p} \rightarrow \text{e}^+\pi^0$ [Nak03] could be deduced. The disagreement with (5.18) rules out the minimal SU(5) model and other groups must be considered.

5.3   The SO(10) model

One such alternative is the SO(10) model [Fri75,Geo75] which contains the SU(5) group as a subgroup. The spinor representation is, in this case, 16-dimensional (see figure 5.3):

$$16_{\text{SO}(10)} = 10_{\text{SU}(5)} \oplus \bar{5}_{\text{SU}(5)} \oplus 1_{\text{SU}(5)}. \tag{5.19}$$

The SU(5) singlet cannot take part in any renormalizable, i.e. gauge SU(5) interaction. This new particle is, therefore, interpreted as the right-handed partner ν_R of the normal neutrino (more accurately, the field ν_L^C is incorporated into the multiplet). ν_R does not take part in any SU(5) interaction and, in particular, does not participate in the normal weak interaction of the GWS model. However, ν_R does participate in interactions mediated by the new SO(10) gauge bosons. Since the SO(10) symmetry contains the SU(5) symmetry, the possibility now exists

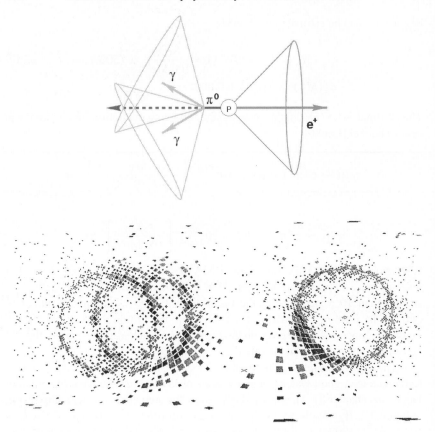

Figure 5.2. Top: Schematic picture of a proton decay $p \to e^+ \pi^0$ and the corresponding Cerenkov cones. Bottom: Monte Carlo simulation of such a proton decay for a water Cerenkov detector like Super-Kamiokande from [Vir99].

that somewhere above M_X the SO(10) symmetry is broken down into the SU(5) symmetry and that it then breaks down further as already discussed:

$$SO(10) \to SU(5) \to SU(3) \otimes SU(2)_L \otimes U(1). \tag{5.20}$$

Other breaking schemes for SO(10) do, however, exist. For example, it can be broken down without any SU(5) phase and even below the breaking scale left–right symmetry remains. Thus, the SO(10) model does represent the simplest left–right symmetrical theory.

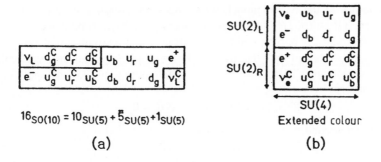

Figure 5.3. (*a*) All fermions of one family can be accommodated in *one* SO(10) multiplet. The 16th element is the as yet unseen right-handed neutrino ν_R or, equivalently, its CP conjugate ν_L^C. The illustrations correspond to different SO(10) breaking schemes. (*b*) The breaking of the SO(10) multiplet according to the $SU(4)_{EC} \otimes SU(2)_L \otimes SU(2)_R$ structure (from [Gro90]).

5.3.1 Left–right symmetric models

In this Pati–Salam model [Pat74] the symmetry breaking happens as follows:

$$SO(10) \rightarrow SU(4)_{EC} \otimes SU(2)_L \otimes SU(2)_R \tag{5.21}$$

where the index EC stands for *extended colour*, an extension of the strong interaction with the leptons as the fourth colour charge. The $SU(2)_R$ factor can be seen as the right-handed equivalent of the left-handed $SU(2)_L$. It describes a completely analogous right-handed weak interaction mediated by right-handed W bosons. Figure 5.3 shows the splitting of the multiplet according to the two symmetry-breaking schemes. The weak Hamiltonian in such a theory has to be extended by the corresponding terms involving right-handed currents:

$$H \sim G_F(j_L J_L + \kappa j_R J_L + \eta j_L J_R + \lambda j_R J_R) \tag{5.22}$$

with the leptonic currents j_i and hadronic currents J_i defined as in chapter 3 and $\kappa, \eta, \lambda \ll 1$. The mass eigenstates of the vector bosons $W_{1,2}^{\pm}$ can be expressed as a mixture of the gauge bosons:

$$W_1^{\pm} = W_L^{\pm} \cos\theta + W_R^{\pm} \sin\theta \tag{5.23}$$
$$W_2^{\pm} = - W_L^{\pm} \sin\theta + W_R^{\pm} \cos\theta \tag{5.24}$$

with $\theta \ll 1$ and $m_2 \gg m_1$. This can be used to rewrite the parameters in (5.22):

$$\eta = \kappa \approx \tan\theta \qquad \lambda \approx (m_1/m_2)^2 + \tan^2\theta \tag{5.25}$$

Lower bounds on the mass of right-handed bosons exist [PDG00]:

$$m_{W_R} > 720 \, \text{GeV} \tag{5.26}$$

In contrast to the SU(5) model, which does not conserve B and L but does conserve $(B - L)$, $(B - L)$ does not necessarily have to be conserved in the SO(10) model. A baryon number as well as a lepton number violation of two units is possible and with that the possibility of not only neutrinoless double β-decay (see chapter 7) but also of neutron–antineutron oscillations opens up. In the first case,

$$\Delta L = 2 \qquad \Delta B = 0 \qquad (5.27)$$

and, in the second,

$$\Delta B = 2 \qquad \Delta L = 0. \qquad (5.28)$$

For more details on the process of neutron–antineutron oscillations see [Kla95, Moh96a]. The SO(10) model can also solve the problem of SU(5) regarding the predictions of the lifetime of the proton. Their predictions lie in the region of 10^{32}–10^{38} yr [Lee95] and prefer other decay channels such as p $\rightarrow \nu K^+$ where the experimental limit is weaker and given by $\tau_p/BR(\text{p} \rightarrow \nu K^+) > 2.2 \times 10^{33}$ yr (90% CL) [Nak03]. It is convenient now to explore another extension of the standard model, which is given by supersymmetry (SUSY). This will also end with a short discussion of SUSY GUT theories.

5.4 Supersymmetry

A theoretical treatment of supersymmetry in all aspects is far beyond the scope of this book. We restrict ourselves to some basic results and applications in particle physics. Several excellent textbooks and reviews exist on this topic for further reading [Dra87, Wes86, 90, Moh86, 92, Nil84, Hab85, Lop96, Tat97, Mar97, Ell98, Oli99, Wei00, Kaz00].

Supersymmetry is a complete symmetry between fermions and bosons [Wes74]. This is a new symmetry and one as fundamental as that between particles and antiparticles. It expands the normal Poincaré algebra for the description of spacetime with extra generators, which changes fermions into bosons and *vice versa*. Let Q be a generator of supersymmetry such that

$$Q|(\text{Fermion})\rangle = |\text{Boson}\rangle \qquad \text{and} \qquad Q|(\text{Boson})\rangle = |\text{Fermion}\rangle.$$

In order to achieve this, Q itself has to have a fermionic character. In principle, there could be several supersymmetric generators Q but we restrict ourselve to one ($N = 1$ supersymmetry). The algebra of the supersymmetry is determined by the following relationships:

$$\{Q_\alpha, Q_\beta\} = 2\gamma_{\alpha\beta}^\mu p_\mu \qquad (5.29)$$
$$[Q_\alpha, p_\mu] = 0 \qquad (5.30)$$

Here p_μ is the 4-momentum operator. Note *that due to the anticommutator relation equation (5.29), internal particle degrees of freedom are connected to the external spacetime degrees of freedom.* This has the consequence that a *local* supersymmetry has to contain gravitation (supergravity theories, SUGRAs). A further generic feature of any supersymmetric theory is that the number of bosons equals that of fermions. A consequence for particle physics is then that the numbers of particles of the standard model are doubled. For every known fermion there is a boson and to each boson a fermion reduced by spin-$\frac{1}{2}$ exists.

One of the most attractive features of supersymmetry with respect to particle physics is an elegant solution to the hierarchy problem. The problem here is to protect the electroweak scale (3.67) from the Planck scale (13.54) which arises from higher order corrections. This is especially dramatic for scalar particles like the Higgs. The Higgs mass receives a correction δm_H via higher orders where [Ell91b, Nil95]

$$\delta m_H^2 \sim g^2 \int^\Lambda \frac{\mathrm{d}^4 k}{(2\pi)^4 k^2} \sim g^2 \Lambda^2. \tag{5.31}$$

If the cut-off scale Λ is set at the GUT scale or even the Planck scale, the lighter Higgs particle would experience corrections of the order M_X or even M_{Pl}. In order to achieve a well-defined theory, it is then necessary to fine tune the parameters in all orders of perturbation theory. With supersymmetry the problem is circumvented by postulating new particles with similar mass and equal couplings. Now corresponding to any boson with mass m_B in the loop there is a fermionic loop with a fermion mass m_F with a relative minus sign. So the total contribution to the 1-loop corrected Higgs mass is

$$\delta m_H^2 \simeq O\left(\frac{\alpha}{4\pi}\right)(\Lambda^2 + m_B^2) - O\left(\frac{\alpha}{4\pi}\right)(\Lambda^2 + m_F^2) = O\left(\frac{\alpha}{4\pi}\right)(m_B^2 - m_F^2). \tag{5.32}$$

When all bosons and fermions have the same mass, the radiative corrections vanish identically. The stability of the hierarchy only requires that the weak scale is preserved, meaning

$$|(m_B^2 - m_F^2)| \leq 1 \text{ TeV}^2. \tag{5.33}$$

Two remarks should be made. If this solution is correct, supersymmetric particles should be observed within the next generation of accelerators, especially the LHC. However, supersymmetry predicts that the masses of particles and their supersymmetric partners are identical. Because no supersymmetric particle has yet been observed, supersymmetry must be a broken symmetry. In the following we restrict our discussion to the *minimal supersymmetric standard model* (MSSM).

5.4.1 The minimal supersymmetric standard model (MSSM)

As already stated, even in the minimal model we have to double the number of particles (introducing a superpartner to each particle) and we have to add another Higgs doublet (and its superpartner). The reason for the second Higgs doublet is given by the fact that there is no way to account for the up and down Yukawa couplings with only one Higgs field. The nomenclature of the supersymmetric partners is as follows: the scalar partners of normal fermions are designated with a preceding 's', so that, for example, the supersymmetric partner of the quark becomes the squark $\tilde{q}$. The super-partners of normal bosons receive the ending '-ino'. The partner of the photon, therefore, becomes the photino $\tilde{\gamma}$.

In the Higgs sector both doublets obtain a vacuum of expectation value (vev):

$$\langle H_1 \rangle = \begin{pmatrix} v_1 \\ 0 \end{pmatrix} \qquad \langle H_2 \rangle = \begin{pmatrix} 0 \\ v_2 \end{pmatrix}. \tag{5.34}$$

Their ratio is often expressed as a parameter of the model:

$$\tan \beta = \frac{v_2}{v_1}. \tag{5.35}$$

Furthermore, in contrast to the SM here one has eight degrees of freedom, three of which can be gauged away as in the SM. The net result is that there are five physical Higgs bosons: two CP-even (scalar) neutrals (h, H), one CP-odd (pseudo-scalar) neutral (A) and two charged Higgses ($H^\pm$).

There are four neutral fermions in the MSSM which receive mass but can mix as well. They are the gauge fermion partners of the B and W^3 gauge bosons (see chapter 3), as well as the partners of the Higgs. They are, in general, called neutralinos or, more specifically, the bino $\tilde{B}$, the wino $\tilde{W}^3$ and the Higgsinos $\tilde{H}_1^0$ and $\tilde{H}_2^0$. The neutralino mass matrix can be written in the ($\tilde{B}, \tilde{W}^3, \tilde{H}_1^0, \tilde{H}_2^0$) basis as

$$\begin{pmatrix} M_1 & 0 & -M_Z s_{\theta_W} \cos \beta & M_Z s_{\theta_W} \sin \beta \\ 0 & M_2 & M_Z c_{\theta_W} \cos \beta & -M_Z c_{\theta_W} \sin \beta \\ -M_Z s_{\theta_W} \cos \beta & M_Z c_{\theta_W} \cos \beta & 0 & -\mu \\ M_Z s_{\theta_W} \sin \beta & -M_Z c_{\theta_W} \sin \beta & -\mu & 0 \end{pmatrix}$$

$$\tag{5.36}$$

where $s_{\theta_W} = \sin \theta_W$ and $c_{\theta_W} = \cos \theta_W$. The eigenstates are determined by diagonalizing the mass matrix. As can be seen, they depend on three parameters M_1 (coming from the bino mass term), M_2 (from the wino mass term) and μ (from the Higgsino mixing term $\frac{1}{2}\mu \tilde{H}_1 \tilde{H}_2$). We also have four charginos coming from $\tilde{W}^\pm$ and $\tilde{H}^\pm$. The chargino mass matrix is composed similar to the neutralino mass matrix.

Using the universality hypothesis that, on the GUT scale, all the gaugino masses (spin-$\frac{1}{2}$ particles) are identical to a common mass $m_{1/2}$ and that all the spin-0 particle masses at this scale are identical to m_0, we end up with μ, $\tan \beta$, m_0, $m_{1/2}$ and A as free parameters. Here A is a soft supersymmetry-breaking parameter (for details see [Oli99]). In total five parameters remain which have to be explored experimentally.

5.4.2 *R*-parity

The MSSM is a model containing the minimal extension of the field contents of the standard model as well as minimal extensions of interactions. Only those required by the standard model and its supersymmetric generalization are considered. It is assumed that R-parity is conserved to guarantee the absence of lepton- and baryon-number-violating terms. R-parity is assigned as follows:

$$R_P = 1 \quad \text{for normal particles}$$
$$R_P = -1 \quad \text{for supersymmetric particles.}$$

R_P is a multiplicative quantum number and is connected to the baryon number B, the lepton number L and the spin S of the particle by

$$R_P = (-1)^{3B+L+2S}. \tag{5.37}$$

Conservation of R-parity has two major consequences:

(i) Supersymmetric particles can only be produced in pairs.
(ii) The lightest supersymmetric particle (LSP) has to be stable.

However, even staying with the minimal particle content and being consistent with all symmetries of the theory, more terms can be written in the superpotential W which violate R-parity given as

$$W_{\not R_p} = \lambda_{ijk} L_i L_j \bar{E}_k + \lambda'_{ijk} L_i Q_j \bar{D}_k + \lambda''_{ijk} U_i \bar{D}_j \bar{D}_k \tag{5.38}$$

where the indices i, j and k denote generations. L, Q denote lepton and quark doublet superfields and $\bar{E}$, $\bar{U}$ and $\bar{D}$ denote lepton and up, down quark singlet superfields respectively. Terms proportional to λ, λ' violate lepton number, those proportional to λ'' violate baryon number. A compilation of existing bounds on the various coupling constants can be found in [Bed99].

After discussing the MSSM as an extension of the standard model and the possibility of R_P violation, it is obvious that one can also construct supersymmetric GUT theories, like SUSY SU(5), SUSY SO(10) and so on, with new experimental consequences. As schematic illustration of unification is shown in figure 5.4. We now want to discuss briefly a few topics of the experimental search.

5.4.3 Experimental search for supersymmetry

Consider, first, the running coupling constants. As already mentioned, new particles change the parameters in the renormalization group equations (5.2). As can be seen in figure 5.5, in contrast to the standard model extrapolation the coupling constants including MSSM now unify and the unified value and scale are given by

$$M_{\text{GUT}} = 10^{15.8\pm0.3\pm0.1} \text{ GeV} \tag{5.39}$$
$$\alpha_{\text{GUT}}^{-1} = 26.3 \pm 1.9 \pm 1.0. \tag{5.40}$$

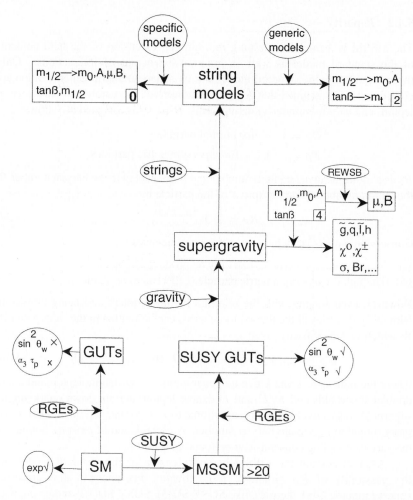

Figure 5.4. Schematic picture of the different steps in grand unification from the Fermi scale to the Planck scale. The numbers indicate the number of new parameters required to describe the corresponding model (from [Lop96]).

Even though this is not a proof that SUSY is correct, it at least gives a hint of its existence. The prediction of the Weinberg angle in supersymmetric models also corresponds better to the experimentally observed value (chapter 3) than those of GUT theories without supersymmetry The predictions of these theories are [Lan93b]:

$$\sin^2 \theta_W(m_Z) = 0.2334 \pm 0.0050 \text{ (MSSM)} \tag{5.41}$$

$$\sin^2 \theta_W(m_Z) = 0.2100 \pm 0.0032 \text{ (SM)}. \tag{5.42}$$

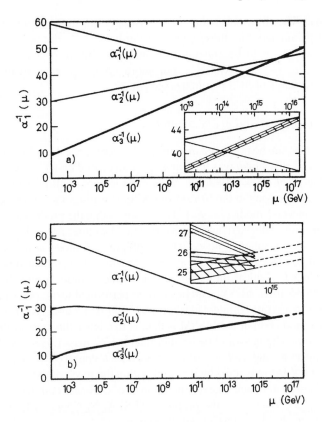

Figure 5.5. Running of the coupling constants. Left: Evolution assuming the SM particle content. Evidently the coupling constants do not meet at the unification scale. Right: Unification is achieved by including the MSSM (from [Ama91]).

The experimental strategies to search for SUSY can be separated into four groups:

- direct production of supersymmetric particles in high-energy accelerators,
- precision measurements,
- search for rare decays and
- dark matter searches.

For the accelerator searches another constraint is applied to work with four free parameters (constrained MSSM, CMSSM). This requires gauge coupling unification at the GUT scale leading to the relation $M_1 = \frac{5}{3}\frac{\alpha_1}{\alpha_2}M_2$ and one can only work with the parameters μ, $\tan\beta$, m_0, $m_{1/2}$. Beside that, as long as R-parity is conserved, the LSP remains stable and acts as a good candidate for dark matter (see chapter 13).

A good example for the second method is a search for an electric dipole moment of the neutron or processes where supersymmetry enters via loop corrections. The third one either uses existing stringent experimental bounds to restrict parameters like those coming from b $\to$ s + γ decay or investigates processes which might be enhanced or modified with respect to the standard model like μ $\to$ 3e [Ays01]. For more comprehensive reviews on the experimental status of SUSY searches see [Kaz00].

5.4.3.1   *SUSY signatures at* e^+e^- *colliders*

SUSY particles can be produced in pairs at e^+e^- colliders. The obvious machine to look at is LEPII, which was running at the end of its data-taking with a centre-of-mass energy of $\sqrt{s}$ = 208 GeV. A common feature of all possible signals as long as we are working in the MSSM or CMSSM is a significant missing energy ($\not{E}_T$) and transverse momenta ($\not{p}_T$). The reason is that the produced stable LSPs escape detection. This signature is accompanied by either jets or leptons. So far all searches have resulted in no evidence and figure 5.6 shows as an example the LSP neutralino mass as a function of tan β. Any mass lighter than about 30 GeV can be excluded. Typical limits for charginos and sleptons are of the order of 100 GeV. In addition, SUSY searches can be performed using other production mechanisms in p$\bar{\text{p}}$ (pp)-colliders like Fermilab Run II and the LHC. The reason is that another prediction of the MSSM is that, at tree level, the mass of the lightest supersymmetric Higgs should be smaller than the Z^0 mass ($m_h < m_Z$). Taking into account first- and second-order corrections, a conservative upper limit of $m_h < 130$ GeV is predicted which is well within the reach of these machines. If SUSY is realized in nature a next generation of e^+e^- linear colliders with higher centre-of-mass energies like the proposed TESLA, NLC and CLIC will have a rich programme in SUSY particle spectroscopy.

5.4.3.2   *SUSY GUTs and proton decay*

Predictions for proton decay are changed within SUSY GUTs. The increased unification scale with respect to the minimal SU(5) results in a bigger M_X mass. This results in a substantially increased lifetime for the proton of about 10^{35} yr, which is compatible with experiment. However, the dominant decay channel (see, e.g., [Moh86, 92]) changes in such models, such that the decays p $\to$ K$^+$ + $\bar{\nu}_\mu$ and n $\to$ K^0 + $\bar{\nu}_\mu$ should dominate. The experimentally determined lower limit [Vir99] of the proton lifetime of $\tau_p > 1.9 \times 10^{33}$ yr for this channel is less restrictive than the p $\to$ π^0 + e$^+$ mode. Recent calculations within SUSY SU(5) and SUSY SO(10) seem to indicate that the upper bound on the theoretical expectation is $\tau_p < 5 \times 10^{33}$ yr which should be well within the reach of longer running Super-K and next-generation experiments like Hyper-Kamiokande, ICARUS, UNO and AQUA-RICH discussed later. Other dominant decay modes might reveal in some left–right symmetric models, which prefer p $\to$ μ^+K^0. The experimental bound here is $\tau_p > 1.2 \times 10^{32}$ yr [Vir99]. For a bound on R_p-violating constants coming from proton decay see [Smi96]. After

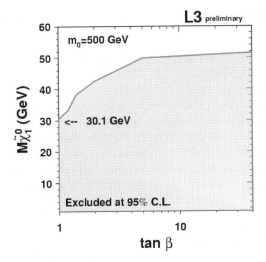

Figure 5.6. Neutralino mass limits as a function of $\tan \beta$ as obtained by the DELPHI and L3 experiments at LEP (from [Kaz00]).

discussing the standard model and its possible extensions we now want to take a look at what type of neutrino mass generation can be realized.

5.5 Neutrino masses

As already stated in chapter 3, neutrino masses are set to zero in the standard model. Therefore, any evidence of a non-vanishing neutrino mass would indicate physics 'beyond the standard model'.[1] A lot of model building has been performed to include neutrino masses in physics, for recent reviews see [Val03, Kin03].

5.5.1 Neutrino masses in the electroweak theory

Neutrino masses can be created in the standard model by extending the particle content of the theory. Dirac mass terms of the form (2.36) and the corresponding Yukawa couplings (3.61) can be written for neutrinos if singlet ν_R are included in the theory as for all other fermions. This would result in (see (3.61))

$$\mathcal{L}_{\text{Yuk}} = -c_\nu \bar{\nu}_R \phi^\dagger \begin{pmatrix} \nu_{e_L} \\ e_L \end{pmatrix} + h.c. \tag{5.43}$$

---

[1] It is a matter of taste what exactly 'beyond the standard model' means. Neutrino masses can be generated within the gauge structure of $SU(3) \otimes SU(2) \times U(1)$ by enlarging the particle content or adding non-renormalizable interactions. Even by adding new particles this sometimes is nevertheless still called 'standard model' because the gauge structure is unchanged.

resulting in terms like (3.61)

$$= -c_\nu \nu \bar{\nu} \nu. \tag{5.44}$$

The smallness of the neutrino mass must then be explained by a correspondingly smaller Yukawa coupling c_ν.

If no additional fermions are included the only possible mass terms are of Majorana type and, therefore, violate lepton number (equivalent to violating $B - L$, which is the only gauge-anomaly-free combination of these quantum numbers). Thus we might introduce new Higgs bosons which can violate $B - L$ in their interactions. Furthermore, the neutrino mass has to be included in a Yukawa coupling. The corresponding fermionic bilinears having a net $B - L$ number and the further requirement of gauge-invariant Yukawa couplings determine the possible Higgs multiplets, which can couple directly to the fermions:

- a triplet Δ and
- a singly charged singlet h^-.

The Higgs triplet is given by

$$\begin{pmatrix} \Delta^0 \\ \Delta^- \\ \Delta^{--} \end{pmatrix} \tag{5.45}$$

and its Yukawa coupling gives neutrinos their mass. The component Δ^0 requires a vacuum expectation value of v_3, which has to be much smaller than the one obtained by the standard Higgs doublet. Because the Higgs potential now contains both multiplets ϕ (3.56) and Δ, both contribute to the mass of the gauge bosons. From that an upper bound on v_3 can already be given:

$$\rho = \frac{m_W^2}{m_Z^2 cos\theta_W} = \frac{1 + 2v_3^2/v^2}{1 + 4v_3^2/v^2} \rightarrow \frac{v_3}{v} < 0.07. \tag{5.46}$$

The second model introducing an SU(2) singlet Higgs h^- has been proposed by Zee [Zee80]. As h^- carries electric charge its vev must vanish and some other sources of $B - L$ violation must be found.

An independent possibility introducing neutrino masses in the standard model would be non-renormalizable operators, also leading to non-standard neutrino interactions. After discussing how by enlarging the particle content of the standard model neutrino masses can be generated, we now want to see what possibilities GUT and SUSY offer.

5.5.2 Neutrino masses in the minimal SU(5) model

In the multiplets given in (5.15) only ν_L with its known two degrees of freedom shows up, allowing only Majorana mass terms for neutrinos. The coupling to the Higgs field Φ has to be of the form $(\nu_L \otimes \nu_L^C)\Phi$. However, $5 \otimes 5$ results in combinations of $10 \oplus 15$ which does not allow us to write SU(5)-invariant mass terms, because with the Higgs, only couplings of 25 and 5 representations are possible. Therefore, in the minimal SU(5) neutrinos remain massless. But, as in the standard model, enlarging the Higgs sector allows us to introduce Majorana mass terms.

5.5.3 Neutrino masses in the SO(10) model and the seesaw mechanism

In the SO(10) model the free singlet can be identified with a right-handed neutrino (see figure 5.3). It is, therefore, possible to produce Dirac mass terms. The corresponding Yukawa couplings have to be made with 10, 120 or 126 dimensional representations of the Higgs. However, as the neutrinos belong to the same multiplet as the remaining fermions, their mass generation is not independent from that of the other fermions and one finds, e.g. by using the 10-dimensional Higgs, that all Dirac mass terms are more or less identical, in strong contradiction to experiments where limits for neutrino masses are much smaller than the corresponding ones on charged leptons and quarks (see chapter 6). This problem can be solved by adding the 126-dimensional representation of the Higgs field and assigning a vev to the SU(5) singlet component. This gives rise to Majorana mass of the right-handed neutrino. This mass term can take on very large values up to M_X. Under these assumptions it is possible to obtain no Majorana mass term for ν_L and a very large term for ν_R. In this case the mass matrix (2.48) has the following form:

$$ M = \begin{pmatrix} 0 & m_D \\ m_D & m_R \end{pmatrix} \tag{5.47} $$

where m_D is of the order of MeV–GeV, while $m_R \gg m_D$. But this is exactly the requirement for a seesaw mechanism as discussed in chapter 2. This means that it is possible for a suitably large Majorana mass m_R in equation (5.47) to reduce the observable masses so far that they are compatible with experiment. This is the *seesaw* mechanism for the production of small neutrino masses [Gel78, Moh80]. If this is taken seriously, a quadratic scaling behaviour of the neutrino masses with the quark masses or charged lepton masses follows (2.63), i.e.

$$ m_{\nu_e} : m_{\nu_\mu} : m_{\nu_\tau} \sim m_u^2 : m_c^2 : m_t^2 \quad \text{or} \quad \sim m_e^2 : m_\mu^2 : m_\tau^2. \tag{5.48} $$

However, several remarks should be made. This relation holds on the GUT scale. By extrapolating down to the electroweak scale using the renormalization group equations, significant factors could disturb the relation. As an example the ratio of the three neutrino masses for two different models is given by [Blu92]

$$ m_1 : m_2 : m_3 = 0.05m_u^2 : 0.09m_c^2 : 0.38m_t^2 \quad \text{SUSY–GUT} \tag{5.49} $$

$$ m_1 : m_2 : m_3 = 0.05m_u^2 : 0.07m_c^2 : 0.18m_t^2 \quad \text{SO(10).} \tag{5.50} $$

Furthermore, it is assumed that the heavy Majorana mass shows no correlation with the Dirac masses. However, if this is the case, a linear seesaw mechanism arises.

5.5.3.1 *Almost degenerated neutrino masses*

If the upper left entry in (5.47) does not vanish exactly, the common seesaw formula might change. The common general seesaw term

$$m_\nu \approx -m_D^T m_R^{-1} m_D \tag{5.51}$$

is modified to

$$m_\nu \approx f \frac{v^2}{v_R} - m_D^T m_R^{-1} m_D \tag{5.52}$$

where the first term includes the vev of the Higgs fields. Clearly if the first term dominates, there will be no hierarchical seesaw but the neutrinos will be more or less degenerated in mass (sometimes called type II seesaw).

5.5.4 Neutrino masses in SUSY and beyond

Including SUSY in various forms like the MSSM, allowing R_p violation and SUSY GUT opens a variety of new possible neutrino mass generations. This can even be extended by including superstring-inspired models or those with extra dimensions. The neutrino mass schemes are driven here mainly by current experimental results such as those described in the following chapters. In the MSSM, neutrinos remain massless as in the standard model, because of lepton and baryon number conservation. For some current models and reviews, see [Die00, Moh01, Alt02, Hir02, Kin03].

5.6 Neutrino mixing

In the following chapters, it will be found that neutrinos have a non-vanishing mass. Then, the weak eigenstates ν_α need not to be identical to the mass eigenstates ν_i. As in the quark sector they could be connected by a unitary matrix U like the CKM matrix (see chapter 3) called the MNS-matrix (Maki–Nakagava–Sakata)[2] [Mak62]:

$$|\nu_\alpha\rangle = U_{\text{MNS}} |\nu_i\rangle \qquad \alpha = e, \mu, \tau; \ i = 1 \ldots 3. \tag{5.53}$$

For three Dirac neutrinos U is given, in analogy to (3.71), as

$$U = \begin{pmatrix} c_{12}c_{13} & s_{12}c_{13} & s_{13}e^{-i\delta} \\ -s_{12}c_{23} - c_{12}s_{23}s_{13}e^{i\delta} & c_{12}c_{23} - s_{12}s_{23}s_{13}e^{i\delta} & s_{23}c_{13} \\ s_{12}s_{23} - c_{12}s_{23}s_{13}e^{i\delta} & -c_{12}s_{23} - s_{12}c_{23}s_{13}e^{i\delta} & c_{23}c_{13} \end{pmatrix} \tag{5.54}$$

where $s_{ij} = \sin\theta_{ij}$, $c_{ij} = \cos\theta_{ij}$ ($i, j = 1, 2, 3$). A graphical illustration of the mixing matrix elements ignoring the CP-phase is shown in figure 5.7. In the Majorana case, the requirement of particle and antiparticle to be identical, restricts

---

[2] It is also often quoted as the Pontecorvo–Maki–Nakagava–Sakata (PMNS) matrix.

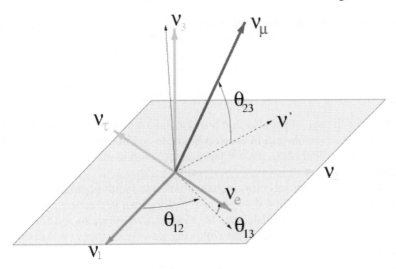

Figure 5.7. Graphical representation of the mixing matrix elements between flavour and mass eigenstates.

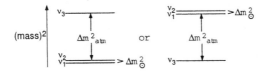

Figure 5.8. Normal and inverted mass hierarchies for three neutrinos. The inverted scheme is characterized by a $\Delta m_{23}^2 = m_3^2 - m_2^2 < 0$.

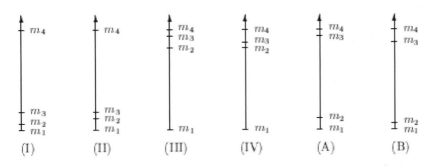

Figure 5.9. Various neutrino mass schemes which can be built on the existence of four different neutrino states to describe current neutrino oscillation evidences. The first four patterns are known as '3 + 1' schemes, because of the one isolated state m_4, while the remaining two are called '2 + 2' schemes.

the freedom to redefine the fundamental fields. The net effect is the appearance of a CP-violating phase already in two flavours. For three flavours two additional phases have to be introduced resulting in a mixing matrix of the form

$$U = U_{\text{MNS}} \operatorname{diag}(1, e^{i\alpha}, e^{i\beta}). \tag{5.55}$$

In the three-flavour scenario several possible mass schemes can still be discussed which will become obvious in chapters 8–10. In addition to normal and inverted mass schemes (figure 5.8), almost degenerated neutrino masses $m_1 \approx m_2 \approx m_3$ are possible.

Further common scenarios include a possible fourth neutrino as shown in figure 5.9. Such a neutrino does not take part in weak interactions and is called a sterile neutrino. Having discussed the theoretical motivations and foundations for a possible neutrino mass in the following we want to focus on experimental searches and evidence.

Chapter 6

Direct neutrino mass searches

In this chapter direct methods for neutrino mass determinations are discussed. The classical way to perform such searches for $\bar{\nu}_e$ is to investigate β-decay. From the historical point of view this process played a major role (see chapter 1), because it was the motivation for W Pauli to introduce the neutrino. Many fundamental properties of weak interactions were discovered by investigating β-decay. For an extensive discussion on weak interactions and β-decay see [Sch66, Sie68, Wu66, Kon66, Mor73, Gro90, Wil01, Wei02].

6.1 Fundamentals of β-decay

Beta-decay is a nuclear transition, where the atomic number Z of the nucleus changes by one unit, while the atomic mass A remains the same.

This results in three possible decay modes:

$$(Z, A) \rightarrow (Z + 1, A) + e^- + \bar{\nu}_e \qquad (\beta^- \text{-decay}) \qquad (6.1)$$

$$(Z, A) \rightarrow (Z - 1, A) + e^+ + \nu_e \qquad (\beta^+ \text{-decay}) \qquad (6.2)$$

$$e^- + (Z, A) \rightarrow (Z - 1, A) + \nu_e \qquad (\text{Electron capture}). \qquad (6.3)$$

The basic underlying mechanism for (6.1) is given by

$$n \rightarrow p + e^- + \bar{\nu}_e \qquad \text{or} \qquad d \rightarrow u + e^- + \bar{\nu}_e \qquad (6.4)$$

on the quark level respectively. The other decay modes can be understood in an analogous way. The corresponding decay energies are given by the following relations, where $m(Z, A)$ denotes the mass of the neutral atom (not the nucleus):

β^--decay:

$$Q_{\beta^-} = [m(Z, A) - Zm_e]c^2 - [(m(Z + 1, A) - (Z + 1)m_e) + m_e]c^2$$
$$= [m(Z, A) - m(Z + 1, A)]c^2. \qquad (6.5)$$

The Q-value corresponds exactly to the mass difference between the mother and daughter atom.

β^+-decay:

$$Q_{\beta^+} = [m(Z, A) - Zm_e]c^2 - [(m(Z - 1, A) - (Z - 1)m_e) + m_e]c^2$$
$$= [m(Z, A) - m(Z - 1, A) - 2m_e]c^2. \qquad (6.6)$$

Because all masses are given for atoms, this decay requires the rest mass of two electrons. Therefore, the mass difference between both has to be larger than $2m_ec^2$ for β^+-decay to occur.

Electron capture:

$$Q_{EC} = [m(Z, A) - Zm_e]c^2 + m_ec^2 - [m(Z - 1, A) - (Z - 1)m_e]c^2$$
$$= [m(Z, A) - m(Z - 1, A)]c^2. \qquad (6.7)$$

As can be expected the Q-values of the last two reactions are related by

$$Q_{\beta^+} = Q_{EC} - 2m_ec^2. \qquad (6.8)$$

If Q is larger than $2m_ec^2$, both electron capture and β^+-decay are competitive processes, because they lead to the same daughter nucleus. For smaller Q-values only electron capture will occur. Obviously, for any of the modes to occur the corresponding Q-value has to be larger than zero.

The way to determine the neutrino mass is related to β^--decay, hence, this mode will be discussed in more detail. More accurately, this method measures the mass of $\bar{\nu}_e$ but CPT-conservation ensures that $m_{\bar{\nu}_e} \equiv m_{\nu_e}$.

The important point is to understand the shape of the observed electron spectrum (see chapter 1) and the impact of a non-vanishing neutrino mass which, for small neutrino masses, shows up only in the endpoint region of the electron spectrum. The following discussion is related to allowed and super-allowed transitions, meaning that the leptons do not carry away any angular momentum ($l = 0$). The transition rate of β-decay to produce an electron in the energy interval between E and $E + \Delta E$ is given by Fermi's Golden Rule:

$$\frac{d^2N}{dt\,dE} = \frac{2\pi}{\hbar}|\langle f|H_{if}|i\rangle|^2\rho(E) \qquad (6.9)$$

where $|\langle f|H_{if}|i\rangle|$ describes the transition matrix element including the weak Hamilton operator H_{if}, $\rho(E)$ denotes the density of final states and E_0 corresponds to the Q-value of the nuclear transition. Neglecting nuclear recoil, the following relation is valid:

$$E_0 = E_\nu + E_e. \qquad (6.10)$$

6.1.1 Matrix elements

Consider first the matrix element given by

$$|\langle f|H_{if}|i\rangle| = \int dV \,\psi_f^* H_{if}\psi_i. \qquad (6.11)$$

The wavefunction ψ_i of the initial state is determined by the nucleons in the mother atom, while the final-state wavefunction ψ_f has to be built by the wavefunction of the daughter as well as the wavefunction of the electron-neutrino field. The interaction between the nucleus and the leptons is weak, thus, in a first approximation wavefunctions normalized to a volume V can be treated as plane waves:

$$\phi_e(r) = \frac{1}{\sqrt{V}}e^{ik_e\cdot r} \qquad (6.12)$$

$$\phi_\nu(r) = \frac{1}{\sqrt{V}}e^{ik_\nu\cdot r}. \qquad (6.13)$$

These wavefunctions can be expanded in a Taylor series around the origin in the form

$$\phi_l(r) = \frac{1}{\sqrt{V}}(1 + ik_l\cdot r + \cdots) \qquad \text{with } l \equiv e, \nu. \qquad (6.14)$$

A comparison of the typical nuclear diameter and the Compton wavelength of the electron and neutrino shows that $k_l r \ll 1$. Therefore, in good approximation, the wavefunctions are

$$\phi_l(r) = \frac{1}{\sqrt{V}} \qquad \text{with } l \equiv e, \nu. \qquad (6.15)$$

The electron wavefunction has to be modified taking into account the electromagnetic interaction of the emitted electron with the Coulomb field of the daughter nucleus $(A, Z + 1)$. For an electron the effect produces an attraction, while for positrons it results in a repulsion (figure 6.4). The correction factor is called the Fermi function $F(Z + 1, E)$ and it is defined as

$$F(Z + 1, E) = \frac{|\phi_e(0)_{\text{Coul}}|^2}{|\phi_e(0)|^2}. \qquad (6.16)$$

In the non-relativistic approach it can be approximated by [Pri68]

$$F(Z + 1, E) = \frac{x}{1 - e^{-x}} \qquad (6.17)$$

with

$$x = \pm\frac{2\pi(Z + 1)\alpha}{\beta} \qquad \text{for } \beta^\mp\text{-decay} \qquad (6.18)$$

and α as the fine structure constant and $\beta = v/c$. An accurate treatment has to take relativistic effects into account and a numerical compilation of Fermi

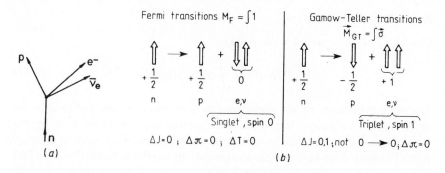

Figure 6.1. Neutron beta decay (*a*) and spin balance (*b*) for Fermi and Gamow–Teller transitions.

functions can be found in [Lan69]. The lepton wavefunctions are practically constant all over the nuclear volume. As a consequence, the term $|\langle f|H_{if}|i\rangle|^2$ will contain a factor $|\phi_e(0)|^2|\phi_\nu(0)|^2 \simeq 1/V^2$. Introducing a coupling constant g to account for the strength of the interaction the matrix element can be written as

$$|\langle f|H_{if}|i\rangle|^2 = g^2 F(E, Z+1)|\phi_e(0)|^2|\phi_\nu(0)|^2|M_{if}|^2$$

$$\simeq \frac{g^2}{V^2} F(E, Z+1)|M_{if}|^2 \tag{6.19}$$

where the so called nuclear matrix element M_{if} is given by

$$M_{if} = \int \mathrm{d}V \, \phi_f^* \mathcal{O}\phi_i. \tag{6.20}$$

This expression now describes the transition between the two nuclear states, where $\mathcal{O}$ is the corresponding operator and, therefore, it is determined by the nuclear structure. Consider again only allowed transitions. In this case two kinds of nuclear transitions can be distinguished depending on whether the emitted leptons form a spin-singlet or spin-triplet state. Assume that the spins of electron and $\bar{\nu}_e$ are antiparallel with a total spin zero. Such transitions are called Fermi transitions (figure 6.1). The transition operator corresponds to the isospin ladder operator τ^- and is given by

$$\mathcal{O}_F = I^- = \sum_{i=1}^{A} \tau^-(i) \tag{6.21}$$

summing over all nucleons. Because the transition neither changes spin J, parity π nor isospin I the following selection rules hold:

$$\Delta I = 0 \qquad \Delta J = 0 \qquad \Delta \pi = 0. \tag{6.22}$$

The second kind of transition is characterized by the fact that both leptons have parallel spins resulting in a total spin 1. Such transitions are called Gamow–Teller transitions and are described by

$$\mathcal{O}_{GT} = \sum_{i=1}^{A} \sigma(i)\tau^-(i) \qquad (6.23)$$

where $\sigma(i)$ are the Pauli spin matrices, which account for the spin flip of the involved nucleons. Also here selection rules are valid:

$$\Delta I = 0, 1$$
$$\Delta J = 0, 1 \qquad \text{no } 0 \to 0$$
$$\Delta \pi = 0. \qquad (6.24)$$

In sum the nuclear matrix element for allowed transitions has the form

$$g^2|M_{if}|^2 = g_V^2|M_F|^2 + g_A^2|M_{GT}|^2 \qquad (6.25)$$

already taking into account the different coupling strength of both transitions by using the vector- and axial vector coupling constants $g_V = G_\beta = G_F \cos \theta_C$ and g_A (see chapter 3). The corresponding matrix elements have to be theoretically calculated. Under the assumptions made, M_{if} does not depend on energy. The overlap between the initial and final wavefunction is especially large for mirror nuclei (the number of protons of one nucleus equals the number of neutrons from the other); therefore, they have a large M_{if}. For super-allowed $0^+ \to 0^+$ transitions $M_{if} = \sqrt{2}$ which results in a single ft value (see section 6.1.3) for such nuclei of about 3100 s (figure 6.2). However, there are nuclei where electrons and neutrinos are emitted with $l \neq 0$ which means that the higher order terms of (6.14) have to be taken into account. The corresponding matrix elements are orders of magnitude smaller and the transitions are called forbidden. For a more extensive discussion on the classification of β-decays see [Sie68, Wu66]. Focusing on allowed transitions, thus the shape of the electron spectrum is determined completely by the density of final states $\rho(E)$, which will be calculated next.

6.1.2 Phase space calculation

The number of different states dn with momentum between p and $p + dp$ in a volume V is

$$dn = \frac{4\pi V p^2 \, dp}{h^3} = \frac{4\pi V p E \, dE}{h^3}. \qquad (6.26)$$

This translates into a density of states per energy interval of

$$\frac{dn}{dE} = \frac{4\pi V p E}{h^3} = \frac{V p E}{2\pi^2 \hbar^3}. \qquad (6.27)$$

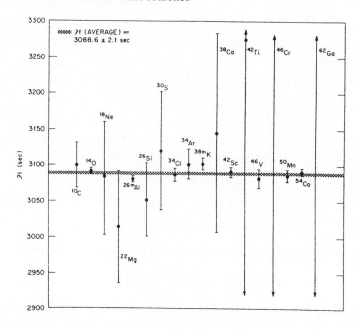

Figure 6.2. Experimental ft values observed in various superallow transitions. As can be seen, they cluster around 3100 s.

Dealing with a three-body decay and a heavy nucleus, the nucleus takes no energy but balances all momenta so the electron and neutrino momenta are not directly correlated and can be treated independently. Thus

$$\rho(E) = \frac{V^2 p_e E_e p_\nu E_\nu}{4\pi^4 \hbar^6}. \tag{6.28}$$

Using (6.10) and including a massive neutrino, the density of states can be expressed in terms of the kinetic energy of the electron E as

$$\rho(E) = \frac{V^2 p_e (E + m)\sqrt{(E_0 - E)^2 - m^2(\nu_e)}(E_0 - E)}{4\pi^4 \hbar^6}. \tag{6.29}$$

Combining this together with (6.9) and (6.25) we get for the β-spectrum of electrons of allowed or super-allowed decays (with $\epsilon = E_0 - E$):

$$\begin{aligned}
\frac{d^2 N}{dt\, dE} &= \frac{g_V^2 |M_F|^2 + g_A^2 |M_{GT}|^2}{2\pi^3 \hbar^7} F(E, Z+1) p_e(E+m) \\
&\quad \times \sqrt{(E_0 - E)^2 - m^2(\nu_e)}(E_0 - E)\theta(E_0 - E - m(\nu_e)) \\
&= A F(E, Z+1) p_e(E+m)\epsilon\sqrt{\epsilon^2 - m^2(\nu_e)}\theta(\epsilon - m(\nu_e)). \tag{6.30}
\end{aligned}$$

As can be seen, the neutrino mass influences the spectral shape only at the upper end below E_0 leading far below the endpoint to a small constant offset proportional to $-m^2(\nu_e)$. Two important modifications might be necessary. First of all, (6.30) only holds for the decay of a bare and infinitely heavy nucleus. In reality, in dealing with atoms or molecules the possible excitation of the electron shell due to a sudden change in the nuclear charge has to be taken into account. The atom or molecule will end in a specific state of excitation energy E_j with a probability P_j. (6.30) will thus be modified into a superposition of β-spectra of amplitude P_j with different endpoint energies $\epsilon_j = E_0 - E_j$:

$$\frac{\mathrm{d}^2 N}{\mathrm{d}t\,\mathrm{d}E} = AF(E, Z+1)p_e(E+m)\sum_j P_j\epsilon_j\sqrt{\epsilon_j^2 - m^2(\nu_e)}\theta(\epsilon_j - m(\nu_e)). \quad (6.31)$$

In addition, in case of neutrino mixing (see chapter 5) the spectrum is a sum of the components of decays into mass eigenstates ν_i:

$$\frac{\mathrm{d}^2 N}{\mathrm{d}t\,\mathrm{d}E} = AF(E, Z + 1)p_e(E + m)\sum_j P_j\epsilon_j$$

$$\times\left(\sum_i |U_{ei}|^2\sqrt{\epsilon_j^2 - m^2(\nu_i)}\theta(\epsilon_j - m(\nu_i))\right). \quad (6.32)$$

As long as the experimental resolution is wider than the mass difference of two neutrino states, the resulting spectrum can be analysed in terms of a single observable—the electron neutrino mass:

$$m^2(\nu_e) = \sum_i |U_{ei}|^2 m^2(\nu_i) \quad (6.33)$$

by using (6.31).

6.1.3 Kurie plot and ft values

The decay constant λ for β-decay can be calculated from (6.30) by integration

$$\lambda = \frac{\ln 2}{T_{1/2}} = \int_0^{p_0} N(p_e)\,\mathrm{d}p_e. \quad (6.34)$$

This results in

$$\lambda = \int_0^{p_0} N(p_e)\,\mathrm{d}p_e = (g_V^2|M_F|^2 + g_A^2|M_{GT}|^2)f(Z + 1, \epsilon_0) \quad (6.35)$$

with

$$f(Z + 1, \epsilon_0) = \int_1^{\epsilon_0} F(Z + 1, \epsilon)\epsilon\sqrt{\epsilon^2 - 1}(\epsilon_0 - \epsilon)^2\,\mathrm{d}\epsilon \quad (6.36)$$

as the so called Fermi integral. ϵ, ϵ_0 are given by

$$\epsilon = \frac{E_e + m_e c^2}{m_e c^2} \qquad \epsilon_0 = \frac{Q}{m_e c^2}. \quad (6.37)$$

Table 6.1. Characterization of β-transitions according to their ft values. Selection rules concern spin I and parity π: $(+)$ means no parity change while $(-)$ implies parity change.

Transition	Selection rule	Log ft	Example	Half-life
Superallowed	$\Delta I = 0, \pm 1, (+)$	3.5 ± 0.2	^{1}n	11.7 min
Allowed	$\Delta I = 0, \pm 1, (+)$	5.7 ± 1.1	^{35}S	87 d
First forbidden	$\Delta I = 0, \pm 1, (-)$	7.5 ± 1.5	^{198}Au	2.7 d
Unique first forbidden	$\Delta I = \pm 2, (-)$	8.5 ± 0.7	^{91}Y	58 d
Second forbidden	$\Delta I = \pm 2, (+)$	12.1 ± 1.0	^{137}Cs	30 yr
Third forbidden	$\Delta I = \pm 3, (-)$	18.2 ± 0.6	^{87}Rb	6×10^{10} yr
Fourth forbidden	$\Delta I = \pm 4, (+)$	22.7 ± 0.5	^{115}In	5×10^{14} yr

The product $f T_{1/2}$, given by

$$f T_{1/2} = \frac{K}{g_V^2 |M_F|^2 + g_A^2 |M_{GT}|^2} \tag{6.38}$$

is called the ft value and can be used to characterize β-transitions (the more accurate log ft is used) as shown in table 6.1. A compilation of ft values of all known β-emitters is shown in figure 6.3. The constant K is given by

$$K = \frac{2\pi^3 \hbar^7}{m_e^5 c^4 \ln 2}. \tag{6.39}$$

It is common in β-decay to plot the spectrum in the form of a so called Kurie plot which is given by

$$\sqrt{\frac{N(p_e)}{p_e^2 F(Z+1, E)}} = A(Q - E_e) \left[1 - \left(\frac{m_\nu c^2}{Q - E_e} \right)^2 \right]^{1/4}. \tag{6.40}$$

Following from this, three important conclusions can be drawn:

(1) For massless neutrinos, the Kurie plot simplifies to

$$\sqrt{\frac{N(p_e)}{p_e^2 F(Z+1, E)}} = A(Q - E_e) \tag{6.41}$$

which is just a straight line intersecting the x-axis at the Q-value.
(2) A light neutrino disturbs the Kurie plot in the region close to the Q-value. This results in an endpoint at $Q - m_\nu c^2$ and the electron spectrum ends perpendicular to the x-axis.

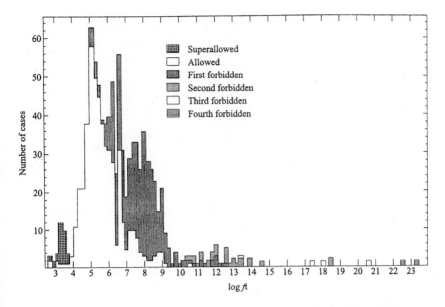

Figure 6.3. Compilation of all known log ft values.

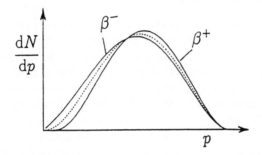

Figure 6.4. Schematic form of an electron beta spectrum. The phase space factor from (6.29) produces a spectrum with a parabolic fall at both ends for $m_\nu - 0$ (dotted line). This is modified by the interaction of the electron/positron with the Coulomb field of the final state nucleus (continuous lines). Taken from [Gro90].

(3) Assuming that there is a difference between the neutrino mass eigenstates and weak eigenstates as mentioned in chapter 5 and discussed in more detail in chapter 8, the Kurie plot is modified to

$$\sqrt{\frac{N(p_e)}{p_e^2 F(Z+1, E)}} = A \sum_i U_{ei}^2 (Q - E_e) \left[1 - \left(\frac{m_i c^2}{Q - E_e} \right)^2 \right]^{1/4}. \quad (6.42)$$

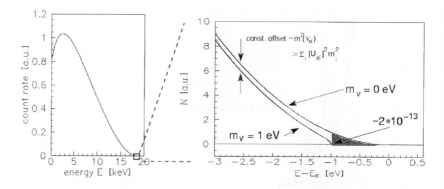

Figure 6.5. Endpoint region of a beta spectrum. The effect of a finite neutrino mass is a reduced endpoint at $Q - m_\nu c^2$ (from [Wei03]).

The result are kinks in the Kurie plot as discussed in section 6.2.4.

6.2 Searches for $m_{\bar{\nu}_e}$

6.2.1 General considerations

As already mentioned, a non-vanishing neutrino mass will reduce the phase space and leads to a change in the shape of the electron spectra, which for small masses can be investigated best near the Q-value of the transition (figure 6.5). First measurements in search of neutrino masses had already been obtained in 1947 resulting in an upper bound of 100 keV. A measurement done in 1952 resulting in a limit of less than 250 eV led to the general assumption of a massless neutrinos [Lan52] which was the motivation to implement massless neutrinos in the standard model (see chapter 3). Several aspects have to be considered before extracting a neutrino mass from a β-decay experiment [Hol92, Ott95, Wil01]:

- the statistics of electrons with an energy close to the endpoint region is small (a small Q-value for the isotope under study is advantageous);
- good energy resolution;
- energy loss within the source;
- atomic and nuclear final state effects, excited state transitions; and
- a theoretical description of the involved wavefunctions.

From all isotopes tritium is the most favoured one. But even in this case with the relatively low endpoint energy of about 18.6 keV only a fraction of 10^{-9} of all electrons lies in a region of 20 eV below the endpoint. A further advantage of tritium is $Z = 1$, making the distortion of the β-spectrum due to Coulomb interactions small and allowing a sufficiently accurate quantum mechanical treatment. Furthermore, the half-life is relatively short ($T_{1/2} = 12.3$ yr) and the involved matrix element is energy independent (the decay is a superallowed $1/2 \rightarrow 1/2$ transition between mirror nuclei). The underlying decay is

$$^3\text{H} \rightarrow {}^3\text{He}^+ + \text{e}^- + \bar{\nu}_\text{e}. \tag{6.43}$$

The $^3\text{H} - {}^3\text{He}$ mass difference has been determined to be $\Delta m = (18.5901 \pm 0.0017)$ keV [Dyc93] and the difference of the atomic binding energies of the shell electrons is $B(^3\text{H}) - B(^3\text{He}) = 65.3$ eV [Ohs94]. In general, ^3H is not used in atomic form but rather in its molecular form H_2. In this case the molecular binding energies have to be considered and for an accurate determination, the small nuclear recoil E_R also has to be included. The result is a Q-value of 18.574 keV. Furthermore, only about 58% of the decays near the endpoint lead to the ground state of the $^3\text{H}\,^3\text{He}^+$ ion, making a detailed treatment of final states necessary. However, in the last 27 eV below the endpoint, there are no molecular excitations.

6.2.2 Searches using spectrometers

While until 1990 magnetic spectrometers were mostly used for the measurements [Hol92, Ott95], the currently running experiments in Mainz and Troitsk use electrostatic retarding spectrometers [Lob85, Pic92]. As an example the Mainz experiment is described in a little more detail. The principal setup is shown in figure 6.6. The tritium source and the detector are located within two solenoids of a $B_S = 2.4$ T maximal magnetic field. This reduces to a minimal field of $B_{\min} \approx 8 \times 10^{-4}$ T in the middle plane of the spectrometer (the analysing plane). The ratio $B_S/B_{\min}$ is 3000. Electrons emitted from the source spiral around the magnetic field lines and will be guided into the spectrometer. By a set of electrodes around the spectrometer a retarding electrostatic potential is created which has its maximum value (a barrier of eU_0 with $U_0 < 0$) in the analysing plane. The emitted electrons will be decelerated by this potential: only those with sufficient energy can pass the potential barrier and will be accelerated and focused on the detector. The main advantage of such a spectrometer is the following: emitted electrons have a longitudinal kinetic energy T_L along the field lines, which is analysed by the spectrometer, and a transverse kinetic energy T_T in the cyclotron motion given by

$$T_T = -\mu \cdot B \qquad \text{with } \mu = \frac{e}{2m_e} L. \tag{6.44}$$

Because of angular momentum conservation, L and, therefore, μ are constants of motion, showing that in an inhomogenous magnetic field T_T changes proportional to B. Thus, the energy in a decreasing field is transformed from $T_T \rightarrow T_L$ and *vice versa* in an increasing field. In the analysing plane all cyclotron energy has been converted into analysable longitudinal energy T_L, except for a small rest between zero (emission under $\theta = 0°$, e.g. $T_T = 0$) and maximal ($\theta = 90°$, e.g.

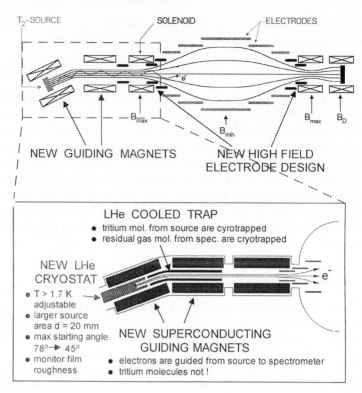

Figure 6.6. Layout of the Mainz electrostatic retarding spectrometer for measuring tritium β-decay.

$T_T = T$). The transmission function has a width of

$$\Delta T = \frac{B_{\min}}{B_S} T = \frac{1}{3000} T = 6 \text{ eV} \qquad \text{(if } T \approx 18 \text{ keV)}. \qquad (6.45)$$

The Mainz filter has a width of only 4.8 eV and guarantees good energy resolution. Figure 6.7 shows the electron spectrum near the endpoint as obtained with the Mainz spectrometer. The main difference between the Mainz and the Troitsk spectrometer is the tritium source. While the Mainz experiment froze a thin film of T_2 onto a substrate, the Troitsk experiment uses a gaseous tritium source. The obtained limits are [Wei03]:

$$m_\nu^2 = -1.2 \pm 2.2(\text{stat.}) \pm 2.1(\text{sys.}) \text{ eV}^2 \to m_{\bar{\nu}_e}$$
$$< 2.2 \text{ eV}(95\% \text{ CL}) \qquad \text{Mainz} \qquad (6.46)$$
$$m_\nu^2 = -2.3 \pm 2.5(\text{stat.}) \pm 2.0(\text{sys.}) \text{ eV}^2 \to m_{\bar{\nu}_e}$$
$$< 2.2 \text{ eV}(95\% \text{ CL}) \qquad \text{Troitsk.} \qquad (6.47)$$

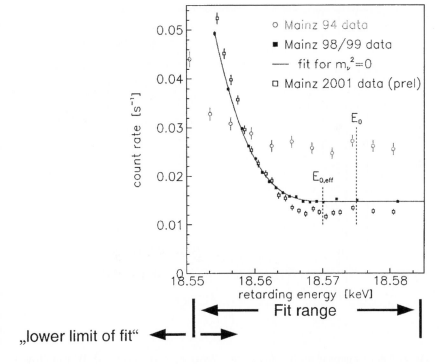

Figure 6.7. Mainz electron spectrum near the endpoint of tritium decay. The signal/background ratio is increased by a factor of 10 in comparison with the 1994 data. The Q-value of 18.574 keV marks to the centre-of-mass of the rotation-vibration excitations of the molecular ground state of the daughter ion $^3\mathrm{HeT}^+$.

The longstanding problem of negative m_ν^2 values (m_ν^2 is a fit parameter to the spectrum and, therefore, can be negative) has finally disappeared (figure 6.8). The Troitsk number is obtained by including an observed anomaly in the analysis. Excess counts have been observed in the region of interest, which can be described by a monoenergetic line just below the endpoint. Even more, a semi-annual modulation of the line position is observed [Lob99]. However, this effect has not been seen by the Mainz experiment, even when measured at the same time as Troitsk [Wei03]. This indicates, most likely, an unknown experimental artefact.

6.2.2.1 *Future spectrometers—KATRIN*

For various physics arguments which become clearer throughout the book, it will be important to improve the sensitivity of neutrino mass searches into a region below 1 eV. However, this requires a new very large spectrometer. The new KATRIN experiment is designed to fulfil this need and probe neutrino masses down to 0.2 eV [Bad01, Osi01, Wei03, Wei03a]. For such a resolution

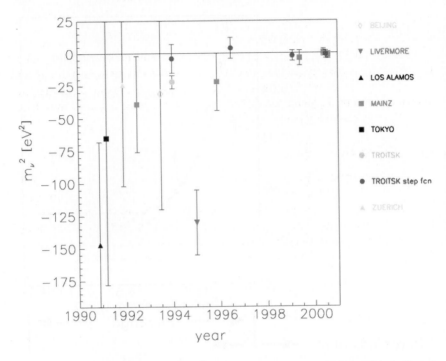

Figure 6.8. Evolution of the fit value m^2 in β-decay as a function of time. With the Troitsk and Mainz experiment the longstanding problem of negative m^2 caused by unknown systematic effects finally disappeared.

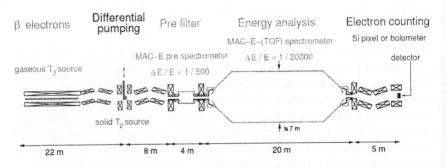

Figure 6.9. Schematic layout of the new proposed KATRIN spectrometer (from [Wei03]).

a transmission window of only 1 eV is neccessary which corresponds to a ratio of $B_{min}/B_S = 5 \times 10^{-5}$. A sketch of the layout is shown in figure 6.9. The main features of the experiment are a windowless gaseous tritium source, minimizing the systematic uncertainties from the source itself, a pre-spectrometer, acting as an energy pre-filter to reject all β-electrons, except the ones in the region of interest close to the endpoint and the main spectrometer. To obtain the required resolution the analysing plane has to have a spectrometer of diameter 10 m. The full spectrometer is 20 m long kept at an ultra-high vacuum below 10^{-11} mbar. One difficulty is the fact that only about 10^{-13} of all electrons from β-decay fall into a region 1 eV below the endpoint. Therefore, the detector has to be shielded, allowing only a background rate of 10^{-2} events s^{-1} and the detectors must have a good energy resolution (less than 600 eV at 18.6 keV).

6.2.3 Cryogenic searches

A complementary strategy to be followed is the use of cryogenic microcalorimeters [Gat01, Fio01]. The idea behind this new detector development is that the released energy is converted, within an absorber, into phonons which leads to a temperature rise. This will be detected by a sensitive thermometer. For this to work, the device has to be cooled down into the mK region. The measurement of the electron energy is related to a temperature rise via the specific heat by

$$\Delta T = \frac{\Delta E}{C_V} \tag{6.48}$$

where the specific heat is given in practical units as [Smi90]

$$C_V \approx 160 \left(\frac{T}{\Theta_D}\right)^3 \text{ J cm}^{-3} \text{ K}^{-1} \approx 1 \times 10^{18} \left(\frac{T}{\Theta_D}\right)^3 \text{ keV cm}^{-3} \text{ K}^{-1} \tag{6.49}$$

with Θ_D as material-dependent Debye temperature. Because these experiments measure the total energy released, the final-state effects are not important. This method allows the investigation of the β-decay of ^{187}Re

$$^{187}\text{Re} \rightarrow {}^{187}\text{Os} + e^- + \bar{\nu}_e \tag{6.50}$$

which has the lowest tabulated Q-value of all β-emitters ($Q = 2.67$ keV) [ToI98]. The associated half-life measurement of the order of 10^{10} yr will be quite important because the ^{187}Re–^{187}Os pair is a well-known cosmochronometer and a more precise half-life measurement would sharpen the dating of events in the early universe such as the formation of the solar system. Cryogenic bolometers have been built from metallic Re as well as AgReO$_4$ crystals with neutron transmutation doped-germanium thermistor readout (figure 6.10). The β-spectra (figure 6.11) were measured successfully [Gat99, Ale99]. The actual measured Q-values of 2481 ± 6 eV and 2460 ± 5(stat.) ± 10(sys.) eV are in agreement with each other but lower than the expected one. A first half-life for ^{187}Re of $T_{1/2} = 43 \pm 4$(stat.) ± 3(sys.) $\times 10^9$ yr is obtained in agreement with measurements using mass spectrometers resulting in $T_{1/2} = 42.3 \pm 1.3 \times 10^9$ yr. Due to the good energy resolution of the devices, for the first time environmental fine structure effects on β-decay could be observed recently [Gat99]. Last but not least the first limits on $m_{\bar{\nu}_e}$ of <22(26) eV are given [Gat01, Arn03]. An upgrade to build large arrays of these detectors to go down to a 1 eV mass sensitivity is forseen.

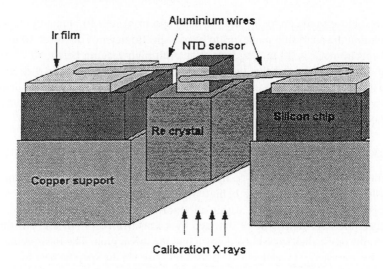

Figure 6.10. Sketch of the rhenium microcalorimeter with an absorbing mass of metallic Re, a neutron-transmutation-doped (NTD) Ge thermistor on top and two aluminium wires for thermal and mechanical connections (from [Meu98]).

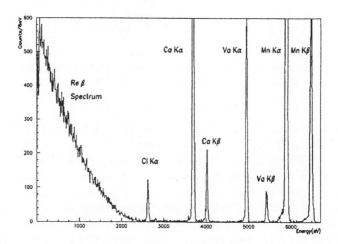

Figure 6.11. ^{187}Re spectrum obtained with a cryogenic bolometer. The big spikes correspond to calibration peaks (from [Meu98]).

6.2.4 Kinks in β-decay

As already stated in chapter 5, the existence of several neutrino mass eigenstates and their mixing might lead to kinks in the Kurie plot of a β-spectrum. This is shown schematically in figure 6.12. The energy range where the Kurie plot shows

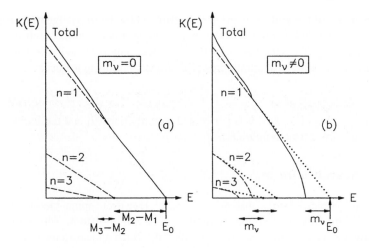

Figure 6.12. Schematic Kurie plot for two massive neutrinos.

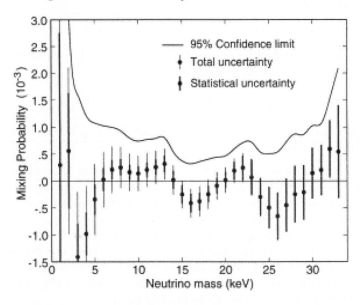

Figure 6.13. Best fit (points) of the mixing probability as a function of assumed neutrino mass in ^{63}Ni decay. The error bars combine statistical and systematic errors. The full line is an upper limit at 95% CL (from [Hol99]).

a kink is small and depends on the mass m_2 and the mixing angle θ:

$$\frac{\Delta K}{K} \simeq \frac{\tan^2 \theta}{2} \left(1 - \frac{m_2^2 c^4}{(E_0 - E_e)^2} \right)^{1/2} \qquad \text{for } E_0 - E_e > m_2 c^2. \qquad (6.51)$$

The position of the kink is, therefore, determined by the mass eigenstate m_2 and the size of the kink is related to the mixing angle θ between the neutrino states (see chapter 8). The searches are performed especially for heavier neutrino mass eigenstates. A search for admixtures of keV neutrinos using the decay

$$^{63}\text{Ni} \rightarrow {}^{63}\text{Cu} + e^- + \bar{\nu}_e \tag{6.52}$$

with a Q-value of 67 keV was performed recently [Hol99] and the limits on the admixture are shown in figure 6.13.

6.3 Searches for m_{ν_e}

CPT invariance ensures that $m_{\bar{\nu}_e} = m_{\nu_e}$. However, some theories beyond the standard model offer the possibility of CPT violation, which makes it worthwhile considering measuring m_{ν_e} directly. Such a measurement of m_{ν_e} has been proposed by [Der81] using the internal bremsstrahlung spectrum in electron capture processes:

$$(A, Z) + e^- \rightarrow (A, Z-1) + \nu_e + \gamma. \tag{6.53}$$

The bremsstrahlung spectrum of photons with energy k for K-shell capture can be given in a form similar to the β-spectrum:

$$N(k)\,dk \propto k(k_0 - k)\sqrt{(k_0 - k)^2 - m_\nu^2} = kE_\nu p_\nu. \tag{6.54}$$

As in the Kurie plot, the endpoint depends on m_ν. Two major problems are associated with this:

- Every state from which electron capture can happen is characterized by its quantum numbers n, l, j and has its own spectrum $N(k)$. Therefore, the measured spectrum is a superposition of these spectra which leads to a smearing in the endpoint region.
- The capture rate is very small. An electron from an $l \neq 0$ state transforms virtually into an intermediate $l = 0$ state via emission of a photon. This state has a non-vanishing wavefunction at the nucleus allowing capture. This effect can be enhanced if the energy of the transition is close to an x-ray transition, which leads to a resonance-like effect.

The most convenient isotope is ^{163}Ho. It has a very low Q-value of about 2.5 keV; therefore, only M-capture and capture from higher shells are possible. Using a source of ^{163}HoF$_3$ and a Si(Li) detector the atomic transition between the $5p \rightarrow 3s$ levels was investigated. Assuming a Q-value of 2.56 keV a limit of

$$m_{\nu_e} < 225 \text{ eV} \qquad (95\% \text{ CL}) \tag{6.55}$$

was obtained [Spr87]. A Q-value of 2.9 keV would worsen this bound to 500 eV. A new attempt using cryogenic microcalorimeters, which measures the total energy and is free of some uncertainties related with the pure x-ray measurements described earlier, has been started. If several lines can be observed, a combined fit for Q and m_ν can be done. First prototypes have been constructed and the obtained Q-value is 2.80 ± 0.05 keV, higher than previously assumed. This method might be useful in the future [Meu98]. Currently, the bounds discussed here are rather weak in comparison with β-decay. Astrophysical limits on m_{ν_e} will be discussed in chapter 13.

6.4 m_{ν_μ} determination from pion-decay

The easiest way to obtain limits on m_{ν_μ} is given by the two-body decay of the π^+. For pion decay at rest the neutrino mass is determined by

$$m_{\nu_\mu}^2 = m_{\pi^+}^2 + m_{\mu^+}^2 - 2m_{\pi^+}\sqrt{p_{\mu^+}^2 + m_{\mu^+}^2}. \tag{6.56}$$

Therefore, a precise measurement of m_{ν_μ} depends on an accurate knowledge of the muon momentum p_μ as well as m_μ and m_π. The pion mass is determined by x-ray measurements in pionic atoms. The measurements lead to two values:

$$m_\pi = 139.567\,82 \pm 0.000\,37 \text{ MeV}$$
$$m_\pi = 139.569\,95 \pm 0.000\,35 \text{ MeV} \tag{6.57}$$

respectively [Jec95] ($\approx$2.5 ppm) but a recent independent measurement supports the higher value by measuring $m_\pi = 139.570\,71 \pm 0.000\,53$ MeV [Len98]. The muon mass is determined by measuring the ratio of the magnetic moments of muons and protons. This results in [PDG02]

$$m_\mu = (105.658\,357 \pm 0.000\,005) \text{ MeV} \qquad (\approx 0.05 \text{ ppm}). \tag{6.58}$$

Latest π-decay measurements were performed at the Paul-Scherrer Institute (PSI) resulting in a muon momentum of [Ass96]

$$p_\mu = (29.792\,00 \pm 0.000\,11) \text{ MeV} \qquad (\approx 4 \text{ ppm}). \tag{6.59}$$

Combining all numbers, a limit of

$$m_{\nu_\mu}^2 = (-0.016 \pm 0.023) \text{ MeV}^2 \rightarrow m_{\nu_\mu} < 190 \text{ keV} \qquad (90\% \text{ CL}) \tag{6.60}$$

could be achieved.

A new experiment (E952) looking for pion decay in flight using the g-2 storage ring at BNL is planned [Car00]. The g-2 ring could act as a high resolution spectrometer and an exploration of m_{ν_μ} down to 8 keV seems feasible.

6.5   Mass of the ν_τ from tau-decay

Before discussing the mass of ν_τ it should be mentioned that the direct detection of ν_τ via CC reactions has been observed only very recently [Kod01]. It was the goal of E872 (DONUT) at Fermilab to detect exactly this reaction (see chapter 4) and they came up with four candidate events.

The present knowledge of the mass of ν_τ stems from measurements with ARGUS (DORIS II) [Alb92], CLEO(CESR) [Cin98], OPAL [Ack98], DELPHI [Pas97] and ALEPH [Bar98] (LEP) all using the reaction $e^+e^- \rightarrow \tau^+\tau^-$. The energy E_τ is given by the different collider centre-of-mass energies $E_\tau = \sqrt{s}/2$. Practically all experiments use the τ-decay into five charged pions:

$$\tau \rightarrow \nu_\tau + 5\pi^\pm(\pi^0) \tag{6.61}$$

with a branching ratio of $BR = (9.7 \pm 0.7) \times 10^{-4}$. To increase the statistics, CLEO, OPAL, DELPHI and ALEPH extended their search by including the three-prong decay mode $\tau \rightarrow \nu_\tau + 3h^\pm$ with $h \equiv \pi, K$. But even with the disfavoured statistics, the five-prong decay is more sensitive, because the mass of the hadronic system m_{had} peaks at about 1.6 GeV, while the effective mass of the three π-system is shaped by the $a_1(1260)$ resonance. While ARGUS and DELPHI obtained their limit by investigating only the invariant mass of the five π-system, ALEPH, CLEO and OPAL performed a two-dimensional analysis by including the energy of the hadronic system E_{had}. In the one dimensional analysis, the maximum energy of the hadronic system is given by

$$m_{had} = m_\tau - m_\nu \tag{6.62}$$

and, therefore, results in an upper bound on m_ν. A bound can also be obtained from the hadronic energy coming from

$$m_\nu < E_\nu = E_\tau - E_{had} \tag{6.63}$$

where E_{had} is given in the rest frame of the τ by

$$E_{had} = \frac{(m_\tau^2 + m_{had}^2 - m_\nu^2)}{2m_\tau} \tag{6.64}$$

which will be boosted in the laboratory frame. A finite neutrino mass leads to a distortion of the edge of the triangle of a plot of the E_{had}–m_{had} plane as shown in figure 6.14. A compilation of the resulting limits is given in table 6.2 with the most stringent one given by ALEPH [Bar98]:

$$m_{\nu_\tau} < 18.2\,\text{MeV} \qquad (95\%\,\text{CL}) \tag{6.65}$$

A combined limit for all four LEP experiments improves this limit only slightly to 15.5 MeV. A chance for improvement might be offered by an investigation of leptonic D_S^+-decays [Pak03].

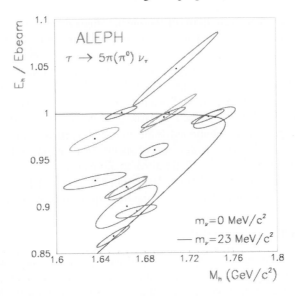

Figure 6.14. Two-dimensional plot of the hadronic energy *versus* the invariant mass of the $5(6)\pi$-system. The error ellipses are positively correlated, because both the hadronic mass and the hadronic energy are determined from the momenta of the particles composing the hadronic system (from [Bar98]).

Table 6.2. Comparision of ν_τ mass limits (95% CL) as measured by various experiments. Numbers with * include also events from 3π-decay. In this case the limit is obtained by the combination of both modes.

Experiment	Number of fitted events	Combined m_{ν_τ} limit (MeV)
ARGUS	20	31
CLEO	60+53	30
CLEO	29 058 (4π)	28
OPAL	2514* + 22	27.6
DELPHI	6534*	27
ALEPH	2939* + 55	18.2

6.6   Electromagnetic properties of neutrinos

Another experimental aspect where a non-vanishing neutrino mass could show up is the search for electromagnetic properties of neutrinos such as their magnetic moments. Even with charge neutrality, neutrinos can participate in electromagnetic interactions by coupling with photons via loop diagrams. As for other particles the electromagnetic properties can be described by form factors (see chapter 4). The hermiticity of the electromagnetic current j_μ allows four independent form factors for Dirac neutrinos, the charge and axial charge form factors $F(Q^2)$ and $G(Q^2)$ and the electric and magnetic dipole moment form factors $D(Q^2)$ and $M(Q^2)$. $F(Q^2)$ and $G(Q^2)$ have to vanish for $Q^2 \to 0$ because of electric charge neutrality. The values of $D(Q^2)$ and $M(Q^2)$ for $Q^2 = 0$ are the electric $D(0) = d_\nu/e$ and magnetic dipole moment $M(0) = \mu_\nu/\mu_B$ of the Dirac neutrinos. CPT and CP invariance make the electric dipole moment vanish. The previously mentioned static moments correspond to the diagonal elements of a matrix. The off-diagonal elements, if the initial and final neutrino flavours are not identical, are called transition moments.

For Majorana neutrinos $F(Q^2)$, $D(Q^2)$ and $M(Q^2)$ vanish, because of their self-conjugate properties. Only $G(Q^2)$ and transition moments are possible.

6.6.1 Electric dipole moments

The Fourier transform of the previously mentioned form factors can be interpreted as spatial distributions of charges and dipole moments. This allows a possible spatial extension of neutrinos to be defined via an effective mean charge radius $\langle r^2 \rangle$ ('effective size of the neutrino') given by

$$\langle r^2 \rangle = 6 \frac{df(Q^2)}{dQ^2}\bigg|_{Q^2=0} \qquad \text{with } f(Q^2) = F(Q^2) + G(Q^2). \qquad (6.66)$$

It can be measured in the elastic νe scattering processes discussed in chapter 4 (replacing $g_V, G_V \to g_V, G_V + 2\delta$), with δ given as

$$\delta = \frac{\sqrt{2}\pi\alpha}{3G_F} \langle r^2 \rangle = 2.38 \times 10^{30} \text{ cm}^{-2} \langle r^2 \rangle. \qquad (6.67)$$

The current limits are:

$$\langle r^2 \rangle(\nu_e) < 5.4 \times 10^{-32} \text{ cm}^2 \qquad \text{(LAMPF [All93])}$$
$$\langle r^2 \rangle(\nu_\mu) < 1.0 \times 10^{-32} \text{ cm}^2 \qquad \text{(CHARM [Dor89])}$$
$$\langle r^2 \rangle(\nu_\mu) < 2.4 \times 10^{-32} \text{ cm}^2 \qquad \text{(E734 [Ahr90])} \qquad (6.68)$$
$$\langle r^2 \rangle(\nu_\mu) < 6.0 \times 10^{-32} \text{ cm}^2 \qquad \text{(CHARM-II [Vil95])}$$

Electric dipole moments have not been observed for any fundamental particle. They always vanish as long as CP or, equivalently, T is conserved as this implies $d_\nu = 0$. However, nothing is known about CP violation in the leptonic sector. This might change with the realization of a neutrino factory. Until then we can use the limits on magnetic dipole moments from νe scattering as bounds, because for not-too-small energies their contribution to the cross section is the same. Bounds of the order of $d_\nu < 10^{-20}$ e cm (ν_e, ν_μ) and $d_\nu < 10^{-17}$ e cm (ν_τ) result. For Majorana neutrinos, CPT invariance ensures that $d_\nu = 0$.

6.6.2 Magnetic dipole moments

Another possibility probing a non-vanishing mass and the neutrino character is the search for its magnetic moment. In the standard model neutrinos have no magnetic moment because they are massless and a magnetic moment would require a coupling of a left-handed state with a right-handed one—the latter does not exist. A simple extension by including right-handed singlets allows for Dirac masses. In this case, it can be shown that due to loop diagrams neutrinos can obtain a magnetic moment which is proportional to their mass and is given by [Lee77, Mar77]

$$\mu_\nu = \frac{3G_F e}{8\sqrt{2}\pi^2} m_\nu = 3.2 \times 10^{-19} \left(\frac{m_\nu}{\text{eV}}\right) \mu_B. \tag{6.69}$$

For neutrino masses in the eV range, this is far too small to be observed and to have any significant effect in astrophysics. Nevertheless, there exist models, which are able to increase the expected magnetic moment [Fuk87, Bab87, Pal92]. However, Majorana neutrinos still have a vanishing static moment because of CPT invariance. This can be seen from the following argument (a more theoretical treatment can be found in [Kim93]). The electromagnetic energy of a neutrino with spin direction σ in an electromagnetic field is given by

$$E_{\text{em}} = -\mu_\nu \sigma \cdot B - d_\nu \sigma \cdot E. \tag{6.70}$$

Applying CPT results in $B \to B$, $E \to E$ and $\sigma \to -\sigma$ which results in $E_{\text{em}} \to -E_{\text{em}}$. However, CPT transforms a Majorana neutrino into itself ($\bar{\nu} = \nu$) which allows no change in E_{em}. Therefore, $E_{\text{em}} = 0$ which is only possible if $\mu_\nu = d_\nu = 0$.

Limits on magnetic moments arise from ν_e e scattering experiments and astrophysical considerations. The differential cross section for ν_e e scattering in the presence of a magnetic moment is given by

$$\frac{d\sigma}{dT} = \frac{G_F^2 m_e}{2\pi} \left[(g_V + x + g_A)^2 + (g_V + x - g_A)^2 \left(1 - \frac{T}{E_\nu}\right)^2 \right.$$
$$\left. + (g_A^2 - (x + g_V)^2) \frac{m_e T}{E_\nu^2} \right] + \frac{\pi \alpha^2 \mu_\nu^2}{m_e^2} \frac{1 - T/E_\nu}{T} \tag{6.71}$$

where T is the kinetic energy of the recoiling electron and x is related to the charge radius $\langle r^2 \rangle$:

$$x = \frac{2m_W^2}{3} \langle r^2 \rangle \sin^2 \theta_W \qquad x \to -x \qquad \text{for } \bar{\nu}_e. \tag{6.72}$$

The contribution associated with the charge radius can be neglected in the case $\mu_\nu \gtrsim 10^{-11} \mu_B$. As can be seen, the largest effect of a magnetic moment can be observed in the low-energy region and because of destructive interference with the

MUNU Time Projection Chamber

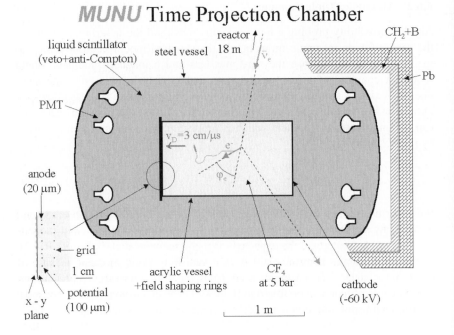

Figure 6.15. Layout of the MUNU TPC to search for magnetic moments at the Bugey reactor (from [Ams97]).

electroweak terms, searches with antineutrinos would be preferred. The obvious sources are, therefore, nuclear reactors.

To improve the experimental situation and, in particular, to check the region relevant for the solar neutrino problem (see chapter 11) new experiments have been performed and planned. The most recent one is the MUNU experiment [Ams97] performed at the Bugey reactor. It consists of a 1 m^3 time projection chamber (TPC) loaded with CF$_4$ under a pressure of 5 bar (figure 6.15). The use of a TPC allows not only the electron energy to be measured but also, for the first time in such experiments, the scattering angle, making the reconstruction of the neutrino energy possible. The neutrino energy spectrum in reactors in the energy region $1.5 < E_\nu < 8$ MeV is known at the 3% level. To suppress background, the TPC is surrounded by 50 cm anti-Compton scintillation detectors as well as a passive shield of lead and polyethylene. If there is no magnetic moment the expected count rate is 9.5 events per day increasing to 13.4 events per day if $\mu_\nu = 10^{-10}\mu_B$ for an energy threshold of 500 keV. The estimated background is six events per day. The expected sensitivity level is down to $\mu_\nu = 3 \times 10^{-11}\mu_B$. A first limit is given as [Dar03]

$$\mu_{\bar{\nu}_e} < 1.0 \times 10^{-10}\mu_B \qquad (90\% \text{ CL}). \qquad (6.73)$$

Another reactor experiment which recently started is TEXONO [Won02], using a 1 kg Ge-detector in combination with 46 kg of CsI(Tl) scintillators. The use of a low background Ge–NaI spectrometer in a shallow depth near a reactor has also been considered [Ded98]. Replacing the nuclear reactor by a strong β-source, low-energy threshold experiments in underground laboratories are also under investigation. Calculations for a scenario of an MCi ^{147}Pm source (an endpoint energy of 234.7 keV) in combination with a 100 kg low-level NaI(Tl) detector with a threshold of about 2 keV can be found in [Bar96].

Astrophysical limits exist and are somewhat more stringent but also more model dependent. Bounds from supernovae will be discussed in section 11. The major constraint on magnetic moments arises from stellar energy-loss arguments. Transverse and longitudinal excitations in a stellar plasma ('plasmons') are both kinematically able to decay into neutrino pairs of sufficiently small mass, namely $2m_\nu < K^2$, where K is the plasmon 4-momentum. In addition, an effective ν–γ coupling is introduced. For $\mu_\nu > 10^{-12}\mu_B$ this process can compete with standard energy-loss mechanisms if the plasma frequency is around 10 keV. The cooling of the hottest white dwarfs will be faster if plasmon decay into neutrinos occurs and, therefore, a suppression of the hottest white dwarfs in the luminosity function might occur. From observations, bounds of the order $\mu_\nu < 10^{-11}\mu_B$ could be obtained [Raf99]. More reliable are globular cluster stars. Here horizontal branch stars and low mass red giants before the He flash would be affected if there is an additional energy loss in form of neutrinos. To prevent the core mass at He ignition from exceeding its standard value by less than 5%, a bound of $\mu_\nu < 3 \times 10^{-12}\mu_B$ has been obtained [Raf90, Raf99].

Measurements based on $\nu_e e \rightarrow \nu_e e$ and $\nu_\mu e \rightarrow \nu_\mu e$ scattering were done at LAMPF and BNL yielding bounds for ν_e and ν_μ of [Kra90] (see also [Dor89, Ahr90, Vil95])

$$\mu_{\nu_e} < 10.8 \times 10^{-10}\mu_B \qquad \text{(if } \mu_{\nu_\mu} = 0) \qquad (6.74)$$

$$\mu_{\nu_\mu} < 7.4 \times 10^{-10}\mu_B \qquad \text{(if } \mu_{\nu_e} = 0). \qquad (6.75)$$

Combining these scattering results and Super-Kamiokande observations (see chapter 9), a limit for the magnetic moment of ν_τ was obtained [Gni00]:

$$\mu_{\nu_\tau} < 1.9 \times 10^{-9}\mu_B. \qquad (6.76)$$

As can be seen, the experimental limits are still orders of magnitude away from the predictions (6.69).

6.7   Neutrino decay

Another physical process which is possible if neutrinos have a non-vanishing rest mass is neutrino decay. Depending on the mass of the heavy neutrino ν_H various

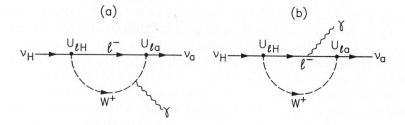

Figure 6.16. Feynman diagrams describing radiative neutrino decay $\nu_H \rightarrow \nu_L + \gamma$.

decay modes into a light neutrino ν_L can be considered, the most common are:

$$\nu_H \rightarrow \nu_L + \gamma$$
$$\nu_H \rightarrow \nu_L + \ell^+ + \ell^- \qquad (\ell \equiv e, \mu) \qquad (6.77)$$
$$\nu_H \rightarrow \nu_L + \nu + \bar{\nu}$$
$$\nu_H \rightarrow \nu_L + \chi$$

The first mode is called radiative neutrino decay and the fourth process is a decay with the emission of a majoron χ, the Goldstone boson of lepton symmetry breaking. Because of the non-detectable majoron the last two modes are often called invisible decays. Note that it is always a mass eigenstate, that decays meaning, e.g., the decay $\nu_\mu \rightarrow \nu_e + \gamma$ is in a two-neutrino mixing scheme caused by the decay $\nu_2 \rightarrow \nu_1 + \gamma$.

6.7.1 Radiative decay $\nu_H \rightarrow \nu_L + \gamma$

The two simplest Feynman graphs for radiative neutrino decay are shown in figure 6.16. The decay rate is given as [Fei88]

$$\Gamma(\nu_H \rightarrow \nu_L + \gamma) = \frac{1}{8\pi} \left[\frac{m_H^2 - m_L^2}{m_H} \right]^3 (|a|^2 + |b|^2) \qquad (6.78)$$

where for Dirac neutrinos the amplitudes a, b are

$$a_D = -\frac{eG_F}{8\sqrt{2}\pi^2}(m_H + m_L) \sum_l U_{lH} U_{lL}^* F(r_l) \qquad (6.79)$$

$$b_D = -\frac{eG_F}{8\sqrt{2}\pi^2}(m_H - m_L) \sum_l U_{lH} U_{lL}^* F(r_l) \qquad (6.80)$$

with U as the corresponding mixing matrix elements and $F(r_l)$ as a smooth function of $r_l = (m_l/m_W)^2$: $F(r_l) \approx 3r/4$ if $r \ll 1$. For Majorana neutrinos $a_M = 0, b_M = 2b_D$ or $a_M = 2a_D, b_M = 0$ depending on the relative CP-phase of the neutrinos ν_H and ν_L. Taking only tau-leptons which dominate the sum in (6.79), one obtains for $m_L \ll m_H$ a decay rate of

$$\Gamma \approx \frac{m_H^5}{30 \text{ eV}} |U_{\tau H} U_{\tau L}^*|^2 \times 10^{-29} \text{ yr}^{-1}. \tag{6.81}$$

This implies very long lifetimes against radiative decays of the order $\tau > 10^{30}$ yr. However, in certain models, like the left–right symmetric models, this can be reduced drastically.

Experimentally the following searches have been performed:

- Search for photons at nuclear reactors by using liquid scintillators. This probes the admixture of ν_H to $\bar{\nu}_e$, therefore it is proportional to $|U_{eH}|^2$. At the Goesgen reactor no difference was observed in the on/off phases of the reactor resulting in [Obe87]

$$\frac{\tau_H}{m_H} > 22(59) \frac{s}{\text{eV}} \qquad \text{for } a = -1(+1) \qquad (68\% \text{ CL}). \tag{6.82}$$

- At LAMPF, using pion and muon decays at rest (therefore looking for $|U_{\mu H}|^2$). No signal was observed and a limit of [Kra91]

$$\frac{\tau_H}{m_H} > 15.4 \frac{s}{\text{eV}} \qquad (90\% \text{ CL}) \tag{6.83}$$

was obtained.
- From the experimental solar x-ray and γ-flux a lower bound was derived as [Raf85]

$$\frac{\tau_H}{m_H} > 7 \times 10^9 \frac{s}{\text{eV}}. \tag{6.84}$$

Observations performed during a solar eclipse to measure only decays between the moon and the Earth have also been performed [Bir97].
- Maybe the most stringent limits come from supernova SN1987A (see chapter 11). There was no excess of the γ-flux measured by the gamma-ray spectrometer (GRS) on the solar maximum mission (SMM) satellite during the time when the neutrino events were detected, which can be converted in lower bounds of [Blu92a, Obe93]

$$\begin{aligned} \tau_H &> 2.8 \times 10^{15} B_\gamma \frac{m_H}{\text{eV}} & m_M &< 50 \text{ eV} \\ \tau_H &> 1.4 \times 10^{17} B_\gamma & 50 \text{ eV} &< m_M < 250 \text{ eV} \\ \tau_H &> 6.0 \times 10^{18} B_\gamma \frac{\text{eV}}{m_H} & m_M &> 250 eV \end{aligned} \tag{6.85}$$

where B_γ is the radiative branching ratio.

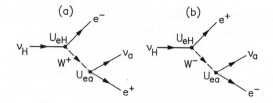

Figure 6.17. Feynman diagrams describing radiative neutrino decay $\nu_H \to \nu_L + e^+ + e^-$.

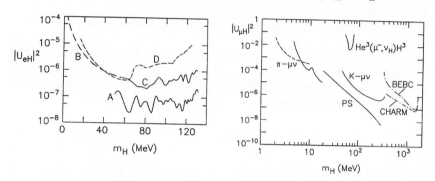

Figure 6.18. Limits on U_{eH} and $U_{\mu H}$ as a function of a heavy neutrino mass m_H. Left: Curves A and B correspond to measurements at TRIUMF [Bri92], curves C and D to earlier experiments from [Bri92]. Right: From [Boe92].

6.7.2 The decay $\nu_H \to \nu_L + e^+ + e^-$

The Feynman graphs for this decay are shown in figure 6.17. Clearly this decay is only possible if $m_H > 2m_e \approx 1$ MeV. The decay rate is given by

$$\Gamma(\nu_H \to \nu_L + e^+ + e^-) = \frac{G_F^2 m_H^5}{192\pi^3}|U_{eH}^2|. \tag{6.86}$$

Here Dirac and Majorana neutrinos result in the same decay rate. Searches are performed with nuclear reactors and high-energy accelerators. The obtained limits on the mixing U_{eH}^2 as well as such on $U_{\mu H}^2$ are shown in figure 6.18.

6.7.3 The decay $\nu_H \to \nu_L + \chi$

To avoid several astrophysical and cosmological problems associated with radiative decays, the invisible decay into a majoron is often considered. Its decay rate is given for highly relativistic neutrinos as [Kim93]

$$\Gamma(\nu_H \to \nu_L + \chi) = \frac{g^2 m_L m_H}{16\pi E_H}\left(\frac{x}{2} - 2 - \frac{2}{x}\ln x + \frac{2}{x^2} - \frac{1}{2x^3}\right) \tag{6.87}$$

with g being an effective coupling constant and $x = m_H/m_L$. Little is known experimentally about this invisible decay.

Matter can enhance the decay rates as discussed in [Kim93]. However, still no neutrino decay has yet been observed.

We now proceed to a further process where neutrino masses can show up and which is generally considered as the gold-plated channel for probing the fundamental character of neutrinos, discussed in chapter 2.

Chapter 7

Double β-decay

A further nuclear decay which is extremely important for neutrino physics is neutrinoless double β-decay. This lepton-number-violating process requires, in addition to a non-vanishing neutrino mass, that neutrinos are Majorana particles. It is, therefore, often regarded as the gold-plated process for probing the fundamental character of neutrinos. For additional literature see [Doi83, Hax84, Doi85, Gro90, Boe92, Moe94, Kla95, Kla95a, Fae99, Eji00, Vog00, Kla01a, Ell02].

7.1 Introduction

Double β-decay is characterized by a nuclear process changing the nuclear charge Z by two units while leaving the atomic mass A unchanged. It is a transition among isobaric isotopes. Using the Weizsäcker mass formula [Wei35] these can be described as

$$m(Z, A = \text{constant}) \propto \text{constant} + \alpha Z + \beta Z^2 + \delta_P \qquad (7.1)$$

with δ_P as the pairing energy, empirically parametrized as [Boh75]

$$\delta_P = \begin{cases} -a_P A^{-1/2} & \text{even–even nuclei} \\ 0 & \text{even–odd and odd–even nuclei} \\ +a_P A^{-1/2} & \text{odd–odd nuclei} \end{cases} \qquad (7.2)$$

with $a_P \approx 12$ MeV. For odd A the pairing energy vanishes resulting in one parabola with one stable isobar, while for even A two parabola separated by $2\delta_P$ exist (figure 7.1). The second case allows for double β-decay and, therefore, all double β-decay emitters are even–even nuclei. It can be understood as two subsequent β-decays via a virtual intermediate state. Thus, a neccessary requirement for double β-decay to occur is

$$m(Z, A) > m(Z + 2, A) \qquad (7.3)$$

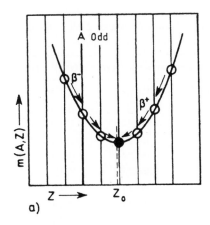 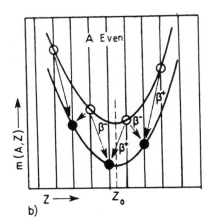

Figure 7.1. Dependence of energy on Z for nuclei with the same mass number A: stable nuclei are denoted by bold circles; left, nuclei with odd mass number A; right, nuclei with even mass number A.

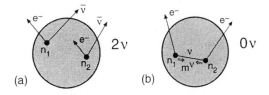

Figure 7.2. Schematic view of double β-decay.

and, for practical purposes, β-decay has to be forbidden:

$$m(Z, A) < m(Z + 1, A) \tag{7.4}$$

or at least strongly suppressed. Such a strong suppression of β-transitions between the involved nuclear states is caused by a large difference ΔL in angular momentum, as in the case of ^{48}Ca (ΔL equal to 5 or 6). Because ground states of even–even nuclei have spin-0 and parity $(+)$, the ground state transitions are characterized as $(0^+ \rightarrow 0^+)$ transitions. Today we know 36 possible double β-decay emitters, the most important of them are listed in table 7.1. A full list can be found in [Boe92].

In the following the two-nucleon mechanism (2n mechanism) is explored in more detail. Discussions of other mechanisms (Δ, π^-) where the same nucleon experiences two successive β-decays can be found in [Mut88]. For $(0^+ \rightarrow 0^+)$ transitions they are forbidden by angular momentum selection rules [Boe92].

Double β-decay was first discussed by M Goeppert-Mayer [Goe35] in the form of

$$(Z, A) \rightarrow (Z + 2, A) + 2e^- + 2\bar{\nu}_e \quad (2\nu\beta\beta\text{-decay}). \tag{7.5}$$

Table 7.1. Compilation of $\beta^-\beta^-$-emitters with a Q-value of at least 2 MeV. Q-values, natural abundances and phase space factors (taken from [Boe92]) are given.

Transition	Q-value (keV)	Nat. ab. (%)	$[G^{0\nu}]^{-1}$ (yr)	$[G^{2\nu}]^{-1}$ (yr)
$^{48}_{20}\text{Ca} \rightarrow {}^{48}_{22}\text{Ti}$	4271	0.187	4.10E24	2.52E16
$^{76}_{32}\text{Ge} \rightarrow {}^{76}_{34}\text{Se}$	2039	7.8	4.09E25	7.66E18
$^{82}_{34}\text{Se} \rightarrow {}^{82}_{36}\text{Kr}$	2995	9.2	9.27E24	9.27E24
$^{96}_{40}\text{Zr} \rightarrow {}^{96}_{42}\text{Mo}$	3350	2.8	4.46E24	5.19E16
$^{100}_{42}\text{Mo} \rightarrow {}^{100}_{44}\text{Ru}$	3034	9.6	5.70E24	1.06E17
$^{110}_{46}\text{Pd} \rightarrow {}^{110}_{48}\text{Cd}$	2013	11.8	1.86E25	2.51E18
$^{116}_{48}\text{Cd} \rightarrow {}^{116}_{50}\text{Sn}$	2802	7.5	5.28E24	5.28E24
$^{124}_{50}\text{Sn} \rightarrow {}^{124}_{52}\text{Te}$	2288	5.64	9.48E24	5.93E17
$^{130}_{52}\text{Te} \rightarrow {}^{130}_{54}\text{Xe}$	2533	34.5	5.89E24	2.08E17
$^{136}_{54}\text{Xe} \rightarrow {}^{136}_{56}\text{Ba}$	2479	8.9	5.52E24	2.07E17
$^{150}_{60}\text{Nd} \rightarrow {}^{150}_{62}\text{Sm}$	3367	5.6	1.25E24	8.41E15

This process can be seen as two simultaneous neutron decays (figure 7.2). This decay mode conserves lepton number and is allowed within the standard model, independently of the nature of the neutrino. This mode is of second-order Fermi theory and, therefore, the lifetime is proportional to $(G_F \cos\theta_C)^{-4}$. Within the GWS model (see chapter 3), this corresponds to a fourth-order process. As double β-decay is a higher-order effect, expected half-lives are long compared to β-decay: rough estimates illustrated in [Wu66, Kla95] result in half-lives of the order of 10^{20} yr and higher. Together with proton decay, this is among the rarest processes envisaged and, therefore, special experimental care has to be taken to observe this process. In contrast to proton decay, it is not easy to build detectors of several kilotons by using water, because one is restricted to the isotope of interest which currently implies typical sample sizes of g to several kg.

Shortly after the classical paper by Majorana [Maj37] discussing a two-component neutrino, Furry discussed another decay mode in form of [Fur39]

$$(Z, A) \rightarrow (Z+2, A) + 2e^- \qquad (0\nu\beta\beta\text{-decay}). \qquad (7.6)$$

Clearly, this process violates lepton number conservation by two units and is forbidden in the standard model. It can be seen as two subsequent steps ('Racah sequence') as shown in figure 7.2:

$$(Z, A) \rightarrow (Z+1, A) + e^- + \bar{\nu}_e$$
$$(Z+1, A) + \nu_e \rightarrow (Z+2, A) + e^-. \qquad (7.7)$$

First a neutron decays under the emission of a right-handed $\bar{\nu}_e$. This has to be absorbed at the second vertex as a left-handed ν_e. To fulfil these conditions, the neutrino and antineutrino have to be identical, i.e. the neutrinos have to be Majorana particles (see chapter 2). Moreover, to allow for helicity matching, a neutrino mass is required. The reason is that the wavefunction describing neutrino mass eigenstates for $m_\nu > 0$ has no fixed helicity and, therefore, besides the dominant left-handed contribution, has an admixture of a right-handed one, which is proportional to m_ν/E.

In principle, V + A weak charged currents could also mediate neutrinoless double β-decay. They could result from left–right symmetric theories like SO(10) (see chapter 5). The left–right symmetry is broken at low energies because the right-handed vector mesons $W_R^\pm$ and Z_R^0 have not yet been observed. Then, in addition to the neutrino mass mechanism, right-handed leptonic and hadronic currents can also contribute. The general Hamiltonian used for $0\nu\beta\beta$-decay rates is then given by

$$H = \frac{G_F \cos\theta_C}{\sqrt{2}} (j_L J_L^\dagger + \kappa j_L J_R^\dagger + \eta j_R J_L^\dagger + \lambda j_R J_R^\dagger) \qquad (7.8)$$

with the left- and right-handed leptonic currents as

$$j_L^\mu = \bar{e}\gamma^\mu(1 - \gamma_5)\nu_e \qquad j_R^\mu = \bar{e}\gamma^\mu(1 + \gamma_5)\nu_{eR}. \qquad (7.9)$$

The hadronic currents J can be expressed in an analogous way by quark currents. Often nucleon currents are used in a non-relativistic approximation treating nucleons within the nucleus as free particles (impulse approximation). The coupling constants κ, η, λ vanish in the GWS model. The mass eigenstates of the vector bosons $W_{1,2}^\pm$ are mixtures of the left- and right-handed gauge bosons

$$W_1^\pm = W_L^\pm \cos\theta + W_R^\pm \sin\theta \qquad (7.10)$$
$$W_2^\pm = -W_L^\pm \sin\theta + W_R^\pm \cos\theta \qquad (7.11)$$

with $\theta \ll 1$ and $M_2 \gg M_1$. Thus, the parameters can be expressed in left–right symmetric GUT models as

$$\eta = \kappa \approx \tan\theta \qquad \lambda \approx (M_1/M_2)^2 + \tan^2\theta. \qquad (7.12)$$

It can be shown that in gauge theories the mass and right-handed current mechanisms are connected and a positive observation of $0\nu\beta\beta$-decay would prove a finite Majorana mass [Sch82, Tak84]. The reason is that, regardless of the mechanism causing $0\nu\beta\beta$-decay, the two emitted electrons together with the two u, d quarks that are involved in the n $\rightarrow$ p transition can be coupled to the two ν_e in such a way that a neutrino–antineutrino transition as in the Majorana mass term occurs (figure 7.3). For an illustrative deduction see [Kay89].

The phase space for neutrinoless double β-decay is about a factor 10^6 larger than for $2\nu\beta\beta$-decay because of a correspondingly larger number of final states. The reason is that the virtual neutrino of process (7.6) is restricted to the volume of

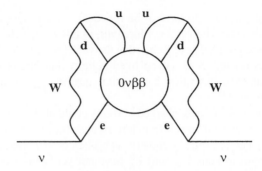

Figure 7.3. Graphical representation of the Schechter–Valle theorem. See text for details.

the nucleus which, according to Heisenberg's uncertainty principle, necessitates taking into account states up to about 100 MeV. In the $2\nu\beta\beta$-decay case, in which real neutrinos are emitted, the number of final states is restricted by the Q-value, which is below 5 MeV.

Of relevance is the alternative process of $\beta^+\beta^+$-decay, which is dominated by right-handed weak currents. This can occur in three variants:

$$(Z, A) \rightarrow (Z - 2, A) + 2e^+(+2\nu_e) \qquad (\beta^+\beta^+) \qquad (7.13)$$

$$e_B^- + (Z, A) \rightarrow (Z - 2, A) + e^+(+2\nu_e) \qquad (\beta^+/EC) \qquad (7.14)$$

$$2e_B^- + (Z, A) \rightarrow (Z - 2, A)(+2\nu_e) \qquad (EC/EC) \qquad (7.15)$$

$\beta^+\beta^+$ is always accompanied by EC/EC or β^+/EC-decay. The Coulomb barrier reduces the Q-value by $4m_ec^2$. The rate for $\beta^+\beta^+$ is, therefore, small and energetically only possible for six nuclides (table 7.2). Predicted half-lives for $2\nu\beta^+\beta^+$ are of the order 10^{26} yr while for β^+/EC (reduction by $Q - 2m_ec^2$) this can be reduced by orders of magnitude down to 10^{22-23} yr making an experimental detection more realistic. The lowest predicted half-life has 2νEC/EC which is the hardest to detect experimentally. A possible 0νEC/EC needs additional particles in the final state because of energy–momentum conservation. Double K-shell capture forbids the emission of a real photon in $0^+ \rightarrow 0^+$ transitions because of angular momentum conservation [Doi92, Doi93]. $\beta^+\beta^+$-decay is currently of minor importance with respect to neutrino physics; however, it might be very important to clarify the underlying mechanism if $0\nu\beta\beta$-decay is ever observed.

To sum up, $0\nu\beta\beta$-decay is only possible if neutrinos are massive Majorana particles and, therefore, the fundamental character of the neutrino can be probed in this process.

Table 7.2. Compilation of the six known $\beta^+\beta^+$ emitters in nature. The Q-values after subtracting $4m_ec^2$, natural abundances and phase space factors (taken from [Boe92]) are given.

Transition	Q-value (keV)	Nat. ab. (%)	$[G^{0\nu}]^{-1}$ (yr)	$[G^{2\nu}]^{-1}$ (yr)
^{78}Kr $\rightarrow$ ^{78}Se	838	0.35	1.8E29	2.56E24
^{96}Ru $\rightarrow$ ^{96}Mo	676	5.5	8.8E29	3.34E25
^{106}Cd $\rightarrow$ ^{106}Pd	738	1.25	7.4E29	1.69E25
^{124}Xe $\rightarrow$ ^{124}Te	822	0.10	5.9E29	7.57E24
^{130}Ba $\rightarrow$ ^{130}Xe	534	0.11	6.4E30	6.92E26
^{136}Ce $\rightarrow$ $^{136}_{48}$Ba	362	0.19	6.1E31	5.15E28

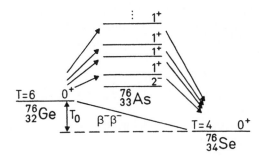

Figure 7.4. Principle of a transition via an intermediate state for $2\nu\beta\beta$-decay. Shown is the transition ^{76}Ge $\rightarrow^{76}$ Se, which can occur via 1^α states in ^{76}As.

7.2 Decay rates

Decay rates can be described analogously to β-decay starting from Fermi's Golden Rule but now the processes under discussion are of second-order perturbation theory. The details of the calculations are rather complex. We refer to the existing literature [Kon66, Doi83, Hax84, Doi85, Mut88, Tom88, Gro90, Boe92, Kla95, Suh98, Fae99, Vog00] and will give only a brief discussion.

7.2.1 The $2\nu\beta\beta$ decay rates

Because ground-state transitions are of the type ($0^+ \rightarrow 0^+$), they can be seen as two subsequent Gamow–Teller transitions and selection rules then require the intermediate states to be 1^+. Fermi transitions are forbidden or at least strongly suppressed by isospin selection rules [Hax84] (figure 7.4).

Using time-dependent perturbation theory, the transition probability W per time from an initial state i to a final state f can be written as (see chapter 6)

$$\frac{\mathrm{d}W}{\mathrm{d}t} = \frac{2\pi}{\hbar}|\langle f|H_{if}|i\rangle|^2\delta(E_f - E_i) \tag{7.16}$$

where the δ-function illustrates the fact that we are dealing with discrete energy levels instead of a density of final states. The corresponding matrix element for double β-decay is one order higher in the pertubation series than single β-decay and is, therefore, given by

$$M_{if} = \sum_m \frac{\langle f|H_{if}|m\rangle\langle m|H_{if}|i\rangle}{E_i - E_m} \tag{7.17}$$

where m characterizes the set of virtual intermediate states and H_{if} is the weak Hamilton operator. As we cannot distinguish the combinations in which the electron–neutrino system appears in the intermediate steps, we have to sum all configurations in (7.17). The energies E_m of the intermediate states are given as

$$E_m = E_{Nm} + E_{e1} + E_{v1} \qquad E_m = E_{Nm} + E_{e2} + E_{v_2} \tag{7.18}$$
$$E_m = E_{Nm} + E_{e1} + E_{v2} \qquad E_m = E_{Nm} + E_{e2} + E_{v_1} \tag{7.19}$$

where E_{Nm} is the energy of the intermediate nucleus. Without an explicit derivation (see [Kon66,Gro90,Boe92] for details), the obtained decay rate is given by

$$\lambda_{2v} = \frac{G_F^4 \cos^4\theta_C}{8\pi^7} \int_{m_e}^{Q+m_e} F(Z, E_{e1})p_{e1}E_{e1}\,\mathrm{d}E_{e1}$$
$$\times \int_{m_e}^{Q+2m_e-E_{e1}} F(Z, E_{e2})p_{e2}E_{e2}\,\mathrm{d}E_{e2}$$
$$\times \int_0^{Q+2m_e-E_{e1}-E_{e2}} E_{v1}^2 E_{v2}^2\,\mathrm{d}E_{v1} \sum_{m,m'} A_{mm'} \tag{7.20}$$

with Q as the nuclear transition energy available to the leptons

$$Q = E_{e1} + E_{e2} + E_{v1} + E_{v2} - 2m_e \tag{7.21}$$

and $F(Z, E)$ the Fermi function (see chapter 6). The quantity $A_{mm'}$ contains the Gamow–Teller nuclear matrix elements and the typical energy denominators from the perturbative calculations

$$A_{mm'} = \langle 0_f^+\|t_-\sigma\|1_j^+\rangle\langle 1_j^+\|t_-\sigma\|0_i^+\rangle\langle 0_f^+\|t_-\sigma\|1_j^+\rangle\langle 1_j^+\|t_-\sigma\|0_i^+\rangle \tag{7.22}$$
$$\times \tfrac{1}{3}(K_m K_{m'} + L_m L_{m'} + \tfrac{1}{2}K_m L_{m'} + \tfrac{1}{2}L_m K_{m'}) \tag{7.23}$$

with t_- as the isospin ladder operator converting a neutron into a proton, σ as spin operator, as already introduced in chapter 6, and

$$K_m = \frac{1}{E_{Nm} + E_{e1} + E_{\nu 1} - E_i} + \frac{1}{E_{Nm} + E_{e2} + E_{\nu 2} - E_i} \qquad (7.24)$$

$$L_m = \frac{1}{E_{Nm} + E_{e1} + E_{\nu 2} - E_i} + \frac{1}{E_{Nm} + E_{e2} + E_{\nu 1} - E_i}. \qquad (7.25)$$

Two more assumptions are good approximations in the case of $0^+ \to 0^+$ transitions. First of all, the lepton energies can be replaced by their corresponding average value, $E_e + E_\nu \approx Q/2 + m_e$ in the denominator of (7.24) and (7.25). This implies that

$$K_m \approx L_m \approx \frac{1}{E_m - E_i + Q/2} = \frac{1}{E_m - (M_i + M_f)/2}. \qquad (7.26)$$

With this approximation the nuclear physics and kinematical parts separate. The second approach is a simplified Fermi function, often called the Primakoff–Rosen approximation [Pri68], given in (6.17). The single-electron spectrum can then be obtained by integrating over $dE_{\nu 1}$ and dE_{e2} in equation (7.20). Then the Primakoff–Rosen approximation allows us to do the integration analytically and this results in a single electron spectrum [Boe92]:

$$\frac{dN}{dT_e} \approx (T_e + 1)^2 (Q - T_e)^6 [(Q - T_e)^2 + 8(Q - T_e) + 28] \qquad (7.27)$$

where T_e is the electron kinetic energy in units of the electron mass. Most experiments measure the sum energy K (also in units of m_e) of both electrons. Here, the spectral form can be obtained by changing to the variables $E_{e1} + E_{e2}$ and $E_{e1} - E_{e2}$ in (7.20) and performing an integration with respect to the latter, resulting in

$$\frac{dN}{dK} \approx K(Q - K)^5 \left(1 + 2K + \frac{4K^2}{3} + \frac{K^3}{3} + \frac{K^4}{30}\right) \qquad (7.28)$$

which shows a maximum at about $0.32 \times Q$. A compilation of expected shapes for all kinds of decay mechanisms is given in [Tre95]. The total rate is obtained by integrating over equations (7.28) and (7.20)

$$\lambda_{2\nu} \approx Q^7 \left(1 + \frac{Q}{2} + \frac{Q^2}{9} + \frac{Q^3}{90} + \frac{Q^4}{1980}\right). \qquad (7.29)$$

The total rate scales with Q^{11}. The decay rate can then be transformed in a half-life which, in its commonly used form, is written as

$$\lambda_{2\nu}/\ln 2 = (T_{1/2}^{2\nu})^{-1} = G^{2\nu}(Q, Z) \left| M_{GT}^{2\nu} + \frac{g_V^2}{g_A^2} M_F^{2\nu}\right|^2 \qquad (7.30)$$

with $G^{2\nu}$ as the phase space and the matrix elements given by

$$M_{GT}^{2\nu} = \sum_j \frac{\langle 0_f^+ \| t_- \sigma \| 1_j^+ \rangle \langle 1_j^+ \| t_- \sigma \| 0_i^+ \rangle}{E_j + Q/2 + m_e - E_i} \qquad (7.31)$$

$$M_F^{2\nu} = \sum_j \frac{\langle 0_f^+ \| t_- \| 1_j^+ \rangle \langle 1_j^+ \| t_- \| 0_i^+ \rangle}{E_j + Q/2 + m_e - E_i}. \qquad (7.32)$$

As already mentioned, Fermi transitions are strongly suppressed.

In earlier times the virtual energies of the intermediate states E_m were replaced by an average energy $\langle E_m \rangle$ and the sum of the intermediate states was taken using $\sum_m |1_m^+\rangle\langle 1_m^+| = 1$ (closure approximation). The advantage was that only the wavefunctions of the initial and final state were required and the complex calculations of the intermediate states could be avoided. However, interference between the different individual terms of the matrix element (7.22) is important and must be considered. Thus, the amplitudes have to be weighted with the correct energy E_m and the closure approximation is not appropriate for estimating $2\nu\beta\beta$-decay rates.

7.2.2 The $0\nu\beta\beta$ decay rates

Now let us consider the neutrinoless case. As stated, beside requiring neutrinos to be Majorana particles, we further have to assume a non-vanishing mass or right-handed (V + A) currents to account for the helicity mismatch. Both mechanisms are associated with different nuclear matrix elements [Doi85, Mut88, Tom91]. A recent formulation of the general problem can be found in [Pae99]. Consider the mass case and no V + A interactions first, a generalization including V + A currents will be given later. The decay rate is then given by [Boe92]

$$\lambda_{0\nu} = 2\pi \sum_{spin} |R_{0\nu}|^2 \delta(E_{e1} + E_{e2} + E_f - M_i) \, d^3 p_{e1} \, d^3 p_{e2} \qquad (7.33)$$

where $R_{0\nu}$ is the transition amplitude containing leptonic and hadronic parts. Because of the complexity, we concentrate on the leptonic part (for details see [Doi85, Gro90, Boe92]). The two electron phase space integral is

$$G^{0\nu} \propto \int_{m_e}^{Q+m_e} F(Z, E_{e1}) F(Z, E_{e2}) p_{e1} p_{e2} E_{e1} E_{e2} \delta(Q - E_{e1} - E_{e2}) \, dE_{e1} \, dE_{e2} \qquad (7.34)$$

with $Q = E_{e1} + E_{e2} - 2m_e$. Using the Primakoff–Rosen approximation (6.18), the decay rate is

$$\lambda_{0\nu} \propto \left(\frac{Q^5}{30} - \frac{2Q^2}{3} + Q - \frac{2}{5} \right). \qquad (7.35)$$

Here, the total rate scales with Q^5 compared to the Q^{11} dependence of $2\nu\beta\beta$-decay. The total decay rate is then

$$\lambda_{0\nu}/\ln 2 = (T_{1/2}^{0\nu})^{-1} = G^{0\nu}(Q, Z)|M_{GT}^{0\nu} - M_F^{0\nu}|^2 \left(\frac{\langle m_{\nu_e}\rangle}{m_e}\right)^2 \tag{7.36}$$

with the matrix elements

$$M_{GT}^{0\nu} = \sum_{m,n}\langle 0_f^+ \| t_{-m}t_{-n}H(r)\sigma_m\sigma_n \| 0_i^+\rangle \tag{7.37}$$

$$M_F^{0\nu} = \sum_{m,n}\langle 0_f^+ \| t_{-m}t_{-n}H(r) \| 0_i^+\rangle \left(\frac{g_V}{g_A}\right)^2 \tag{7.38}$$

with $r = |r_m - r_n|$. Beside the transition operator there is now also a neutrino potential $H(r)$ acting on the nuclear wavefunctions describing the exchange of the virtual neutrino. Because of this propagator Fermi-transitions can also occur as explained in [Mut88]. The dependence of the lifetime on the neutrino mass arises from the leptonic part of $|R_{0\nu}|$.

The measuring quantity $\langle m_{\nu_e}\rangle$, called the effective Majorana neutrino mass, which can be deduced from the half-life measurement, is of course the one of great interest for neutrino physics. It is given by

$$\langle m_{\nu_e}\rangle = \left|\sum_i U_{ei}^2 m_i\right| = \left|\sum_i |U_{ei}|^2 e^{2i\alpha_i} m_i\right| \tag{7.39}$$

with U_{ei} as the mixing matrix elements, m_i as the corresponding mass eigenvalues and the CP phases $\alpha_i/2$. If CP is conserved then $\alpha_i = k\pi$. A Dirac neutrino is a pair of degenerate Majorana neutrinos with $\alpha_i = \pm 1$, whose contributions exactly cancel. In addition, CP-violating phases can already occur for two generations because transformations of the form $\nu_i \to \nu_i' = e^{i\alpha_i}\nu_i$ cannot be performed, because they would violate the self-conjugation property. Note the fact of possible interference among the different terms contributing to the sum in (7.39) in contrast to single β-decay. A general direct comparison with β-decay results can only be made under certain assumptions. Both results should be treated complementarily. Anyhow, limits on $\langle m_{\nu_e}\rangle$ are only valid for Majorana neutrinos.

If right-handed currents are included, expression (7.36) can be generalized to

$$(T_{1/2}^{0\nu})^{-1} = C_{mm}\left(\frac{\langle m_{\nu_e}\rangle}{m_e}\right)^2 + C_{\eta\eta}\langle\eta\rangle^2 + C_{\lambda\lambda}\langle\lambda\rangle^2 \tag{7.40}$$

$$+ C_{m\eta}(\frac{\langle m_{\nu_e}\rangle}{m_e})\langle\eta\rangle + C_{m\lambda}\left(\frac{\langle m_{\nu_e}\rangle}{m_e}\right)\langle\lambda\rangle + C_{\eta\lambda}\langle\eta\rangle\langle\lambda\rangle \tag{7.41}$$

where the coefficients C contain the phase space factors and the matrix elements and the effective quantities are

$$\langle\eta\rangle = \eta\sum_j U_{ej}V_{ej} \qquad \langle\lambda\rangle = \lambda\sum_j U_{ej}V_{ej} \tag{7.42}$$

with V_{ej} as the mixing matrix elements among the right-handed neutrino states. Equation (7.40) reduces to (7.36) when $\langle \eta \rangle$, $\langle \lambda \rangle = 0$. For example the element C_{mm} is given by

$$C_{mm} = |M_{GT}^{0\nu} - M_F^{0\nu}|^2 G^{0\nu}(Q, Z). \tag{7.43}$$

The ratio $R = \langle \lambda \rangle / \langle \eta \rangle$, being independent of V_{ej}, is, under certain assumptions, a simple function of $K = (m_{W_L}/m_{W_R})^2$ and of the mixing angle θ introduced in (7.10) [Suh93].

The signature for the sum energy spectrum of both electrons in $0\nu\beta\beta$-decay is outstanding, namely a peak at the Q-value of the transition. The single electron spectrum is given in the used approximation by

$$\frac{dN}{dT_e} \propto (T_e + 1)^2(Q + 1 - T_e)^2. \tag{7.44}$$

7.2.3 Majoron accompanied double β-decay

A completely new class of decays emerges in connection with the emission of a majoron χ [Doi88]

$$(Z, A) \rightarrow (Z + 2, A) + 2e^- + \chi. \tag{7.45}$$

Majorana mass terms violate lepton number by two units and, therefore, also $(B - L)$ symmetry, which is the only anomaly-free combination of both quantum numbers. A breaking can be achieved in basically three ways:

- explicit $(B-L)$ breaking, meaning the Lagrangian contains $(B-L)$ breaking terms,
- spontaneous breaking of a local $(B - L)$ symmetry and
- spontaneous breaking of a global $(B - L)$ symmetry.

Associated with the last method is the existence of a Goldstone boson, which is called the majoron χ. Depending on its transformation properties under weak isospin, singlet [Chi80], doublet [San88] and triplet [Gel81] models exist. The triplet and pure doublet model are excluded by the measurements of the Z-width at LEP because such majorons would contribute the analogue of 2 (triplet) or 0.5 (doublet) neutrino flavours. Several new majoron models have evolved in recent years [Bur94, Hir96c].

A consequence for experiments is a different sum electron spectrum. The predicted spectral shapes are analogous to (7.28) as

$$\frac{dN}{dK} \propto (Q - K)^n \left(1 + 2K + \frac{4K^2}{3} + \frac{K^3}{3} + \frac{K^4}{30}\right) \tag{7.46}$$

where the spectral index n is now 1 for the triplet majoron, 3 for lepton-number-carrying majorons and 7 for various other majoron models. The different shape allows discrimination with respect to $2\nu\beta\beta$-decay, where $n = 5$. It should be noted that supersymmetric Zino-exchange allows the emission of two majorons, which also results in a $n = 3$ type spectrum, but a possible bound on a Zino-mass is less stringent than the one from direct accelerator experiments [Moh88]. In the $n = 1$ model, the effective neutrino–majoron coupling $\langle g_{\nu\chi} \rangle$ can be deduced from

$$(T_{1/2}^{0\nu\chi})^{-1} = |M_{GT} - M_F|^2 G^{0\nu\chi} |\langle g_{\nu\chi} \rangle|^2 \qquad (7.47)$$

where $\langle g_{\nu\chi} \rangle$ is given by

$$\langle g_{\nu\chi} \rangle = \sum_{i,j} g_{\nu\chi} U_{ei} U_{ej}. \qquad (7.48)$$

7.3   Nuclear structure effects on matrix elements

A main uncertainty in extracting a bound or a value on $\langle m_{\nu_e} \rangle$ from experimental half-life limits is the nuclear matrix element involved. Different questions are associated with the various decay modes. $2\nu\beta\beta$-decay is basically a study of Gamow–Teller (GT) amplitudes. $0\nu\beta\beta$-decay with the exchange of light Majorana neutrinos does not have selection rules on multipoles but the role of nucleon correlations and the sensitivity to nuclear models is important. The exchange of heavy neutrinos is basically dominated by the physics of nucleon–nucleon states at short distances. Two basic strategies are followed in the calculations: either the nuclear shell model approach or the quasi random phase approximation (QRPA). All calculations are quite complex and beyond the scope of this book. Detailed treatmeants can be found in [Hax84, Doi85, Sta90, Boe92, Mut88, Gro90, Suh98, Fae99].

   $2\nu\beta\beta$-decay is a standard weak process and does not involve any uncertainty from particle physics aspects. Its rate is governed by (7.17). The first factor in the numerator is identical to the β^+ or (n,p) amplitude for the final state nucleus, the second factor is equivalent to the β^- or (p,n) amplitude of the initial nucleus. In principle, all GT amplitudes including their signs have to be used. The difficulty is that the 2ν matrix elements only exhaust a small fraction (10^{-5}–10^{-7}) of the double GT sum rule [Vog88, Mut92] and, hence, it is sensitive to details of the nuclear structure. Various approaches have been done, with QRPA being the most common. A compilation is given in [Suh98]. The main ingredients are a repulsive particle–hole spin–isospin interaction and an attractive pp interaction. They play a decisive role in concentrating the β^- strength in the GT resonance and for the relative suppression of β^+ strength and its concentration at low excitation energies. The calculations typically show a strong dependence on the strength of a particle–particle force g_{PP}, which for realistic values is often close to its critical value ('collapse'). This indicates a rearrangement of the nuclear ground state but QRPA is meant to describe small deviations from the unperturbed ground state and, thus, is not fully applicable near the point of collapse. QRPA and

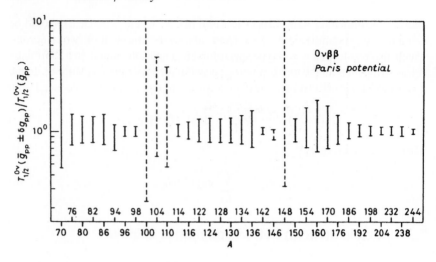

Figure 7.5. The uncertainty of $0\nu\beta\beta$-decay half-lives calculated using QRPA, resulting from limited knowledge of the particle–particle force (g_{pp}) for potential double-β emitters (from [Sta90]).

its various extensions are typically able to explain the experimental values by adjusting only one parameter. It could also be shown [Eri94] that a few low-lying states account for the whole matrix element, i.e. it is sufficient to describe correctly the β^+ and β^- amplitudes of low-lying states and include everything else in the overall renormalization (quenching) of the GT strength. In recent years nuclear shell model methods have become capable of handling much larger configuration spaces than before and can be used for descriptions as well. They avoid the above difficulties of QRPA and can also be tested with other data from nuclear spectroscopy.

In $0\nu\beta\beta$-decay mediated by light virtual Majorana neutrinos several new features arise. According to Heisenberg's uncertainty relation the virtual neutrino can have a momentum up to $q \simeq 1/r_{nn} \simeq 50\text{–}100$ MeV where r_{nn} is the distance between the decaying nucleons. Therefore, the dependence on the energy of the intermediate state is small and the closure approximation can be applied. Also, because $qR > 1$ (R being the radius of the nucleus), the expansion in multipoles does not converge and all multipoles contribute by comparable amounts (figure 7.6). Finally the neutrino propagator results in a long-range neutrino potential. A half-life calculation for all isotopes and the involved uncertainties is shown in figure 7.5. For a more detailed discussion and the treatment of heavy Majorana neutrinos, see [Mut88, Boe92, Suh98, Fae99, Vog00]. A compilation of representative calculations is shown in figure 7.7.

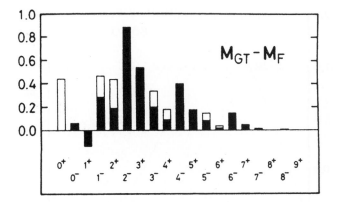

Figure 7.6. Decomposition of the nuclear matrix element $M_{GT} - M_F$ into contributions of the intermediate states with spin and parity I^π for the $0\nu\beta\beta$-decay of ^{76}Ge. Open and filled histograms describe the contributions of $-M_F$ and M_{GT} respectively (from [Mut89]).

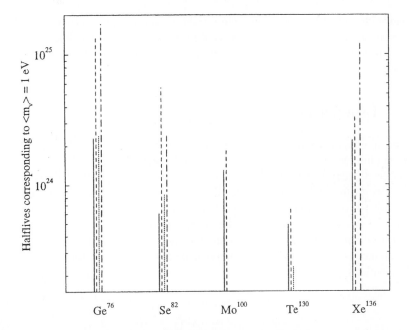

Figure 7.7. Comparison of representative half-life calculations of nuclear matrix elements for various isotopes using $\langle m_{\nu_e} \rangle = 1$ eV. Full curves are QRPA from [Sta90] and broken curves QRPA from [Eng88] (recalculated for $g_A = 1.25$ and $\alpha' = -390$ MeV fm^3), dotted [Hax84] and dot-and-dashed [Cau96] lines are shell model calculations (from [PDG02]).

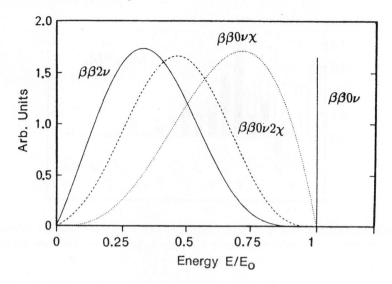

Figure 7.8. Different spectral shapes of observable sum energy spectra of emitted electrons in double β-decay. The $n = 1, 3, \lambda$ forms (dashed lines) correspond to different majoron accompanied modes, $n = 5$ (solid line) is the $2\nu\beta\beta$-decay and the $0\nu\beta\beta$-decay results in a peak.

7.4   Experiments

Typical energies for double β-decay are in the region of a few MeV distributed among the four leptons which are emitted as s-waves. The signal for neutrinoless double β-decay is a peak in the sum energy spectrum of both electrons at the Q-value of the transition, while for the $2\nu\beta\beta$-decay a continous spectrum with the form given (7.28) can be expected (figure 7.8). In tracking experiments where both electrons can be measured separately angular distributions can be used to distinguish among the various transitions and underlying processes. In the 2n mechanism, the main transitions can be described by the following angular distributions:

$$P(\theta_{12}) \propto 1 - \beta_1\beta_2 \cos\theta_{12} \qquad (0^+ \rightarrow 0^+) \qquad (7.49)$$

$$P(\theta_{12}) \propto 1 + \tfrac{1}{3}\beta_1\beta_2 \cos\theta_{12} \qquad (0^+ \rightarrow 2^+) \qquad (7.50)$$

with θ_{12} the angle between both electrons and $\beta_{1,2} = p_{1,2}/E_{1,2}$ their velocity. For a compilation of angular distributions of additional decay modes, see [Tre95]. Being a nuclear decay, the actual measured quantity is a half-life, whose value can be determined from the radioactive decay law assuming $T_{1/2} \gg t$:

$$T_{1/2}^{0\nu} = \ln 2ma N_A / N_{\beta\beta} \qquad (7.51)$$

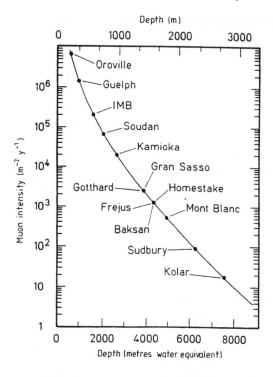

Figure 7.9. Muon intensity *versus* depth of some of the most important underground laboratories. Their shielding depth is given in metres of water equivalent (mwe) together with the attenuation of the atmospheric muon flux.

with m the used mass, a the isotopical abundance, N_A the Avogadro constant and $N_{\beta\beta}$ the number of events, which has to be taken from the experiment. If no peak is observed and the constant background scales linearly with time, $N_{\beta\beta}$ is often estimated at the 1σ level as a possible fluctuation of the background events $N_{\beta\beta} = \sqrt{N_B}$. The $0\nu\beta\beta$ half-life limit can then be estimated from experimental quantities to be

$$T_{1/2}^{0\nu} \propto a\sqrt{\frac{M \times t}{B \times \Delta E}} \tag{7.52}$$

where ΔE is the energy resolution at the peak position and B the background index normally given in counts/year/kg/keV. In addition, $\langle m_{\nu_e}\rangle$ scales with the square root of the half-life (7.36). With zero background the $\langle m_{\nu_e}\rangle$ sensitivity itself already scales with $\sqrt{Mt}$ [Moe91a, Ell02, Cre03].

7.4.1 Practical considerations in low-level counting

For a fair chance of detection, isotopes with large phase space factors (high Q-value) and large nuclear matrix elements should be considered. A significant amount of source material should be available, which is acquired in second-generation double β-decay experiments by using isotopical enriched materials. Being an extremely rare process the use of low-level counting techniques is necessary. The main concern is, therefore, background. Some of the most common background sources follow:

- Atmospheric muons and their interactions in the surrounding producing neutrons. They can be avoided by going underground (figure 7.9). The shielding depth is usually given in metre water equivalent (mwe). Neutron shielding might be required.
- Natural radioactive decay chains (U, Th). With the most energetic natural γ-line at 2.614 MeV (from ^{208}Tl decay) it is beyond most of the Q-values of double β-decay and, hence, disturb these experiments. Other prominent background components coming from these chains are ^{210}Pb producing β electrons with energies up to 1.1 MeV, ^{214}Bi β-decay up to 3 MeV and ^{222}Rn being parent to ^{214}Bi and ^{208}Tl.
- Man-made activities. In particular, ^{137}Cs (prominent γ-line at 662 keV) should be mentioned.
- Cosmogenic activation. Production of radionuclides by cosmic ray spallation in the materials during their stay on the Earth's surface.
- ^{40}K (γ-line at 1.461 MeV)

These background components not only influence double β-decay experiments but underground neutrino experiments in general. However, there might be additional background components which are more specific to a certain experiment. For more details on low-level counting techniques see [Heu95].

All direct experiments focus on electron detection and can be either active or passive. The advantage of active detectors are that the source and detector are identical but often only measure the sum energy of both electrons. However, passive detectors allow us to get more information (e.g. they measure energy and tracks of both electrons separately) but they usually have a smaller source strength. Some experiments will now be described in a little more detail.

7.4.2 Direct counting experiments

7.4.2.1 Semiconductor experiments

In this type of experiment, first done by a group from Milan [Fio67], Ge diodes are used. The source and detector are identical, the isotope under investigation is ^{76}Ge with a Q-value of 2039 keV. The big advantage is the excellent energy resolution of Ge semiconductors (typically about 3–4 keV at 2 MeV). However, the technique only allows the measurement of the sum energy of the two

Figure 7.10.   Photograph of the installation of enriched detectors in the Heidelberg–Moscow experiment (with kind permission of H V Klapdor-Kleingrothaus).

electrons. A big step forward was taken by using enriched germanium (the natural abundance of ^{76}Ge is 7.8%).

The Heidelberg–Moscow experiment. The Heidelberg–Moscow experiment [Gue97] in the Gran Sasso Laboratory uses 11 kg of Ge enriched to about 86% in ^{76}Ge in the form of five HP Ge detectors (figure 7.10). After 53.9 kg × y of data-taking the peak region reveals no signal (figure 7.11). A background as low as 0.2 counts/yr/kg/keV at the peak position has been achieved. The obtained half-life limit after 53.9 kg × yr is [Kla01]

$$T_{1/2}^{0\nu} > 1.9 \times 10^{25} \text{ yr} \qquad (90\% \text{ CL}) \qquad (7.53)$$

which can be converted using (7.36) and the matrix elements given in [Sta90] to an upper bound of

$$\langle m_{\nu_e} \rangle < 0.35 \text{ eV}. \qquad (7.54)$$

This is currently the best available bound coming from double β-decay. Quite controversial is the discussion of a recently claimed evidence for $0\nu\beta\beta$-decay, for details see [Kla01, Aal02, Fer02, Kla02, Har02, Zde02].

    A $2\nu\beta\beta$-decay half-life was obtained by carefully subtracting all identified background sources. The resulting spectrum is shown in figure 7.12. The obtained

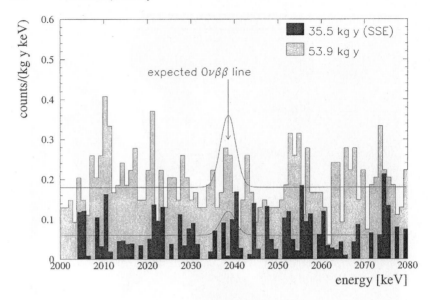

Figure 7.11. Observed spectrum around the expected $0\nu\beta\beta$-decay peak of the Heidelberg–Moscow collaboration. No signal can be seen. The two different spectra correspond to measuring periods with (black) and without (grey) pulse shape discrimination (from [Kla01]).

half-life is [Kla01]

$$T_{1/2}^{2\nu} = 1.55 \pm 0.01^{+0.19}_{-0.15} \times 10^{21} \text{ yr.} \tag{7.55}$$

The total amount of $2\nu\beta\beta$-decay events corresponds to more than 50 000 events. Comparing this number with the 36 events of the discovery in 1987 [Ell87] demonstrates the progress in the field.

A second experiment using Ge in the form of enriched detectors is the IGEX collaboration. After 5.7kg × y the obtained half-life limit is [Aal99, Aal02]

$$T_{1/2}^{0\nu} > 1.6 \times 10^{25} \text{ yr} \quad (90\% \text{ CL).} \tag{7.56}$$

Moreover, there is always the possibility of depositing a double β-decay emitter near a semiconductor detector to study its decay but then only transitions to excited states can be observed by detecting the corresponding gamma rays. Searches for $\beta^+\beta^+$-decay and β^+/EC-decay were also done in this way searching for the 511 keV photons. This has been widely used in the past.

COBRA. A new approach to take advantage of the good energy resolution of semiconductors is COBRA [Zub01]. The idea here is to use CdTe or CdZnTe detectors, mainly to explore ^{116}Cd and ^{130}Te decays. In total, there are seven

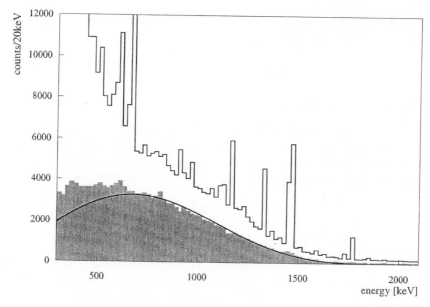

Figure 7.12. Measured spectrum from the Heidelberg–Moscow experiment and the contribution from $2\nu\beta\beta$-decay(from [Kla01]).

(nine in case of CdZnTe) double β-emitters within the detector including those of $\beta^+\beta^+$-decay. In pixelized detectors this might even offer tracking possibilities. First results can be found in [Kie03].

7.4.2.2 Scintillator experiments

Some double β-decay isotopes can be used in scintillators and also follow the idea that source and detector are identical. However, the energy resolution in scintillation counters is worse than in semiconductors. The first of this type of experiment was done with ^{48}Ca using CaF_2 crystals. The latest of these approaches used 37.4 kg of CaF_2 (containing 43 g of ^{48}Ca) in a coal mine near Beijing and obtained a limit for the neutrinoless double β-decay of [You91]

$$T_{1/2}^{0\nu}(^{48}Ca) > 9.5 \times 10^{21} \text{ yr} \qquad (76\% \text{ CL}). \qquad (7.57)$$

Another isotope used is ^{116}Cd in form of $CdWO_4$ [Geo95]. Four detectors of this kind, enriched in ^{116}Cd, are installed in the Solotvino salt mine in the Ukraine. They obtained a limit of [Dan00]

$$T_{1/2}^{0\nu}(^{116}Cd) > 2.9 \times 10^{22} \text{ yr} \qquad (90\% \text{ CL}). \qquad (7.58)$$

A cerium-doped Gd silicate crystal (Gd_2SiO_5:Ce) has also been used giving a half-life of ^{160}Gd of [Dan01]:

$$T_{1/2}^{0\nu}(^{160}\text{Gd}) > 1.3 \times 10^{21} \text{ yr} \quad (90\% \text{ CL}). \quad (7.59)$$

7.4.2.3 Cryogenic detectors

A new technique which might become more important in the future are bolometers running at very low temperature (mK) (see chapter 6). In dielectric materials the specific heat $C(T)$ at such temperatures scales according to (6.50). Therefore, the energy deposition, ΔE, of double β-decay would lead to a temperature rise ΔT of

$$\Delta T = \frac{\Delta E}{C(T)M}. \quad (7.60)$$

Such detectors normally have a very good energy resolution of a few keV at 2 MeV. Currently, only one such experiment (MIBETA) is running using twenty 334 g TeO_2 crystals at 8 mK to search for the ^{130}Te decay [Ale97]. As detectors NTD Ge thermistors are used. The obtained half-life limit corresponds to [Cre03]

$$T_{1/2}^{0\nu}(^{130}\text{Te}) > 2.1 \times 10^{23} \text{ yr} \quad (90\% \text{ CL}). \quad (7.61)$$

A larger version consisting of 62 crystals (CUORICINO) has recently started data-taking.

7.4.2.4 Ionization experiments

These passive experiments are mostly built in the form of time projection chambers (TPCs) where the emitter is either the filling gas (e.g. ^{136}Xe) or is included in thin foils. The advantage is that energy measurements as well as tracking of the two electrons is possible. The disadvantages are the worse energy resolution and, in the case of thin foils, the limited source strength. It was a device such as this which first gave evidence for $2\nu\beta\beta$-decay in a direct counting experiment using ^{82}Se [Ell87]. The experiment used a 14 g selenium source, enriched to 97% in ^{82}Se, in the form of a thin foil installed in the centre of a TPC (figure 7.13). The TPC was shielded against cosmic rays by a veto system. After 7960 hr of measuring time and background subtraction, 36 events remained which, if attributed to $2\nu\beta\beta$-decay, resulted in a half-life of

$$T_{1/2}^{2\nu}(^{82}\text{Se}) = (1.1^{+0.8}_{-0.3}) \times 10^{20} \text{ yr}. \quad (7.62)$$

The TPC at UC Irvine was further used to study the decays of ^{82}Se, ^{100}Mo and ^{150}Nd [Des97]. A limit of

$$T_{1/2}^{0\nu}(^{150}\text{Nd}) > 1.22 \times 10^{21} \text{ yr} \quad (7.63)$$

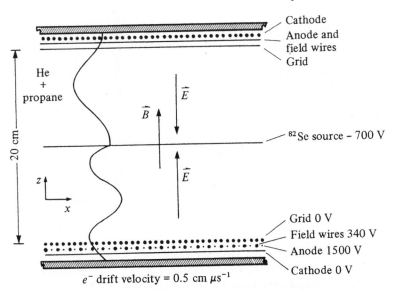

Figure 7.13. Schematic view of the setup of the TPC at UC Irvine, showing the wires, direction of the fields and the ^{82}Se source. A sample electron trajectory is shown on the left-hand side.

has been achieved. An experiment using a TPC with an active volume of 180 l filled with Xe (enriched to 62.5% in ^{136}Xe which corresponds to 3.3 kg) under a pressure of 5 atm was run at the Gotthard tunnel [Lue98]. They obtained a lower bound of

$$T^{0\nu}_{1/2}(^{136}\text{Xe}) > 4.4 \times 10^{23} \text{ yr} \qquad (90\% \text{ CL}). \qquad (7.64)$$

Another recently started experiment is NEMOIII in the Frejus Underground Laboratory. It is a passive source detector using thin foils made out of double beta elements. It consists of a tracking (wire chambers) and a calometric (plastic scintillators) device put into a 30 G magnetic field. The total source strength is about 10 kg which, in a first run, is dominated by using enriched ^{100}Mo foils.

A combination of drift chambers, plastic scintillators and NaI detectors was used in the ELEGANT V detector to investigate samples of the order of 100 g enriched in ^{100}Mo [Eji01] and ^{116}Cd [Eji97].

A compilation of some obtained double β results is shown in table 7.3.

7.4.3 Geochemical experiments

The geochemical approach is to use old ores, which could have accumulated a significant amount of nuclei due to double β-decay over geological time scales. This would lead to an isotopic anomaly which could be measured by mass spectrometry. Clearly the advantage of such experiments is the long exposure

Table 7.3. Compilation of obtained limits for $0\nu\beta\beta$-decay.

$^{48}_{20}\mathrm{Ca} \rightarrow {}^{48}_{22}\mathrm{Ti}$	$> 9.5 \times 10^{21}\,(76\%)$	< 8.3	(76%)
$^{76}_{32}\mathrm{Ge} \rightarrow {}^{76}_{34}\mathrm{Se}$	$> 1.9 \times 10^{25}\,(90\%)$	< 0.35	(90%)
$^{82}_{34}\mathrm{Se} \rightarrow {}^{82}_{36}\mathrm{Kr}$	$> 2.7 \times 10^{22}\,(68\%)$	< 5.0	(68%)
$^{100}_{42}\mathrm{Mo} \rightarrow {}^{100}_{44}\mathrm{Ru}$	$> 5.5 \times 10^{22}\,(90\%)$	< 2.1	(90%)
$^{116}_{48}\mathrm{Cd} \rightarrow {}^{116}_{50}\mathrm{Sn}$	$> 7 \times 10^{22}\ \ (90\%)$	< 2.6	(90%)
$^{128}_{52}\mathrm{Te} \rightarrow {}^{128}_{54}\mathrm{Xe}$	$> 7.7 \times 10^{24}\,(68\%)$	< 1.1	(68%)
$^{130}_{52}\mathrm{Te} \rightarrow {}^{130}_{54}\mathrm{Xe}$	$> 2.1 \times 10^{23}\,(90\%)$	$< 0.85\text{–}2.1$	(90%)
$^{136}_{54}\mathrm{Xe} \rightarrow {}^{136}_{56}\mathrm{Ba}$	$> 4.4 \times 10^{23}\,(90\%)$	< 2.3	(90%)
$^{150}_{60}\mathrm{Nd} \rightarrow {}^{150}_{62}\mathrm{Sm}$	$> 2.1 \times 10^{21}\,(90\%)$	< 4.1	(90%)

time of up to billions of years. Using the age T of the ore, and measuring the abundance of the mother $N(Z, A)$ and daughter $N(Z \pm 2, A)$ isotopes, the decay rate can be determined from the exponential decay law ($t \ll T_{1/2}$)

$$\lambda \simeq \frac{N(Z \pm 2, A)}{N(Z, A)} \times \frac{1}{T}. \tag{7.65}$$

As only the total amount of the daughter is observed, this type of measurement does not allow us to differentiate between the production mechanisms; therefore, the measured decay rate is

$$\lambda = \lambda_{2\nu} + \lambda_{0\nu}. \tag{7.66}$$

To be useful, several requirements and uncertainties have to be taken into account if applying this method. The isotope of interest should be present in a high concentration within the ore. In addition, a high initial concentration of the daughter should be avoided if possible. Other external effects which could influence the daughter concentration should be excluded. Last but not least, an accurate age determination of the ore is necessary. From all these considerations, only Se and Te ores are usable in practice. ^{82}Se, ^{128}Te and ^{130}Te decay to inert noble gases (^{82}Kr, ^{128,130}Xe). The noble gas concentration during crystallization and ore formation will be small. The detection of the small expected isotopical anomaly is made possible due to the large sensitivity of noble gas mass spectrometry [Kir86]. Although experiments of this type were initially performed in 1949, real convincing evidence for double β-decay was observed later in experiments using selenium and tellurium ores [Kir67, Kir68, Kir86]. More recent measurements can be found in [Kir86, Lin88, Ber92]. Comparing the decay rates of the two Te isotopes, phase space arguments ($2\nu\beta\beta$-decay scales with Q^{11}, while $0\nu\beta\beta$-decay scales with Q^5) and the assumption of almost identical matrix elements show that the observed half-life for ^{130}Te can be attributed to $2\nu\beta\beta$-decay [Mut88]. However, the obtained $2\nu\beta\beta$-decay half-life of ^{128}Te and the corresponding $0\nu\beta\beta$-decay limit is still the best beside ^{76}Ge.

A different approach using thermal ionization mass spectrometry allowed the determination of the $2\nu\beta\beta$-decay half-life of ^{96}Zr to be about 10^{19} yr [Kaw93, Wie01], a measurement also performed by NEMO II [Arn99].

7.4.4 Radiochemical experiments

This method takes advantage of the radioactive decay of the daughter nuclei, allowing a shorter 'measuring' time than geochemical experiments ('milking experiments'). It is also independent of some uncertainties in the latter, e.g. the geological age of the sample, original concentration of the daughter and possible diffusion effects of noble gases in geochemical samples. No information on the decay mode can be obtained—only the total concentration of daughter nuclei is measured.

Two possible candidates are the decays ^{232}Th $\rightarrow$ ^{232}U and ^{238}U $\rightarrow$ ^{238}Pu with Q-values of 850 keV (^{232}Th) and 1.15 MeV (^{238}U) respectively. Both daughters are unstable against α-decay with half-lives of 70 yr (^{232}Th) and 87.7 yr (^{238}U). For the detection of the ^{238}U $\rightarrow$ ^{238}Pu decay, the emission of a 5.5 MeV α-particle from the ^{238}Pu decay is used as a signal. The first such experiment was originally performed in 1950 using a six-year-old UO$_3$ sample. From the non-observation of the 5.51 MeV α-particles, a lower limit of

$$T_{1/2}^{0\nu}(^{238}\text{U}) > 6 \times 10^{18} \text{ yr} \qquad (7.67)$$

was deduced. Recently, a sample of 8.47 kg of uranium nitrate, which was purified in 1956 and analysed in 1989, was investigated, and a half-life of

$$T_{1/2}^{2\nu}(^{238}\text{U}) = (2.0 \pm 0.6) \times 10^{21} \text{ yr} \qquad (7.68)$$

was obtained [Tur92]. Both geo- and radio-chemical methods measure only the total decay rate by examining the concentration of the daughter nuclei. Because they are not able to distinguish between the different decay modes, their sensitivity is finally limited by $2\nu\beta\beta$-decay. This makes it almost impossible to establish real positive evidence for the neutrinoless mode by these methods.

Observation of $2\nu\beta\beta$-decay has been quoted now for nine isotopes. A compilation of measured half-lives is given in table 7.4. A complete list of all experimental results obtained until end of 2001 can be found in [Tre02].

7.5 Interpretation of the obtained results

As already stated, the best limit for $0\nu\beta\beta$-decay has been obtained with ^{76}Ge by the Heidelberg–Moscow experiment giving an upper bound of 0.35 eV for $\langle m_{\nu_e} \rangle$. If right-handed currents are included, $\langle m_{\nu_e} \rangle$ is fixed by an ellipsoid which is shown in figure 7.14. The weakest mass limit allowed occurs for $\langle \lambda \rangle$, $\langle \eta \rangle \neq 0$. In this

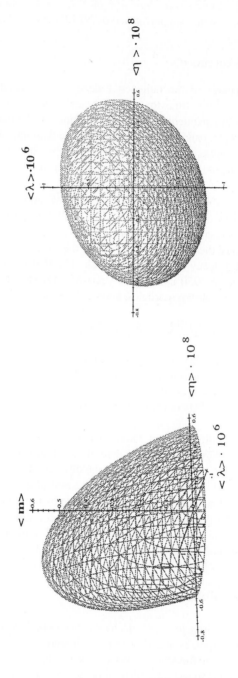

Figure 7.14. Obtained neutrino mass $\langle m_{\nu_e} \rangle$ as a function of the two right-handed current parameters $\langle \lambda \rangle$, $\langle \eta \rangle$. Left: Three dimensional ellipsoid. Right: Projection on the $\langle \eta \rangle - \langle \lambda \rangle$ plane. Usually limits on the neutrino mass are presented assuming $\langle \eta \rangle = \langle \lambda \rangle = 0$.

Table 7.4. Compilation of obtained half-lives for $2\nu\beta\beta$-decay and the deduced matrix element.

Isotope	Experiment	$T_{1/2}$ (10^{20} yr)	$M_{GT}^{2\nu}$ (MeV^{-1})
^{48}Ca	Calt.-KIAE	$0.43^{+0.24}_{-0.11} \pm 0.14$	0.05
^{76}Ge	MPIK-KIAE	$15.5 \pm 0.1^{+1.9}_{-1.5}\, 0.15$	
^{76}Ge	IGEX	11 ± 1.5	0.15
^{82}Se	NEMO 2	$0.89 \pm 0.10 \pm 0.10$	0.10
^{100}Mo	ELEGANT V	$0.115^{+0.03}_{-0.02}$	
^{100}Mo	NEMO 2	$0.095 \pm 0.004 \pm 0.009$	
^{100}Mo	UCI	$0.0682^{+0.0038}_{-0.0053} \pm 0.0068$	0.22
^{116}Cd	NEMO 2	$0.375 \pm 0.035 \pm 0.021$	0.12
^{116}Cd	ELEGANT V	$0.26^{+0.09}_{-0.05}$	
^{116}Cd	ELEGANT V	$0.26^{+0.07}_{-0.04}$	
^{128}Te*	Wash. Uni-Tata	77000 ± 4000	0.025
^{150}Nd	ITEP/INR	$0.188^{+0.066}_{-0.039} \pm 0.019$	
^{150}Nd	UCI	$0.0675^{+0.0037}_{-0.0042} \pm 0.0068$	0.07
^{238}U		20 ± 6	0.05

case the half-life of (7.53) corresponds to limits of

$$\langle m_{\nu_e} \rangle < 0.56\, eV \tag{7.69}$$

$$\langle \eta \rangle < 6.5 \times 10^{-9} \tag{7.70}$$

$$\langle \lambda \rangle < 8.2 \times 10^{-7}. \tag{7.71}$$

The obtained half-life limit also sets bounds on other physical quantities because the intermediate transition can be realized by other mechanisms. Among these are double charged Higgs bosons, right-handed weak currents, R-parity-violating SUSY and leptoquarks (see [Kla99]).

7.5.1 Effects of MeV neutrinos

Equation (7.39) has to be modified for heavy neutrinos ($m_\nu \geq 1$ MeV). Now the neutrino mass in the propagator can no longer be neglected with respect to the neutrino momentum. This results in a change in the radial shape of the used neutrino potential $H(r)$ from

$$H(r) \propto \frac{1}{r} \text{ light neutrinos} \rightarrow H(r) \propto \frac{\exp(-m_h r)}{r} \text{ heavy neutrinos.} \quad (7.72)$$

The change in H(r) can be accommodated by introducing an additional factor $F(m_h, A)$ into (7.36) resulting in an atomic mass A dependent contribution:

$$\langle m_{\nu_e} \rangle = \left| \sum_{i=1,\text{light}}^{N} U_{ei}^2 m_i + \sum_{h=1,\text{heavy}}^{M} F(m_h, A) U_{eh}^2 m_h \right|. \quad (7.73)$$

By comparing the $\langle m_{\nu_e} \rangle$ obtained for different isotopes, interesting limits on the mixing angles for an MeV neutrino can be deduced [Hal83, Zub97].

7.5.2 Transitions to excited states

From the point of view of right-handed currents, investigating transitions to the first excited 2^+ state is important, because here the contribution of the mass term vanishes in first order. The phase space for this transition is smaller (the Q-value is correspondingly lower) but the de-excitation photon might allow a good experimental signal. For a compilation of existing bounds on transitions to excited states see [Bar96a, Tre02]. Typical half-life limits obtained are in the order of 10^{19}–10^{21} yr. As long as no signal is seen, bounds on $\langle \eta \rangle$ and $\langle \lambda \rangle$ from ground-state transitions are much more stringent.

7.5.3 Majoron accompanied decays

Present half-life limits for the decay mode ($n = 1$) are of the order 10^{21}–10^{22} yr resulting in a deduced upper limit on the coupling constant of

$$\langle g_{\nu\chi} \rangle \lesssim 10^{-4}. \quad (7.74)$$

Recent compilations of available limits can be found in [Zub98, Ell02]. A first half-life limit for the $n = 3$ mode was obtained with ^{76}Ge [Zub92]. Ninety percent CL limits of additional modes obtained by the Heidelberg–Moscow experiment with a statistical significance of 4.84 kg × yr are

$$T_{1/2}^{0\nu\chi} > 5.85 \times 10^{21} \text{ yr} \quad (n = 3) \quad (7.75)$$

$$T_{1/2}^{0\nu\chi} > 6.64 \times 10^{21} \text{ yr} \quad (n = 7). \quad (7.76)$$

7.5.4 Decay rates for SUSY-induced $0\nu\beta\beta$-decay

Double β-decay can also proceed via R_P-violating SUSY graphs [Moh86, Hir95, Hir96]: the dominant ones are shown in figure 7.15. The obtainable half-life is

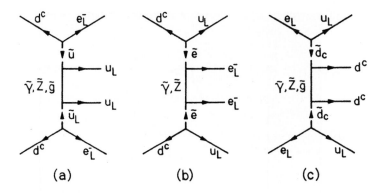

Figure 7.15. Dominant Feynman graphs from R-parity-violating SUSY contributing to double β-decay (from [Moh86]).

given by

$$(T_{1/2}^{0\nu}(0^+ \to 0^+))^{-1} \propto G \left(\frac{\lambda'_{111}}{m_{\tilde{q},\tilde{e}}^4 m_{\tilde{g},\chi}} M \right)^2 \tag{7.77}$$

with G and M the corresponding phase space factor and nuclear matrix element, λ'_{111} the strength of the R-parity violation and $m_{\tilde{q},\tilde{e},\tilde{g},\chi}$ as the mass of the involved squarks, selectrons, gluinos and neutralinos (see chapter 5). The bound on λ'_{111} is shown in figure 7.16.

Other mechanisms for $0\nu\beta\beta$-decay such as double charged Higgs bosons and leptoquarks have been discussed. They will not be discussed here and the reader is referred to [Kla99].

7.6 The future

Several upgrades are planned or proposed to improve some of the existing half-life limits. The best way of improving sensitivity can be achieved by isotopical enrichment, which is expensive, and by trying to make the experiment background free. Using the neutrino oscillation results described in the next chapters, it is a common goal to reach a mass region of around 50 meV or below. This offers the potential for a discovery or at least putting stringent bounds on the various neutrino mass models currently available. However, a factor of at least an order of magnitude in mass sensitivity implies more than two orders of magnitude improvement in half-life. This implies large scale (several hundred kgs of material) experiments and the $2\nu\beta\beta$-decay now becomes prominent as an irreducible background component. The projects can basically be separated into two groups: first, experiments which explore and improve already existing technologies (like GENIUS, MAJORANA, CUORE); and, second, really new experimental ideas (like COBRA, EXO, XMASS, MOON). The variety

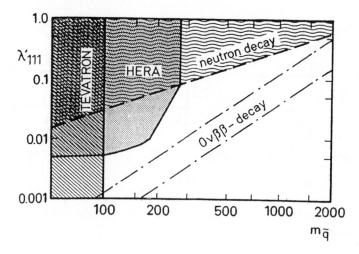

Figure 7.16. Bounds on the Yukawa coupling λ'_{111} as a function of the assumed squark mass in R-parity violating supersymmetric theories. Parts on the left of the curves are excluded. The dashed-dotted lines represent the limits from double beta decay, the upper (lower) line corresponding to a mass of the gluino of 1 TeV (100 GeV) respectively (from [Hir95]).

of experimental proposals is compiled in table 7.5. For recent overviews see [Ell02, Cre03].

7.7 $\beta^+\beta^+$-decay

The interest in this process is dominated by the fact that it is driven by right-handed currents. The experimental signatures of the decay modes involving positrons, (7.13) and (7.14), in the final state are promising because of two or four 511 keV photons. Experimentally more challenging is the EC/EC mode. In an excited state transition a characteristic gamma can be used in association with x-ray emission. In the 0ν mode, because of energy and momentum conservation, additional particles must be emitted such as an e^+e^- pair or internal bremsstrahlung photons [Doi93]. Current half-life limits are of the order of 10^{20} yr obtained with ^{106}Cd and ^{78}Kr for the modes involving positrons [Tre02]. The proposed COBRA experiment has the chance of simultaneously measuring five different isotopes for this decay channel [Zub01]. As the decay is intrinsic to the CdTe detectors there is a good chance of observing the 2νEC/EC and for the positron-emitting modes coincidences among the crystals can be used. Improved results have already been obtained [Kie03].

Table 7.5. Experiments that are planned, proposed or in construction and their proposed half-lives (after [Cre03]). NEMO3 and CUORICINO have recently started data taking.

Experiment	Isotope	Detector	Prop. half-life (yr)
COBRA	^{116}Cd, ^{130}Te	10 kg CdTe semiconductor	1×10^{24}
CUORICINO	^{130}Te	40 kg TeO$_2$ bolometers	1.5×10^{25}
NEMO3	^{100}Mo	10 kg foils with TPC	4×10^{24}
CUORE	^{130}Te	760 kg TeO$_2$ bolometers	7×10^{26}
EXO	^{136}Xe	1 t enriched Xe TPC	8×10^{26}
GEM	^{76}Ge	1 t enriched Ge in LN$_2$	7×10^{27}
GENIUS	^{76}Ge	1 t enriched Ge in LN$_2$	1×10^{28}
MAJORANA	^{76}Ge	0.5 t enriched Ge segmented diodes	4×10^{27}
CAMEO	^{116}Cd	1 t CdWO$_4$ crystals in liquid scintillator	$> 10^{26}$
CANDLES	^{48}Ca	several tons of CaF$_2$ in liquid scintillator	1×10^{26}
GSO	^{160}Gd	2 t Gd$_2$SiO$_5$:Ce crystal scintillator in liquid scintillator	2×10^{26}
MOON	^{100}Mo	34 t nat. Mo sheets between plastic scintillators	1×10^{27}
Xe	^{136}Xe	1.56 t enriched Xe in liquid scintillator	5×10^{26}
XMASS	^{136}Xe	10 t of liquid Xe	3×10^{26}

7.8 *C P* phases and double β-decay

As already mentioned, additional phases exist in the case of Majorana neutrinos. The neutrino mixing matrix (5.54) for three flavours can be written in the form

$$U = U_{\text{MNS}} \, \text{diag}(1, e^{i\alpha}, e^{i\beta}) \qquad (7.78)$$

where U_{MNS} is given in chapter 5 and α, β are the new phases associated with Majorana neutrinos. Neutrino oscillations (see chapter 8) can only probe δ because it violates flavour lepton number but conserves total lepton number. Double β-decay is unique in a sense for probing the additional Majorana phases. The effective Majorana mass can be written in the three flavour scenario as

$$\langle m_{\nu_e} \rangle = ||U_{e1}|^2 m_1 + |U_{e2}|^2 e^{i\alpha} m_2 + |U_{e3}|^2 e^{i\beta} m_3|. \qquad (7.79)$$

C P conserved cases are given for α, $\beta = k\pi$ with $k = 0, 1, 2, \ldots$. Investigations on the effect of Majorana phases can be found in [Rod01a, Pas02]. They might

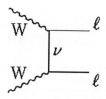

Figure 7.17. General Feynman diagram for $\Delta L = 2$ processes mediated by a virtual Majorana neutrino.

play a crucial role in creating a baryon asymmetry in the early Universe via leptogenesis (see chapter 13).

7.9 Generalization to three flavours

In general, there is a 3×3 matrix of effective Majorana masses, the elements being

$$\langle m_{\alpha\beta} \rangle = \left| \sum m_m \eta_m^{CP} U_{\alpha m} U_{\beta m} \right| \quad \text{with } \alpha, \beta \equiv e, \mu, \tau. \quad (7.80)$$

Double β-decay measures the element $\langle m_{\nu_e} \rangle = \langle m_{ee} \rangle$. In contrast to $0\nu\beta\beta$-decay, little is known about the other matrix elements.

7.9.1 General considerations

The underlying Feynman graph for all these $\Delta L = 2$ processes mediated by a virtual massive Majorana neutrino is shown in figure 7.17. The general behaviour can be described by

$$\sigma \propto \frac{m_i^2}{(q^2 - m_i^2)^2} \rightarrow \begin{cases} m_i^2 & \text{for } m_i^2 \ll q^2 \\ m_i^{-2} & \text{for } m_i^2 \gg q^2. \end{cases} \quad (7.81)$$

As long as an experimental bound does not intersect the cross section prediction a limit on $\langle m_{\alpha\beta} \rangle$ can, in principle, be obtained by linearly extrapolating the low-energy part. However, such a limit is unphysical and should only give a rough estimate of how far away from actually becoming meaningful the result still is. What physical processes can explore the remaining eight terms? It should already be mentioned here that all following bounds are unphysical, because the experimental limits are currently not strong enough.

7.9.1.1 *Muon–positron conversion on nuclei*

$$\mu^- + (A, Z) \rightarrow e^+ + (A, Z - 2) \quad (7.82)$$

is a process closely related to double β-decay and, within the context discussed here, measures $\langle m_{e\mu} \rangle$. The current best bound comes from SINDRUMII and is given by [Kau98]

$$\frac{\Gamma(\text{Ti} + \mu^- \rightarrow \text{Ca}^{GS} + e^+)}{\Gamma(\text{Ti} + \mu^- \rightarrow \text{Sc} + \nu_\mu)} < 1.7 \times 10^{-12} \quad (90\% \text{ CL}) \quad (7.83)$$

which can be converted into a new limit of $\langle m_{e\mu} \rangle < 17(82)$ MeV depending on whether the proton pairs in the final state are in a spin singlet or triplet state [Doi85]. A recent calculation [Sim01] comes to a cross section ten orders of magnitude smaller, which will worsen the bound by five orders of magnitude. Clearly this has to be better understood. Note that a process like $\mu \rightarrow e\gamma$ does not give direct bounds on the quantities discussed here, because it measures $m_{e\mu} = \sqrt{\sum_i U_{ei} U_{\mu i} m_i^2}$. Therefore, without specifying a neutrino-mixing and mass scheme, the quantities are rather difficult to compare. However, if this can be done, these indirect bounds are more stringent.

7.9.1.2 *Processes investigating* $\langle m_{\mu\mu} \rangle$

Three different kind of searches can be considered. One process under study is muon lepton-number-violating ($\Delta L_\mu = 2$) trimuon production in neutrino–nucleon scattering via charged current (CC) reactions

$$\nu_\mu N \rightarrow \mu^- \mu^+ \mu^+ X \quad (7.84)$$

where X denotes the hadronic final state. Detailed calculations can be found in [Fla00]. Taking the fact that, in past experiments, no excess events of this type were observed on the level of 10^{-5} of CC events, a limit of $\langle m_{\mu\mu} \rangle \lesssim 10^4$ GeV can be deduced.

A further possibility for probing $\langle m_{\mu\mu} \rangle$ is to explore rare meson decays such as the rare kaon decay

$$K^+ \rightarrow \pi^- \mu^+ \mu^+. \quad (7.85)$$

A new upper limit on the branching ratio of

$$\frac{\Gamma(K^+ \rightarrow \pi^- \mu^+ \mu^+)}{\Gamma(K^+ \rightarrow \text{all})} < 3 \times 10^{-9} \quad (90\% \text{ CL}) \quad (7.86)$$

could be deduced [App00] resulting in a bound of $\langle m_{\mu\mu} \rangle \lesssim 500$ MeV [Zub00a]. Other rare meson decays can be envisaged, the current status of some decays is shown in table 7.6. A full compilation is given in [Zub02].

A realistic chance to bring $\langle m_{\mu\mu} \rangle$ at least into the physical region by improving both methods and especially using trimuon production will be given by a neutrino factory [Rod01]. However, this would require a muon beam energy of at least 500 GeV, currently not a favoured option.

Table 7.6. Branching ratios of $\Delta L = 2$ decays of rare mesons, which can be described by the same Feynman graph as double β-decay.

Decay mode	Limit on branching ratio
$K^+ \to \pi^- e^+ e^+$	6.4×10^{-10}
$K^+ \to \pi^- \mu^+ \mu^+$	3.0×10^{-9}
$K^+ \to \pi^- e^+ \mu^+$	5.0×10^{-10}
$D^+ \to \pi^- e^+ e^+$	9.6×10^{-5}
$D^+ \to \pi^- \mu^+ \mu^+$	4.8×10^{-6}
$D^+ \to \pi^- e^+ \mu^+$	5.0×10^{-5}
$D_s^+ \to \pi^- e^+ e^+$	6.9×10^{-4}
$D_s^+ \to \pi^- \mu^+ \mu^+$	2.9×10^{-5}
$D_s^+ \to \pi^- e^+ \mu^+$	7.3×10^{-4}
$B^+ \to \pi^- e^+ e^+$	1.0×10^{-6}
$B^+ \to \pi^- \mu^+ \mu^+$	1.8×10^{-6}
$B^+ \to \pi^- e^+ \mu^+$	2.0×10^{-6}

Probably the closest analogy for performing a measurement on nuclear scales would be μ^- capture by nuclei with a μ^+ in the final state as discussed in [Mis94]. No such experiment has yet been performed, probably because of the requirement to use radioactive targets due to energy conservation arguments. The ratio with respect to standard muon capture can be given in the case of the favoured ^{44}Ti and a light neutrino exchange ($m_m \ll q^2$) as

$$R = \frac{\Gamma(\mu^- + \text{Ti} \to \mu^+ + \text{Ca})}{\Gamma(\mu^- + \text{Ti} \to \nu_\mu + \text{Sc})} \simeq 5 \times 10^{-24} \left(\frac{\langle m_{\mu\mu}\rangle}{250\,\text{keV}}\right)^2. \tag{7.87}$$

many orders of magnitude smaller than current μe conversion experiments.

7.9.1.3 *Limits on $\langle m_{\tau\tau}\rangle$ from CC events at HERA*

Limits for mass terms involving the τ-sector were obtained by using HERA data [Fla00a]. The process studied is

$$e^\pm p \to \overset{(-)}{\nu_e} l^\pm l'^\pm X \qquad \text{with } (ll') = (e\tau),\ (\mu\tau),\ (\mu\mu) \text{ and } (\tau\tau). \tag{7.88}$$

Such a process has a spectacular signature with large missing transverse momentum ($\not{p}_T$) and two like-sign leptons, isolated from the hadronic remnants. In addition, the fact that one of the leptons escapes in the beam pipe, which would look like the excess events recently observed with H1 [Adl98], are explored [Rod00].

Unfortunately, all the bounds given except for $\langle m_{ee} \rangle$ are still without physical meaning and currently only the advent of a neutrino factory might change the situation. Nevertheless, it is worthwhile considering these additional processes because, as in the case of $0\nu\beta\beta$-decay, they might provide stringent bounds on other quantities such as those coming from R-parity-violating SUSY.

After discussing only the limits for a possible neutrino mass, we now come to neutrino oscillations where evidence for a non-vanishing rest mass are found.

Chapter 8

Neutrino oscillations

In the case of a non-vanishing rest mass of the neutrino the weak and mass eigenstates are not necessarily identical, a fact well known in the quark sector where both types of states are connected by the CKM matrix (see section 3.3.2). This allows for the phenomenon of neutrino oscillations, a kind of flavour oscillation which is already known in other particle systems. It can be described by pure quantum mechanics. They are observable as long as the neutrino wave packets form a coherent superposition of states. Such oscillations among the different neutrino flavours do not conserve individual lepton numbers only total lepton number. We start with the most general case first, before turning to the more common two- and three-flavour scenarios. For additional literature see [Bil78, Bil87, Kay81, Kay89, Boe92, Kim93, Gro97, Sch97, Bil99, Lip99].

8.1 General formalism

Let us assume that there is an arbitrary number of n orthonormal eigenstates. The n flavour eigenstates $|\nu_\alpha\rangle$ with $\langle\nu_\beta|\nu_\alpha\rangle = \delta_{\alpha\beta}$ are connected to the n mass eigenstates $|\nu_i\rangle$ with $\langle\nu_i|\nu_j\rangle = \delta_{ij}$ via a unitary mixing matrix U:

$$|\nu_\alpha\rangle = \sum_i U_{\alpha i}|\nu_i\rangle \qquad |\nu_i\rangle = \sum_\alpha (U^\dagger)_{i\alpha}|\nu_\alpha\rangle = \sum_\alpha U_{\alpha i}^*|\nu_\alpha\rangle \qquad (8.1)$$

with

$$U^\dagger U = 1 \qquad \sum_i U_{\alpha i} U_{\beta i}^* = \delta_{\alpha\beta} \qquad \sum_\alpha U_{\alpha i} U_{\alpha j}^* = \delta_{ij}. \qquad (8.2)$$

In the case of antineutrinos, i.e. $U_{\alpha i}$ has to be replaced by $U_{\alpha i}^*$:

$$|\bar\nu_\alpha\rangle = \sum_i U_{\alpha i}^*|\bar\nu_i\rangle. \qquad (8.3)$$

The number of parameters in an $n \times n$ unitary matrix is n^2. The $2n - 1$ relative phases of the $2n$ neutrino states can be fixed in such a way that $(n - 1)^2$

independent parameters remain. It is convenient to write them as $\frac{1}{2}n(n-1)$ weak mixing angles of an n-dimensional rotational matrix together with $\frac{1}{2}(n-1)(n-2)$ CP-violating phases.

The mass eigenstates $|\nu_i\rangle$ are stationary states and show a time dependence according to

$$|\nu_i(x,t)\rangle = e^{-iE_it}|\nu_i(x,0)\rangle \tag{8.4}$$

assuming neutrinos with momentum p emitted by a source positioned at $x=0$ ($t=0$)

$$|\nu_i(x,0)\rangle = e^{ipx}|\nu_i\rangle \tag{8.5}$$

and being relativistic

$$E_i = \sqrt{m_i^2 + p_i^2} \simeq p_i + \frac{m_i^2}{2p_i} \simeq E + \frac{m_i^2}{2E} \tag{8.6}$$

for $p \gg m_i$ and $E \approx p$ as neutrino energy. Assume that the difference in mass between two neutrino states with different mass $\Delta m_{ij}^2 = m_i^2 - m_j^2$ cannot be resolved. Then the flavour neutrino is a coherent superposition of neutrino states with definite mass.[1] Neutrinos are produced and detected as flavour states. Therefore, neutrinos with flavour $|\nu_\alpha\rangle$ emitted by a source at $t=0$ develop with time into a state

$$|\nu(x,t)\rangle = \sum_i U_{\alpha i} e^{-iE_it}|\nu_i\rangle = \sum_{i,\beta} U_{\alpha i} U_{\beta i}^* e^{ipx} e^{-iE_it}|\nu_\beta\rangle. \tag{8.7}$$

Different neutrino masses imply that the phase factor in (8.7) is different. This means that the flavour content of the final state differs from the initial one. At macroscopic distances this effect can be large in spite of small differences in neutrino masses. The time-dependent transition amplitude for a flavour conversion $\nu_\alpha \to \nu_\beta$ is then given by

$$A(\alpha \to \beta)(t) = \langle \nu_\beta|\nu(x,t)\rangle = \sum_i U_{\beta i}^* U_{\alpha i} e^{ipx} e^{-iE_it}. \tag{8.8}$$

Using (8.6) this can be written as

$$A(\alpha \to \beta)(t) = \langle \nu_\beta|\nu(x,t)\rangle = \sum_i U_{\beta i}^* U_{\alpha i} \exp\left(-i\frac{m_i^2}{2}\frac{L}{E}\right) = A(\alpha \to \beta)(L) \tag{8.9}$$

with $L = x = ct$ being the distance between source and detector. In an analogous way, the amplitude for antineutrino transitions is obtained:

$$A(\bar\alpha \to \bar\beta)(t) = \sum_i U_{\beta i} U_{\alpha i}^* e^{-iE_it}. \tag{8.10}$$

[1] This is identical to the kaon system. The states K^0 and $\bar K^0$ are states of definite strangeness which are related to K_S^0 and K_L^0 as states with definite masses and widths.

The transition probability P can be obtained from the transition amplitude A:

$$P(\alpha \to \beta)(t) = |A(\alpha \to \beta)|^2 = \sum_i \sum_j U_{\alpha i} U_{\alpha j}^* U_{\beta i}^* U_{\beta j} e^{-i(E_i - E_j)t}$$

$$= \sum_i |U_{\alpha i} U_{\beta i}^*|^2 + 2\,\mathrm{Re} \sum_{j>i} U_{\alpha i} U_{\alpha j}^* U_{\beta i}^* U_{\beta j} \exp\left(-i\frac{\Delta m^2}{2}\right)\frac{L}{E}$$

$$(8.11)$$

with

$$\Delta m_{ij}^2 = m_i^2 - m_j^2. \tag{8.12}$$

The second term in (8.11) describes the time- (or spatial-) dependent neutrino oscillations. The first one is an average transition probability, which also can be written as

$$\langle P_{\alpha \to \beta} \rangle = \sum_i |U_{\alpha i} U_{\beta i}^*|^2 = \sum_i |U_{\alpha i}^* U_{\beta i}|^2 = \langle P_{\beta \to \alpha} \rangle. \tag{8.13}$$

Using CP invariance ($U_{\alpha i}$ real), this can be simplified to

$$P(\alpha \to \beta)(t) = \sum_i U_{\alpha i}^2 U_{\beta i}^2 + 2 \sum_{j>i} U_{\alpha i} U_{\alpha j} U_{\beta i} U_{\beta j} \cos\left(\frac{\Delta m_{ij}^2}{2}\frac{L}{E}\right)$$

$$= \delta_{\alpha\beta} - 4 \sum_{j>i} U_{\alpha i} U_{\alpha j} U_{\beta i} U_{\beta j} \sin^2\left(\frac{\Delta m_{ij}^2}{4}\frac{L}{E}\right). \tag{8.14}$$

Evidently, the probability of finding the original flavour is given by

$$P(\alpha \to \alpha) = 1 - \sum_{\alpha \neq \beta} P(\alpha \to \beta). \tag{8.15}$$

As can be seen from (8.11) there will be oscillatory behaviour as long as at least one neutrino mass eigenstate is different from zero and if there is a mixing (non-diagonal terms in U) among the flavours. In addition, the observation of oscillations allows no absolute mass measurement, oscillations are only sensitive to Δm^2. Last but not least neutrino masses should not be exactly degenerated. Another important feature is the dependence of the oscillation probability on L/E. Majorana phases as described in chapter 7 are unobservable in oscillations because the form given in (7.28) and implemented in (8.11) shows that the diagonal matrix containing the phases always results in the identity matrix [Bil80]. The same results for oscillation probabilities are also obtained by performing a more sophisticated wavepacket treatment [Kay81].

The result can also be obtained from very general arguments [Gro97, Lip99], which show that such flavour oscillations are completely determined by the propagation dynamics and the boundary condition that the probability of observing the wrong flavour at the position of the source at any time must vanish. The propagation in free space for each state is given in (8.4). The expansion of the neutrino wavefunction in energy eigenstates is

$$\psi = \int g(E)\,\mathrm{d}E\,\mathrm{e}^{-\mathrm{i}Et} \sum_{i=1}^{3} c_i \mathrm{e}^{\mathrm{i}px} |v_i\rangle \tag{8.16}$$

with the energy-independent coefficients c_i. The function $g(E)$ describing the exact form of the energy wavepacket is irrelevant at this stage. Each energy eigenstate has three terms, one for each mass eigenstate, if three generations are assumed. The boundary condition for creating a v_e and only a v_e at the source (or at $t = 0$) then requires

$$\sum_{i=1}^{3} c_i \langle v_i | v_\mu \rangle = \sum_{i=1}^{3} c_i \langle v_i | v_\tau \rangle = 0. \tag{8.17}$$

The momentum of each of the three components is determined by the energy and the neutrino masses. The propagation of this energy eigenstate, the relative phases of its three mass components and its flavour mixture at the detector are completely determined by the energy–momentum kinematics of the three mass eigenstates and lead to the same oscillation formula as described before.

8.2   *CP* and *T* violation in neutrino oscillations

Comparison of (8.8) with (8.10) yields a relation between neutrinos and antineutrinos transitions:

$$A(\bar{\alpha} \to \bar{\beta})(t) = A(\alpha \to \beta)(t) \neq A(\beta \to \alpha)(t). \tag{8.18}$$

This relation is a direct consequence of the *CPT* theorem. *CP* violation manifests itself if the oscillation probabilities of $v_\alpha \to v_\beta$ is different from its *CP* conjugate process $\bar{v}_\alpha \to \bar{v}_\beta$. So an observable would be

$$\Delta P_{\alpha\beta}^{CP} = P(v_\alpha \to v_\beta) - P(\bar{v}_\alpha \to \bar{v}_\beta) \neq 0 \qquad \alpha \neq \beta. \tag{8.19}$$

This has to be done with the proposed neutrino superbeams and neutrino factories (see section 8.10). Similarly, *T* violation can be tested if the probabilities of $v_\alpha \to v_\beta$ are different from the *T* conjugate process $v_\beta \to v_\alpha$. Here, the observable is

$$\Delta P_{\alpha\beta}^{T} = P(v_\alpha \to v_\beta) - P(v_\beta \to v_\alpha) \neq 0 \qquad \alpha \neq \beta. \tag{8.20}$$

If CPT conservation holds, which is the case for neutrino oscillations in vacuum, violation of T is equivalent to that of CP. Using U_{MNS} it can be shown explicitly that in vacuum $\Delta P_{\alpha\beta}^{CP}$ and $\Delta P_{\alpha\beta}^{T}$ are equal and given by

$$\Delta P_{\alpha\beta}^{CP} = \Delta P_{\alpha\beta}^{T} = -16 J_{\alpha\beta} \sin\left(\frac{\Delta m_{12}^2}{4E}L\right) \sin\left(\frac{\Delta m_{23}^2}{4E}L\right) \sin\left(\frac{\Delta m_{13}^2}{4E}L\right)$$

(8.21)

where

$$J_{2\beta} \equiv \text{Im}[U_{\alpha 1}U_{\alpha 2}^*U_{\beta 1}^*U_{\beta 2}] = \pm c_{12}s_{12}c_{23}s_{23}c_{13}^2 s_{13}\sin\delta \qquad (8.22)$$

with $+(-)$ sign denoting cyclic (anticyclic) permutation of (α, β) $=$ $(e, \mu), (\mu, \tau), (\tau, e)$. Note that for CP or T violation effects to be present, all the angles must be non-zero and, therefore, three-flavour mixing is essential. To be a bit more specific we now consider the case of two flavour oscillations.

8.3   Oscillations with two neutrino flavours

This is still by far the most common case used in data analysis. In this case the relation between the neutrino states is described by one mixing angle θ and one mass difference $\Delta m^2 = m_2^2 - m_1^2$. The unitary transformation (8.1) is analogous to the Cabibbo matrix given by (taking ν_e and ν_μ):

$$\begin{pmatrix} \nu_e \\ \nu_\mu \end{pmatrix} = \begin{pmatrix} \cos\theta & \sin\theta \\ -\sin\theta & \cos\theta \end{pmatrix} \begin{pmatrix} \nu_1 \\ \nu_2 \end{pmatrix}. \qquad (8.23)$$

Using the formulae from the previous section, the corresponding transition probability is

$$P(\nu_e \to \nu_\mu) = P(\nu_\mu \to \nu_e) = P(\bar\nu_e \to \bar\nu_\mu) = P(\bar\nu_\mu \to \bar\nu_e)$$

$$= \sin^2 2\theta \times \sin^2 \frac{\Delta m^2}{4} \times \frac{L}{E} = 1 - P(\nu_e \to \nu_e). \quad (8.24)$$

This formula explicitly shows that oscillations only occur if both θ and Δm^2 are non-vanishing. All two-flavour oscillations probabilities can be characterized by these two quantities because $P(\nu_\alpha \to \nu_\alpha) = P(\nu_\beta \to \nu_\beta)$. The phase factor can be rewritten as

$$\frac{E_i - E_j}{\hbar}t = \frac{1}{2\hbar c}\Delta m_{ij}^2\frac{L}{E} = 2.534\frac{\Delta m_{ij}^2}{\text{eV}^2}\frac{L/m}{E/\text{MeV}} \qquad (8.25)$$

where in the last step some practical units were used. The oscillatory term can then be expressed as

$$\sin^2\left(\frac{\Delta m_{ij}^2}{4}\frac{L}{E}\right) = \sin^2\pi\frac{L}{L_0} \qquad \text{with } L_0 = 4\pi\hbar c\frac{E}{\Delta m^2} = 2.48\frac{E/\text{MeV}}{\Delta m^2/\text{eV}^2}m.$$

(8.26)

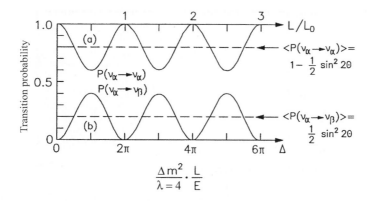

Figure 8.1. Example of neutrino oscillations in the two-flavour scheme: upper curve, $P(\nu_\alpha \rightarrow \nu_\alpha)$ (disappearance); lower curve, $P(\nu_\alpha \rightarrow \nu_\beta)$ (appearance) as a function of $L/L_0 = \Delta m^2/\lambda = 4\pi$ for $\sin^2 2\theta = 0.4$. The dashed lines show the average oscillation probabilities.

In the last step the oscillation length L_0, describing the period of one full oscillation cycle, is introduced (figure 8.1). It becomes larger with higher energies and smaller Δm^2. The mixing angle $\sin^2 2\theta$ determines the amplitude of the oscillation while Δm^2 influences the oscillation length. Both unknown parameters are typically drawn in a double logarithmic plot as shown in figure 8.2. The time average over many oscillations results in ($\langle\sin^2 \frac{\Delta m^2}{2}\rangle = \frac{1}{2}$) according to (8.13). The relative phase of the two neutrino states at a position x is (see (8.16))

$$\delta\phi(x) = (p_1 - p_2)x + \frac{(p_1^2 - p_2^2)}{(p_1 + p_2)}x = \frac{\Delta m^2}{(p_1 + p_2)}x. \tag{8.27}$$

Since the neutrino mass difference is small compared to all momenta $|m_1 - m_2| \ll p \equiv (1/2)(p_1 + p_2)$, this can be rewritten in first order in Δm^2 as

$$\delta\phi(x) = \frac{\Delta m^2}{2p}x \tag{8.28}$$

identical to (8.24) with $x = L$ and $p = E$.

8.4 The case for three flavours

A probably more realistic scenario to consider is that of three known neutrino flavours. The mixing matrix U_{MNS} is given in chapter 5. Note that now more Δm^2 quantities are involved both in magnitude and sign: although in a two-flavour oscillation in vacuum the sign does not enter, in three-flavour oscillation, which includes both matter effects (see section 8.9) and CP violation, the signs of the

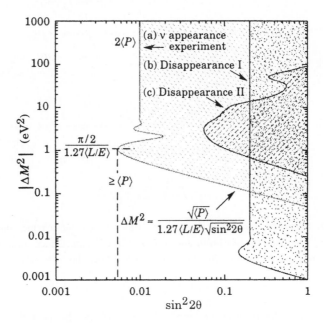

Figure 8.2. Standard double logarithmic plot of Δm^2 *versus* $\sin^2 2\theta$. The excluded parameter ranges of hypothetical appearance and disappearance experiments are shown. At low Δm^2 the experiment loses sensitivity being too close to the source, so the oscillation barely develops. This implies a slope of -2 until one reaches maximal sensitivity in the first oscillation maximum. At very high Δm^2 the oscillation itself can no longer be observed, only an average transition probability (after [PDG00]).

Δm^2 quantities enter and can, in principle, be measured. In the absence of any matter effect, the probability is given by

$$P(\nu_\alpha \to \nu_\beta) = \delta_{\alpha\beta} - 4 \sum_{i>j=1}^{3} \text{Re}(K_{\alpha\beta,ij}) \sin^2 \left(\frac{\Delta m_{ij}^2 L}{4E} \right)$$
$$+ 4 \sum_{i>j=1}^{3} \text{Im}(K_{\alpha\beta,ij}) \sin \left(\frac{\Delta m_{ij}^2 L}{4E} \right) \cos \left(\frac{\Delta m_{ij}^2 L}{4E} \right) \quad (8.29)$$

where

$$K_{\alpha\beta,ij} = U_{\alpha i} U_{\beta i}^* U_{\alpha j}^* U_{\beta j}. \quad (8.30)$$

The general formulae in the three-flavour scenario are quite complex; therefore, the following assumption is made: in most cases only one mass scale is relevant, i.e. $\Delta m_{\text{atm}}^2 \sim \text{few} \times 10^{-3} \text{ eV}^2$, which is discussed in more detail in chapter 9. Furthermore, one possible neutrino mass spectrum such as the hierarchical one is taken:

$$\Delta m_{21}^2 = \Delta m_{\text{sol}}^2 \ll \Delta m_{31}^2 \approx \Delta m_{32}^2 = \Delta m_{\text{atm}}^2. \qquad (8.31)$$

Then the expressions for the specific oscillation transitions are:

$$P(\nu_\mu \to \nu_\tau) = 4|U_{33}|^2|U_{23}|^2 \sin^2\left(\frac{\Delta m_{\text{atm}}^2 L}{4E}\right)$$

$$= \sin^2(2\theta_{23})\cos^2(\theta_{13})\sin^2\left(\frac{\Delta m_{\text{atm}}^2 L}{4E}\right) \qquad (8.32)$$

$$P(\nu_e \to \nu_\mu) = 4|U_{13}|^2|U_{23}|^2 \sin^2\left(\frac{\Delta m_{\text{atm}}^2 L}{4E}\right)$$

$$= \sin^2(2\theta_{13})\sin^2(\theta_{23})\sin^2\left(\frac{\Delta m_{\text{atm}}^2 L}{4E}\right) \qquad (8.33)$$

$$P(\nu_e \to \nu_\tau) = 4|U_{33}|^2|U_{13}|^2 \sin^2\left(\frac{\Delta m_{\text{atm}}^2 L}{4E}\right)$$

$$= \sin^2(2\theta_{13})\cos^2(\theta_{23})\sin^2\left(\frac{\Delta m_{\text{atm}}^2 L}{4E}\right). \qquad (8.34)$$

8.5 Experimental considerations

The search for neutrino oscillations can be performed in two different ways—an appearance or disappearance mode. In the latter case one explores whether less than the expected number of neutrinos of a produced flavour arrive at a detector or whether the spectral shape changes if observed at various distances from a source. This method is not able to determine the new neutrino flavour. An appearance experiment searches for possible new flavours, which do not exist in the original beam or produce an enhancement of an existing neutrino flavour. The identification of the various flavours relies on the detection of the corresponding charged lepton produced in their charged current interactions

$$\nu_l + N \to l^- + X \qquad \text{with } l \equiv e, \mu, \tau \qquad (8.35)$$

where X denotes the hadronic final state.

Several neutrino sources can be used to search for oscillations which will be discussed in this and the following chapters more extensively. The most important are:

- nuclear power plants ($\bar{\nu}_e$),
- accelerators ($\nu_e, \nu_\mu, \bar{\nu}_e, \bar{\nu}_\mu$),
- the atmosphere ($\nu_e, \nu_\mu, \bar{\nu}_e, \bar{\nu}_\mu$) and
- the Sun (ν_e).

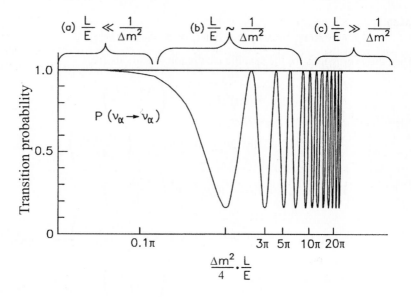

Figure 8.3. Logarithmic plot of the oscillation probability $P(\nu_\alpha \to \nu_\alpha)$ as a function of L/E for $\sin^2 2\theta = 0.83$. The brackets denote three possible cases: (a) no oscillations ($L/E \ll 1/\Delta m^2$); (b) oscillation $L/E \approx 1/\Delta m^2$; and (c) average oscillations for $L/E \gg 1/\Delta m^2$.

Which part of the Δm^2–$\sin^2 2\theta$ parameter space is explored depends on the ratio L/E. The relation

$$\Delta m^2 \propto E/L \tag{8.36}$$

shows that the various mentioned sources sometimes cannot probe each other, i.e. high-energy accelerators ($E \approx 1$–100 GeV, $L \approx 1$ km) are not able to check the solar neutrino data ($E \approx 1$ MeV, $L \approx 10^8$ km). Equation (8.36) also defines the minimal Δm^2 which can be explored. Three cases have to be considered with respect to a possible observation of oscillations (figure 8.3):

- $L/E \ll \frac{4}{\Delta m^2}$, i.e. $L \ll L_0$. Here, the experiment is too close to the source and the oscillations have no time to develop.
- $L/E \gtrsim \frac{4}{\Delta m^2}$, i.e. $L/E \gtrsim \frac{1}{\Delta m^2}$. This is a necessary condition to observe oscillations and it is the most sensitive region.
- $L/E \gg \frac{4}{\Delta m^2}$, i.e. $L \gg L_0$. Several oscillations have happened between the source and the detector. Normally, experiments do then measure L/E not precisely enough to resolve the oscillation pattern but measure only an average transition probability.

Two more points which influence the experimental sensitivity to and the observation of oscillations have to be considered. First of all, L is often not well defined. This is the case when dealing with an extended source (Sun, atmosphere, decay tunnels). Alternatively, E might not be known exactly. This might be the case if the neutrino source has an energy spectrum $N(E)$ and E will not be measured in a detector. Last but not least, for some experiments there is no chance to vary L and/or E because it is fixed (e.g. in the case of the Sun); therefore, the explorable Δm^2 region is constrained by nature.

8.6　Nuclear reactor experiments

Nuclear reactors are the strongest terrestrial neutrino source, coming from the β-decays of unstable neutron-rich products of ^{235}U, ^{238}U, ^{239}Pu and ^{241}Pu fission. The average yield is about $6\bar{\nu}_e$/fission. The flux density is given by

$$\Phi_\nu = 1.5 \times 10^{12} \frac{P/\text{MW}}{L^2/\text{m}^2} \ \text{cm}^{-2} \ \text{s}^{-1} \tag{8.37}$$

where P is the thermal power (in MW) of the reactor and L (in m) is the distance from the reactor core. The total isotropic flux of emitted $\bar{\nu}_e$ is then ($F = 4\pi L^2$)

$$F\Phi_\nu = 1.9 \times 10^{17} \frac{P}{\text{MW}} \ \text{s}^{-1}. \tag{8.38}$$

Reactor experiments are disappearance experiments looking for $\bar{\nu}_e \rightarrow \bar{\nu}_X$, because the energy is far below the threshold for μ, τ production. The spectrum peaks around 2–3 MeV and extends up to about 8 MeV. Experiments typically try to measure the positron spectrum which can be deduced from the $\bar{\nu}_e$ spectrum and either compare it directly to the theoretical predictions or measure it at several distances from the reactor and search for spectral changes. Both types of experiments have been performed in the past. However, the first approach requires a detailed theoretical understanding of the fission processes as well as a good knowledge of the operational parameters of the reactor during a duty cycle which changes the relative contributions of the fission products.

The detection reaction used mostly is

$$\bar{\nu}_e + p \rightarrow e^+ + n \tag{8.39}$$

with an energy threshold of 1.804 MeV. The $\bar{\nu}_e$ energy can be obtained by measuring the positron energy spectrum as

$$E_{\bar{\nu}_e} = E_{e^+} + m_n - m_p = E_{e^+} + 1.293 \ \text{MeV} = T_{e^+} + 1.804 \ \text{MeV} \tag{8.40}$$

neglecting the small neutron recoil energy (≈ 20 keV). The cross section for (8.39) is given by

$$\sigma(\bar{\nu}_e + p \rightarrow e^+ + n) = \sigma(\nu_e + n \rightarrow e^- + p)$$

$$= \frac{G_{F^2} E_{\nu^2}}{\pi} |\cos\theta_c|^2 \left(1 + 3\left(\frac{g_A}{g_\nu}\right)^2\right)$$

$$= 9.23 \times 10^{-42} \left(\frac{E_\nu}{10 \ \text{MeV}}\right)^2 \ \text{cm}^2. \tag{8.41}$$

Table 8.1. List of finished 'short-baseline' ($\leq$ 300 m) reactor experiments. The power of the reactors and the distance of the experiments with respect to the reactor are given.

Reactor	Thermal power [MW]	Distance [m]
ILL-Grenoble (F)	57	8.75
Bugey (F)	2800	13.6, 18.3
Rovno (USSR)	1400	18.0, 25.0
Savannah River (USA)	2300	18.5, 23.8
Gösgen (CH)	2800	37.9, 45.9, 64.7
Krasnojarsk (Russia)	?	57.0, 57.6, 231.4
Bugey III (F)	2800	15.0, 40.0, 95.0

Normally, coincidence techniques are used for detection between the annihilation photons and the neutrons which diffuse and thermalize within 10–100 μs.

Sometimes the reactions

$$\bar{\nu}_e + D \rightarrow e^+ + n + n \quad (E_{Thr} = 4.0 \, \text{MeV}) \quad (CC) \quad (8.42)$$

$$\bar{\nu}_e + D \rightarrow \bar{\nu}_e + p + n \quad (E_{Thr} = 2.2 \, \text{MeV}) \quad (NC) \quad (8.43)$$

were used.

The main backgrounds in reactor neutrino experiments originate from uncorrelated cosmic-ray hits in coincidence with natural radioactivity and correlated events from cosmic-ray muons and induced neutrons [Boe00, Bem02].

8.6.1 Experimental status

Several reactor experiments have been performed in the past (see table 8.1). All these experiments had a fiducial mass of less than 0.5 t and the distance to the reactor was never more than 250 m. Two new reactor experiments performed recently were CHOOZ and Palo Verde, which will be discussed in a little more detail. Both were motivated by the fact that the ν_e might participate in the atmospheric neutrino anomaly, discussed in more detail in chapter 9. The fact that the testable Δm^2 region is between 10^{-2}–10^{-3} eV^2 requires a distance of about 1 km to the reactors.

8.6.1.1   CHOOZ

The CHOOZ experiment in France [Apo98, Apo99] was performed between April 1997 and July 1998. It had some advantages with respect to previous experiments. The detector was located 1115 m and 998 m away from two 4.2 GW reactors, more than a factor four in comparison to previous experiments. In addition, the detector was located underground with a shielding of 300 mwe, reducing

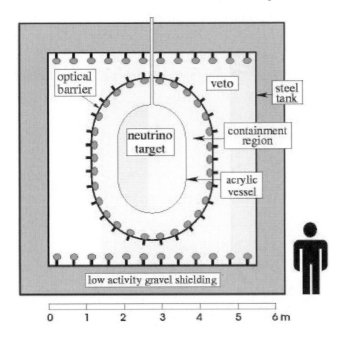

Figure 8.4. Schematic drawing of the CHOOZ detector.

the background due to atmospheric muons by a factor of 300. This allowed the construction of a homogeneous detector (figure 8.4). The main target was about 4.8 t and was, therefore, much larger than those used before. It consisted of a specially developed scintillator loaded with 0.1% Gd within an acrylic vessel. This inner detector was surrounded by an additional detector containing 17 t of scintillator without Gd and 90 t of scintillator as an outer veto. The signal is the detection of the annihilation photons in coincidence with n-capture on Gd, the latter producing gammas with a total sum of 8 MeV. The typical neutron capture time was about 30.5 μs. The published positron spectrum [Apo98] is shown in figure 8.5 and shows no hints for oscillation. The measured energy averaged ratio between expected and observed events is

$$R = 1.01 \pm 2.8\%(\text{stat.}) \pm 2.7\%(\text{sys.}). \tag{8.44}$$

This result is in perfect agreement with the absence of any oscillations, leading to the exclusion plot shown in figure 8.6. This limits any mixing angle with electrons (also θ_{13}) to

$$\sin^2 2\theta < 0.12(90\% \text{ CL}) \qquad \text{at } \Delta m^2 \approx 3 \times 10^{-3} \text{ eV}^2. \tag{8.45}$$

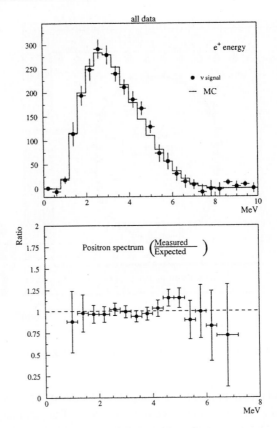

Figure 8.5. Background subtracted positron energy spectrum CHOOZ data. Error bars represents statistical errors only. The filled histogram represents the expectations for the case of no oscillations (from [Apo98]).

8.6.1.2   Palo Verde

The Palo Verde experiment [Boe01, Bem02] was performed near Phoenix, AZ (USA) and took data from October 1998 to July 2000. The total thermal power of the three reactors used was 11.6 GW and two of them were located 890 m and one 750 m away from the detector. The detector consisted of 12 t of a liquid scintillator also loaded with 0.1% Gd. Because of its rather shallow depth, with a shielding of only about 32 mwe, the detector had to be designed in a modular way. The scintillator was filled in 66 modules each 9 m long and with 12.7 cm × 25.4 cm cross section, which were arranged in an 11 × 6 array. The detector was surrounded by a 1m water shield to moderate background neutrons and an additional veto system against cosmic muons using 32 large scintillation counters. The space and time coincidence of three modules coming from two 511

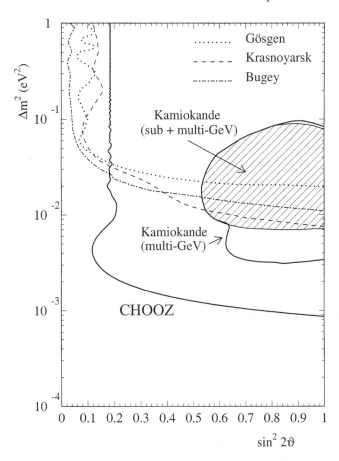

Figure 8.6. Exclusion plot of ν_e–ν_μ oscillations for various reactor experiments. Also shown are the parameter ranges which describe the atmospheric neutrino anomaly (see chapter 9). As can be seen CHOOZ excludes the ν_e–ν_μ oscillation channel as a possible explanation (from [Apo98]).

keV photons together with the neutron capture served as a signal.

Also in Palo Verde no evidence for oscillation was seen and a ratio of

$$R = 1.01 \pm 2.4\%\,(\text{stat.}) \pm 5.3\%\,(\text{sys.}) \tag{8.46}$$

is given. The resulting exclusion plot is shown in figure 8.7. Therefore, it can be concluded that ν_μ–ν_e oscillations play only a minor role in the atmospheric neutrino anomaly.

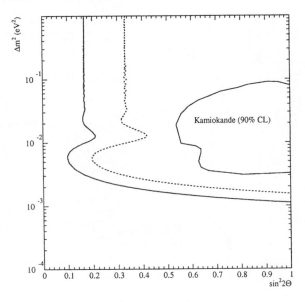

Figure 8.7. Same as figure 8.6 but showing the Palo Verde exlcusion region. The two curves correspond to two different analyses using different background subtraction. The main result here disfavours the ν_e–ν_μ oscillation channel as a possible explanation (from [Boe01]).

8.6.1.3 *KamLAND*

Future activities are motivated by the current solar neutrino data (chapter 10) implying going to even larger baselines. As discussed in chapter 10 the preferred solar solution suggests a region of $\Delta m^2 \approx 10^{-5}$ eV^2 with a large mixing angle $\sin^2 2\theta$. Using (8.24) and the fact that reactor and solar neutrino energies are about the same, this requires a baseline for searches of at least 100 km, two orders of magnitude larger than ever before.

An experiment designed for this goal is KamLAND in Japan installed in the Kamioka mine [Pie01]. Close to the mine, 16 commercial nuclear power plants are delivering a total of 130 GW. Taking also reactors from South Korea into account there is a total flux of $\bar{\nu}_e$ at Kamioka of about 4×10^6 cm^{-2} s^{-1} (or 1.3×10^6 cm^{-2} s^{-1} for $E_{\bar{\nu}} > 1.8$ MeV). Of this flux 80% derives from reactors in a distance between 140 and 210 km. The detector itself consists of 1000 t of liquid scintillator contained within a sphere. The scintillator is based on mineral oil and pseudocumene, designed to achieve a sufficiently light yield and n–γ discrimination by pulse shape analysis. This inner balloon is surrounded by 2.5 m of non-scintillating fluid as shielding. Both are contained and mechanically supported by a spherical stainless steel vessel. On this vessel 1280 phototubes for readout of the fiducial volume are also mounted. The signal is obtained by

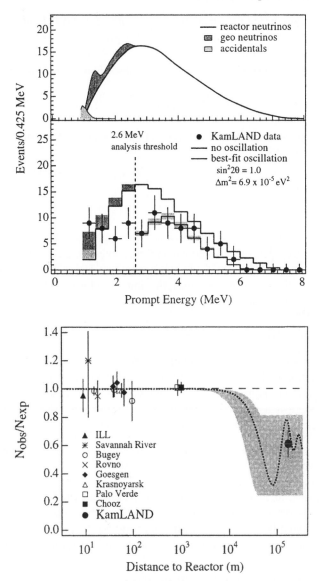

Figure 8.8. First results from KamLAND. Top: The measured positron spectrum. The deviation from the expected spectral shape can be clearly seen. Bottom: Ratio as a function of L/E. A clear reduction with respect to short baseline reactor experiments is seen. For comparison a theoretical oscillation curve is included (from [Egu03]).

a delayed coincidence of the prompt photons from positron annihilation and the 2.2 MeV photons from p(n, γ)d capture.

First results based on a measuring time of 145.1 days [Egu03] have been recently obtained. A cut on the energy of the prompt photon to be larger than 2.6 MeV has been applied for the analysis. Fifty-four $\bar{\nu}_e$ events were observed while the expectation had been 86.8 ± 5.6 events (figure 8.8). The obtained ratio is

$$\frac{N_{\text{obs}} - N_{BG}}{N_{\text{exp}}} = 0.611 \pm 0.085 \pm 0.041. \tag{8.47}$$

The implications of this result with respect to the solar neutrino problem will be discussed in chapter 10.

8.6.2 Future

8.6.2.1 Borexino

Another experiment might be the solar neutrino experiment Borexino in the Gran Sasso Laboratory described in more detail in chapter 10. Because of the absence of nuclear power plants in Italy the baseline is even larger than for KamLAND, typically more than 600 km. However, this implies a lower $\bar{\nu}_e$ flux and taking the fact of having only 300 t fiducial volume, a smaller signal is expected.

8.6.2.2 Measuring θ_{13} at reactors

An important quantity for future neutrino activities especially within the context of CP violation is the mixing angle θ_{13}. Its value has to be non-zero to allow a search for CP violation and $\sin^2 \theta_{13}$ should be larger than about 0.01, because otherwise there is a drastic change in the CP sensitivity. Reactor experiments perform disappearance searches, where the probability is given by

$$P(\bar{\nu}_e \to \bar{\nu}_e) = 1 - \sin^2 \theta_{13} \sin^2 \frac{\Delta m_{13}^2 L}{4E}. \tag{8.48}$$

This complements planned accelerator searches, which are appearance searches for ν_e in a ν_μ beam. To be sensitive enough, a number of nuclear power plants producing a high flux combined with a near and far detector to observe spectral distortions have to be used. Current ideas can be found in [Mik02, Sue03, Sha03].

8.7 Accelerator-based oscillation experiments

High-energy accelerators offer the chance for both appearance and disappearance searches. Both were and are still commonly used. Having typically much higher beam energies than reactors they probe normally higher Δm^2 regions. However, because of the intensity of the beam, the event rate can be much higher allowing smaller mixing angles $\sin^2 2\theta$ to be probed. Future long-baseline ($L \gg 100$ km) experiments will be able to extend the accelerator searches down to Δm^2 regions relevant for atmospheric neutrino studies and will be discussed in chapter 9.

Figure 8.9. Photograph of the interior of the LSND detector (with kind permission of the LSND collaboration).

A large number of searches have been performed in the past. Therefore, we will focus on the more recent ones and start with medium-energy experiments, namely LSND and KARMEN.

8.7.1 LSND

The LSND experiment [Ath97] at LANL was a 167 t mineral-oil-based liquid scintillation detector using scintillation and Cerenkov light for detection. It consisted of an approximately cylindrical tank 8.3 m long and 5.7 m in diameter (figure 8.9). The neutrino beam was produced by a proton beam with 800 MeV kinetic energy hitting a 30 cm long water target located about 1 m upstream of a copper beam stop. The experiment was about 30 m away from the beam stop under an angle of $12°$ with respect to the proton beam direction and can be called a short-baseline experiment. Most of the π^+ are stopped in the target and decay into muons which come to rest and decay in the target as well. The expected neutrino

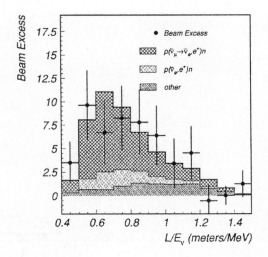

Figure 8.10. The L/E distribution for events with $20 < E_e < 60$ MeV. The data (points) as well as two background components and a fit accounting for the oscillation (hatched area) are shown (from [Agu01]).

spectrum has already been shown in figure 4.27. The decay at rest (DAR) of the positively charged muons allows $\bar{\nu}_\mu \to \bar{\nu}_e$ oscillation to be investigated. A small fraction of the positively charged pions (about 3%) decays in flight in the 1 m long space between target and beam stop and was used for the study of $\nu_\mu \to \nu_e$ oscillations. Note that the beam contamination of $\bar{\nu}_e$ is only of the order 10^{-4}, because negative pions are captured by nuclei before decay. LSND took data from 1993 to 1998.

For the DAR analysis in the channel $\bar{\nu}_\mu \to \bar{\nu}_e$, the signal reaction was

$$\bar{\nu}_e + p \to e^+ + n. \tag{8.49}$$

As experimental signature, a positron within the energy range $20 < E_e < 60$ MeV together with a time and spatial correlated delayed 2.2 MeV photon from $p(n, \gamma)d$ are required. LSND is not able to distinguish between positron and electron. After background subtraction an excess of $87.9 \pm 22.4 \pm 6.0$ events was indeed observed (figure 8.10) [Agu01]. Interpreting these as oscillations would correspond to a transition probability of $P(\bar{\nu}_\mu \to \bar{\nu}_e) = 2.64 \pm 0.67 \pm 0.45 \times 10^{-3}$. The analysis, therefore, ends up as evidence for oscillations in the region shown in figure 8.11.

The DIF analysis is looking for isolated electrons in the region $60 < E_e < 200$ MeV coming from $^{12}C(\nu_e, e^-)\,^{12}N_{gs}$ reactions. The lower bound of 60 MeV is well above 52.8 MeV, the endpoint of the electron spectrum from muon decay at rest. Here, a total excess of $8.1 \pm 12.2 \pm 1.7$ events was observed showing no clear effect of oscillations [Ath98].

8.7.2 KARMEN

The KARMEN experiment [Dre94] was operated at the neutron spallation source ISIS at Rutherford Appleton Laboratory from 1990 to 2001. KARMEN also used a 800 MeV proton beam but took advantage of the time structure of the beam. It was a pulsed beam having a repetition rate of 50 Hz and consisted of two pulses of 100 ns each with a separation of 325 ns. This time structure of the neutrino beam was important for identifying of neutrino-induced reactions and an effective suppression of the cosmic-ray background. The spectral shape of the neutrino beam is identical to the one described for LSND DAR. The detector was installed 18 m away from the target. In order to improve the sensitivity to oscillations by reducing the neutron background, an additional veto shield against atmospheric muons was constructed in 1996 which has been in operation since February 1997 (KARMEN2) and which surrounded the whole detector. The total shielding consisted out of 7000 t steel and a system of two layers of active veto counters. The detector itself consisted out of 56 t of liquid scintillator. The central scintillation calorimeter was segmented into 512 optically individual modules. The neutron capture detection was done with Gd_2O_3-coated paper within the module walls.

The $\bar{\nu}_\mu \rightarrow \bar{\nu}_e$ analysis again used reaction (8.48). Because of the pulsed beam, positrons were expected within a few μs after beam on target. The signature for detection is a spatially correlated delayed coincidence of a positron with energy up to 51 MeV together with γ-emission from either p(n, γ)d or Gd(n, γ)Gd reactions. The first one results in 2.2 MeV photons while the latter results in photons with a total energy of 8 MeV. The time difference between annihilation and neutron capture is given by thermalization, diffusion and capture of neutrons and is about 110 μs. After analysis of the 1997–2001 dataset 15 candidates remain with a total expected background of 15.8 events [Arm02]. There is no visible evidence for oscillation and the excluded region is shown in figure 8.11 with limits given as $\sin^2 2\theta < 1.7 \times 10^{-3}$ for $\Delta m^2 > 100$ eV^2. Obviously, in the large Δm^2 region ($\Delta m^2 > 10$ eV^2) both experiments are not in agreement; however, in the low-energy region there is still some allowed parameter space for LSND which is not covered by KARMEN. To what extent both experiment are in agreement or not is a severe statistical problem of handling both datasets. Such a combined analysis has been performed [Chu02]. The result is shown in figure 8.12.

8.7.3 Future test of the LSND evidence—MiniBooNE

The next step to test the full LSND evidence will be the MiniBooNE experiment at Fermilab looking for $\nu_\mu \rightarrow \nu_e$ oscillation [Baz01, Tay03]. The neutrino beam will be produced by the Fermilab Booster, sending a high-intensity pulsed proton beam of 8 GeV to a Be target. The positively charged secondaries, mostly pions, will be focused by a magnetic horn and brought into a decay tunnel. This results

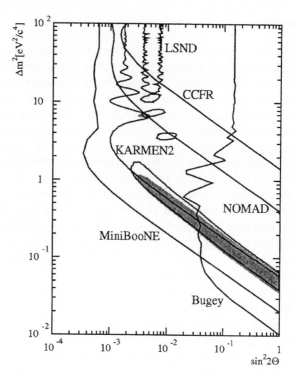

Figure 8.11. Δm^2 *versus* $\sin^2 2\theta$ plot for ν_e–ν_μ oscillations. The parameter ranges describing the LSND evidence as well as the exclusion curves of KARMEN, NOMAD, CCFR and the Bugey reactor experiment are shown (from [Ast03]).

in an almost pure ν_μ beam (ν_e contamination less than 0.3%). The detector itself is installed about 500 m away from the end of the decay tunnel. It consists of 800 t of pure mineral oil, contained in a 12.2 m diameter spherical tank. A support structure carries about 1550 phototubes for detection of Cerenkov and scintillation light. Data-taking started in August 2002.

8.8   Searches at higher neutrino energy

Short-baseline oscillation searches were recently performed at higher energies. The main motivation was the search for ν_μ–ν_τ oscillations assuming that ν_τ might have a mass in the eV range and would be a good candidate for hot dark matter (see chapter 13). The two experiments at CERN performing this search were NOMAD and CHORUS, both described in more detail in chapter 4. Therefore, here only the complementary search strategies are discussed.

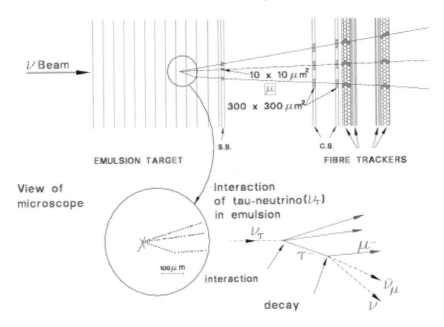

Figure 8.12. Same as figure 8.12 but showing the region of parameters (grey area) from a combined analysis assuming statistical compatibility of KARMEN2 and LSND. Also shown is the envisaged sensitivity of the MiniBooNE experiment (from [Chu02]).

8.8.1 CHORUS and NOMAD

CHORUS took advantage of the excellent spatial resolution of a few μm of emulsions. The dominant ν_μ beam produced ν_μ CC interactions. An oscillation event would result in a ν_τ CC interaction. Using the average beam energy of about 25 GeV, a produced τ travels about 1 mm before it decays. Such a track is clearly visible in the emulsion and, furthermore, the corresponding kink from the decay can be seen as well (figure 8.13). After data-taking the emulsions were scanned with automated microscopes equipped with CCD cameras and fast processors. Data were collected from 1994 to 1997 and all 0μ and 1-prong events were analysed. No signal was found and an upper limit of 2.4 τ-decays is given [Lud01]. This can be converted to an oscillation probability of

$$P(\nu_\mu \to \nu_\tau) \leq 3.4 \times 10^{-4}. \tag{8.50}$$

NOMAD, in contrast, uses kinematical criteria to search for ν_τ. The kinematic situation is shown in figure 8.15. As can be seen for ν_μ CC, the outgoing lepton is in the plane transverse to the beam more or less back to back to the hadronic final state (momentum conservation) and the missing momentum is rather small. In ν_μ NC events there is large missing momentum and no lepton at about 180

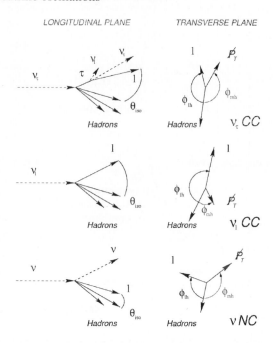

Figure 8.13. ν_τ detection principle used by the CHORUS experiment at CERN. A τ-lepton produced in a CC interaction produces a track of a few hundred μm before its decay. Focusing on the decay into muons the signature results in a kink. The short τ track and the kink are clearly visible because of the excellent spatial resolution of nuclear emulsions.

degrees is expected. The signal, namely a ν_τ CC interaction, is somewhere in between. The τ lepton, unlike in CHORUS, is invisible in NOMAD and can only be detected via some of its decay products. They follow the original τ direction resulting in a more back-to-back-like signature. However, in the τ-decay at least one neutrino is produced resulting in significant missing momentum. The analysis now proceeds in a way to find optimal experimental variables for the momentum imbalance and lepton isolation to discriminate between these backgrounds and the signal. This is done on the basis of likelihood functions performed as a 'blind box' analysis. Also NOMAD did not observe any oscillation signal (55 candidates observed, 58 background events expected) and gives an upper limit for the oscillation probability of [Ast01, Esk00]

$$P(\nu_\mu \rightarrow \nu_\tau) \leq 2 \times 10^{-4}. \tag{8.51}$$

The exclusion plots of both experiments together with former experiments are shown in figure 8.15. Using the beam contamination of ν_e, both could also produce limits on $\nu_e \rightarrow \nu_\tau$ oscillations [Ast01]. They are of the order of $P(\nu_e \rightarrow \nu_\tau) < 10^{-2}$ and are also shown in figure 8.15. Recently, NOMAD

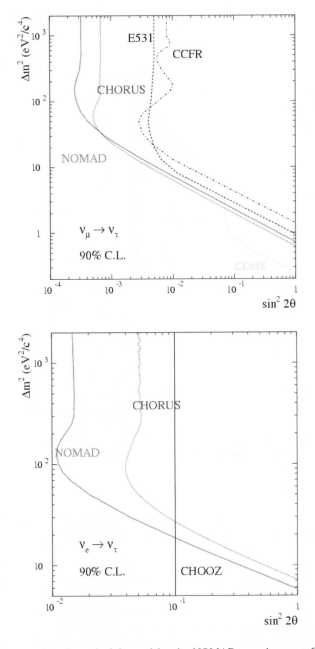

Figure 8.14. ν_τ detection principle used by the NOMAD experiment at CERN. The analysis is based on the kinematics of CC and NC interactions using momentum imbalance and isolation of charged tracks as main criteria, because the τ-lepton cannot be observed directly.

Wrong-Sign Muon Measurements

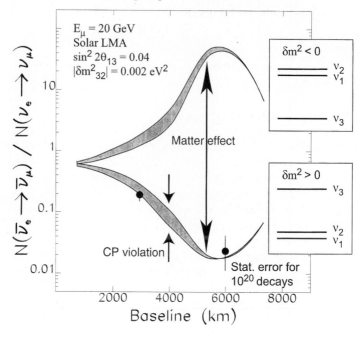

Figure 8.15. Top: ν_μ–ν_τ exclusion plot with the final result of NOMAD and the current result of CHORUS. Both experiments lead to an improvement to about one order of magnitude with respect to the former E531 and CCFR experiments at Fermilab. Bottom: ν_e–ν_τ exclusion plot showing the NOMAD and CHORUS result. This is based on the impurity of the used beam containing about 1% ν_e. Also shown is the CHOOZ limit. Note the different Δm^2 region in comparison with figure 8.6 (from [Ast01]).

published their results on ν_e–ν_μ oscillations [Ast03]. Like KARMEN they did not see any evidence and their exclusion curve is also shown in figure 8.11.

Currently, no further short-baseline experiment is planned. One of the reasons is that the atmospheric neutrino anomaly (see chapter 9) points towards a parameter region of $\Delta m^2 \approx 3 \times 10^{-3}$ eV^2 and large mixing $\sin^2 2\theta \approx 1$. Taking a 1 GeV beam this would correspond to an oscillation length of about 500 km, which requires long-baseline experiments. These will be discussed in chapter 9. Such beams have to cross a significant amount of matter on their way through the Earth and now we want to discuss how matter, in general, might affect neutrino oscillations. This is not only of importance for long-baseline experiments, which will be discussed in the next chapter, but also for solar and supernova neutrinos as well.

Table 8.2. The list of matter densities relevant for two-neutrino oscillations.

	$\nu_e \to \nu_{\mu,\tau}$	$\nu_e \to \nu_s$	$\nu_\mu \to \nu_\tau$	$\nu_{\mu,\tau} \to \nu_s$
$\dfrac{A}{2\sqrt{2}EG_F}$	N_e	$N_e - \frac{1}{2}N_n$	0	$-\frac{1}{2}N_n$

8.9 Neutrino oscillations in matter

Matter effects can occur if the neutrinos under consideration experience different interactions by passing through matter. In the Sun and the Earth ν_e can have NC and CC interactions with leptons because of the existence of electrons, while for ν_μ and ν_τ only NC reactions are possible. In addition, for a ν_μ beam traversing the Earth, in the case of the existence of sterile neutrinos ν_S, there is a difference between weak reactions (ν_μ) and no weak interactions at all (ν_S), see also [Kuo89, Kim93, Sch97, Bil99].

Starting from the weak interaction Lagrangian (3.48) one gets for low-energy neutrino interactions of flavour ℓ with the background matter

$$-\mathcal{L}_{\nu_\ell} = \frac{G_F}{\sqrt{2}} \nu_\ell^\dagger (1 - \gamma_5) \nu_\ell \sum_f N_f (\delta_{\ell f} + T_{3f_L} - 2\sin^2\theta_W Q_f) \tag{8.52}$$

where G_F is the Fermi coupling constant, θ_W the Weinberg angle, T_{3f_L} the eigenvalue of the fermion field f_L of the third component of the weak isospin and Q_f is the charge of f. In the matter Lagrangian (8.52), the CC interaction is represented by the Kronecker symbol $\delta_{\ell f}$ which states that for neutrinos of flavour ℓ the charged current only contributes if background matter containing charged leptons of the same flavour is present. For real matter with electrons, protons and neutrons which is electrically neutral, i.e. $N_e = N_p$, we have $T_{3e_L} = -T_{3p_L} = T_{3n_L} = -1/2$ and $Q_e = -Q_p = -1$, $Q_n = 0$ for electrons, protons and neutrons, respectively. To discuss two-neutrino oscillations in matter two useful definitions are:

$$N(\nu_\alpha) \equiv \delta_{\alpha e} N_e - \frac{1}{2}N_n \qquad (\alpha \equiv e, \mu, \tau) \qquad N(\nu_s) \equiv 0 \tag{8.53}$$

following directly from (8.52) and

$$A \equiv 2\sqrt{2}G_F E(N(\nu_\alpha) - N(\nu_\beta)). \tag{8.54}$$

The list of all possible matter densities which determine A and occur in the different oscillation channels is given in table 8.2. We start with the vacuum case again. The time dependence of mass eigenstates is given by (8.4). Neglecting the common phase by differentiation, we obtain the equation of motion (Schrödinger equation)

$$i\frac{d\nu_i(t)}{dt} = \frac{m_i^2}{2E}\nu_i(t) \tag{8.55}$$

which can be written in matrix notation as follows:

$$i\frac{d\nu(t)}{dt} = H^i \nu(t)$$

with

$$\nu = \begin{pmatrix} \nu_1 \\ . \\ . \\ . \\ \nu_n \end{pmatrix} \tag{8.56}$$

and

$$H_{ij}^i = \frac{m_i^2}{2E}\delta_{ij}.$$

H^i is the Hamilton matrix ('mass matrix') in the ν_i representation and it is diagonal meaning that the mass eigenstates in vacuum are eigenstates of H. By applying the unitary transformation

$$\nu = U^\dagger \nu' \quad \text{with } \nu' = \begin{pmatrix} \nu_\alpha \\ . \\ . \\ . \end{pmatrix} \tag{8.57}$$

and the mixing matrix U, the equation of motion and the Hamilton matrix H^α can be written in the representation of flavour eigenstates ν_α:

$$i\frac{d\nu'(t)}{dt} = H^\alpha \nu'(t) \quad \text{with } H^\alpha = U H^i U^\dagger. \tag{8.58}$$

Consider the case of two neutrinos (ν_e, ν_μ): the Hamilton matrix can be written in both representations as

$$H^i = \frac{1}{2E}\begin{pmatrix} m_1^2 & 0 \\ 0 & m_2^2 \end{pmatrix}$$

$$H^\alpha = \frac{1}{2E}\begin{pmatrix} m_{ee}^2 & m_{e\mu}^2 \\ m_{e\mu}^2 & m_{\mu\mu}^2 \end{pmatrix}$$

$$= \frac{1}{2E}\begin{pmatrix} m_1^2\cos^2\theta + m_2^2\sin^2\theta & (m_2^2 - m_1^2)\sin\theta\cos\theta \\ (m_2^2 - m_1^2)\sin\theta\cos\theta & m_1^2\sin^2\theta + m_2^2\cos^2\theta \end{pmatrix}$$

$$= \frac{1}{4E}\Sigma\begin{pmatrix} 1 & 0 \\ 0 & 1 \end{pmatrix} + \frac{1}{4E}\Delta m^2\begin{pmatrix} -\cos 2\theta & \sin 2\theta \\ \sin 2\theta & \cos 2\theta \end{pmatrix} \tag{8.59}$$

with $\Sigma = m_2^2 + m_1^2$ and $\Delta m^2 = m_2^2 - m_1^2$. How does the behaviour change in matter? As already stated, the ν_e mass is modified in matter according to (using ν_e and ν_μ as examples)

$$m_{ee}^2 \rightarrow m_{eem}^2 = m_{ee}^2 + A \quad \text{with } A = 2\sqrt{2}G_F E N_e \tag{8.60}$$

the latter following directly from (8.54). The Hamilton matrix H_m^α in matter is, therefore, given in the flavour representation as

$$H_m^\alpha = H^\alpha + \frac{1}{2E} \begin{pmatrix} A & 0 \\ 0 & 0 \end{pmatrix} = \frac{1}{2E} \begin{pmatrix} m_{ee}^2 + A & m_{e\mu}^2 \\ m_{e\mu}^2 & m_{\mu\mu}^2 \end{pmatrix}$$

$$= \frac{1}{4E} (\Sigma + A) \begin{pmatrix} 1 & 0 \\ 0 & 1 \end{pmatrix}$$

$$+ \frac{1}{4E} \begin{pmatrix} A - \Delta m^2 \cos 2\theta & \Delta m^2 \sin 2\theta \\ \Delta m^2 \sin 2\theta & -A + \Delta m^2 \cos 2\theta \end{pmatrix}. \quad (8.61)$$

The same relations hold for antineutrinos with the exchange $A \rightarrow -A$. Transforming this matrix back into the (ν_1, ν_2) representation results in

$$H_m^i = U^\dagger H_m^\alpha U = U^\dagger H^\alpha U + \frac{1}{2E} U^\dagger \begin{pmatrix} A & 0 \\ 0 & 0 \end{pmatrix} U$$

$$= H^i + \frac{1}{2E} U^\dagger \begin{pmatrix} A & 0 \\ 0 & 0 \end{pmatrix} U$$

$$= \frac{1}{2E} \begin{pmatrix} m_1^2 + A \cos^2 \theta & A \cos \theta \sin \theta \\ A \cos \theta \sin \theta & m_2^2 + A \sin^2 \theta \end{pmatrix}. \quad (8.62)$$

The matrix now contains non-diagonal terms, meaning that the mass eigenstates of the vacuum are no longer eigenstates in matter. To obtain the mass eigenstates (ν_{1m}, ν_{2m}) in matter and the corresponding mass eigenvalues (m_{1m}^2, m_{2m}^2) (effective masses) H_m^i must be diagonalized. This results in mass eigenstates of

$$m_{1m,2m}^2 = \frac{1}{2} \left[(\Sigma + A) \mp \sqrt{(A - \Delta m^2 \cos 2\theta)^2 + (\Delta m^2)^2 \sin^2 2\theta} \right]. \quad (8.63)$$

For $A \rightarrow 0$, it follows that $m_{1m,2m}^2 \rightarrow m_{1,2}^2$. Considering now a mixing matrix U_m connecting the mass eigenstates in matter $m_{1m,2m}$ with the flavour eigenstates (ν_e, ν_μ) the corresponding mixing angle θ_m is given by

$$\tan 2\theta_m = \frac{\sin 2\theta}{\cos 2\theta - A/\Delta m^2} \qquad \sin 2\theta_m = \frac{\sin 2\theta}{\sqrt{(A/\Delta m^2 - \cos 2\theta)^2 + \sin^2 2\theta}}. \quad (8.64)$$

Here again, for $A \rightarrow 0$, it follows that $\theta_m \rightarrow \theta$. Using the relation

$$\Delta m_m^2 = m_{2m}^2 - m_{1m}^2 = \Delta m^2 \sqrt{\left(\frac{A}{\Delta m^2} - \cos 2\theta\right)^2 + \sin^2 2\theta} \quad (8.65)$$

the oscillation probabilities in matter can be written analogously to those of the vacuum:

$$P_m(\nu_e \rightarrow \nu_\mu) = \sin^2 2\theta_m \times \sin^2 \frac{\Delta m_m^2}{4} \times \frac{L}{E} \quad (8.66)$$

$$P_m(\nu_e \rightarrow \nu_e) = 1 - P_m(\nu_e \rightarrow \nu_\mu) \quad (8.67)$$

with a corresponding oscillation length in matter:

$$L_m = \frac{4\pi E}{\Delta m_m^2} = \frac{L_0}{\sqrt{\left(\frac{A}{\Delta m^2} - \cos 2\theta\right)^2 + \sin^2 2\theta}} = \frac{\sin 2\theta_m}{\sin 2\theta} L_0. \qquad (8.68)$$

Note already here that (8.64) allows the possibility of maximal mixing in matter, $\sin 2\theta_m \approx 1$, even for small $\sin\theta$ because of the resonance type form. This will be of importance when discussing the MSW effect on solar neutrinos in chapter 10.

A further scenario where the matter effect can be prominent is in very-long-baseline experiments like the planned neutrino factory.

8.10 CP and T violation in matter

In matter, the measurement of CP violation can become more complicated, because of the fact that the oscillation probabilities for neutrinos and antineutrinos are, in general, different in matter, even if $\delta = 0$. Indeed, the matter effect can either contaminate or enhance the effect of an intrinsic CP violation effect coming from δ [Ara97, Min98, Min00, Min02]. For the case of T violation, the situation is different. If $\Delta P_{\alpha\beta}^T \neq 0$ for $\alpha \neq \beta$ would be established, then this implies $\delta \neq 0$ even in the presence of matter. The reason is that the oscillation probability is invariant under time reversal even in the presence of matter. Similar to the case of CP violation, T violation effects can either be enhanced or suppressed in matter [Par01]. However, a measurement of T violation is experimentally more difficult to perform because there is a need for a non-muon neutrino beam, like a beta beam.

An additional problem arises in the form of parameter degeneracy. Assuming that all mixing parameters except θ_{13} and δ are known, and a precise measurement of $P(\nu_\mu \rightarrow \nu_e)$ and $P(\bar{\nu}_\mu \rightarrow \bar{\nu}_e)$ has been performed, there is still a situation where you find four different solutions (two for CP-even, two for CP-odd) [Bur01, Bar02]. The only chance to remove the ambiguities is to perform either an experiment at two different energies or baselines or to combine two different experiments. A compilation of expected matter effects and CP violation is shown in figure 8.16.

8.11   Possible future beams

Driven by the recent evidences for oscillations and facing the three angles and one phase in the MNS matrix, the idea of building new beams with very high intensity has been pushed forward. One of the main goals besides the matter effects is the observation of CP violation in the leptonic sector. However, this requires a non-vanishing θ_{13} which might be measured at reactors or in 'off-axis' experiments.

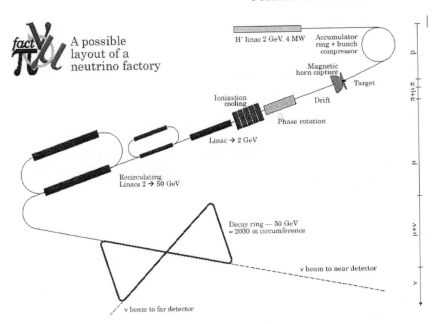

A possible layout of a neutrino factory

Figure 8.16. Possibilities for observing CP violation and matter effects using beams from a neutrino factory by using wrong sign muons. Matter effect start to significantly split in two bands if a detector is at least 1000 km away from the source. The two bands correspond to normal and inverted mass hierarchies. The width of the band gives the size of the possible CP violation using the parameters stated. It will only be observable if the LMA solution of the solar neutrino problem is correct (see chapter 10) and if the angle θ_{13} in the MNS matrix is different from zero.

8.11.1 Off-axis beams and experiments

A search for θ_{13} in a parasitic mode of already existing beam-lines such as NuMI is possible due to the pion decay kinematics. The goal is to obtain a high-intensity NBB. The ν_μ momentum in the laboratory frame is given by

$$p_L = \gamma(p^* \cos\theta^* + \beta p^*) \tag{8.69}$$
$$p_T = p^* \sin\theta^* \tag{8.70}$$

with $p^* = 0.03$ GeV/c as the neutrino momentum and θ^* as the polar angle of neutrino emission with respect to the pion direction of flight, both given in the pion rest frame. In the laboratory frame, θ is given by

$$\theta = \frac{R}{L} = \frac{1}{\gamma}\frac{\sin\theta^*}{1 + \cos\theta^*} \tag{8.71}$$

with L as the baseline and R as the distance of the detector from the beam centre. If the neutrino emission in the pion rest frame is perpendicular to the pion flight direction ($\theta^* = 90°$), then

$$\theta = \frac{1}{\gamma}. \tag{8.72}$$

The neutrino energy E_ν as a function is given by

$$E_\nu(R) = \frac{2\gamma p^*}{1 + (\gamma R/L)^2} \tag{8.73}$$

which is half of the energy at beam centre for $\theta = 1/\gamma$. However, the most important kinematic property is that, at this angle, the neutrino energy is, in first order, independent of the energy of the parent pion

$$\frac{\partial E_\nu}{\partial \gamma} = 0. \tag{8.74}$$

This opens a way for an NBB with a high intensity. The idea is to measure ν_e appearance in a ν_μ beam. The oscillation probability is directly proportional to $\sin^2 \theta_{13}$. Various experiments try to use this advantage for a measurement [Ito01, Ayr02, Dyd03].

8.11.2 Beta beams

The idea is to accelerate β-unstable isotopes [Zuc02] to energies of a few 100 MeV using ion accelerators like ISOLDE at CERN. This would give a clearly defined beam of ν_e or $\bar{\nu}_e$. Among the favoured isotopes discussed are ^{6}He in the case of a $\bar{\nu}_e$ beam and ^{18}Ne in the case of a ν_e beam.

8.11.3 Superbeams

Conventional neutrino beams in the GeV range run into systematics when investigating oscillations involving ν_μ and ν_e because of the beam contaminations of ν_e from K_{e3} decays (see chapter 4). To reduce this component, lower energy beams with high intensity are proposed. Here, quasi-elastic interactions are dominant. A first realization could be from the Japanese Hadron Facility (JHF) in Tohai, in its first phase producing a 0.77 MW beam of protons with 50 GeV on a target and using Super-Kamiokande as the far detector [Aok03]. The baseline corresponds to 225 km. This could be updated in a second phase to 4 MW and also a 1 Mt detector (Hyper-K). A similar idea exists at CERN to use the proposed SPL making a high-intensity beam to Modane (130 km away). Such experiments would allow $\sin^2 2\theta_{23}$, Δm_{23}^2 to be measured and might discover $\sin^2 2\theta_{13}$.

8.11.4 Muon storage rings—neutrino factories

In recent years the idea to use muon storage rings to obtain high-intensity neutrino beams has become very popular [Gee98, Aut99, Alb00, Als02, Apo02]. The two

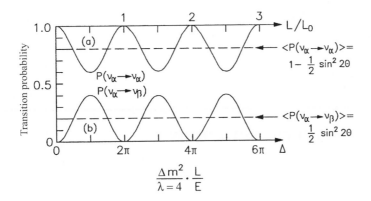

Figure 8.17. Proposed layout for a neutrino factory. The main ingredients are: a high intensity proton linac, a target able to survive the deposited energy and giving a good yield of pions, a cooling device for the decay muons, an accelerator for the muons and a storage ring allowing for muon decay and therefore neutrino beams.

main advantages are the precisely known neutrino beam composition and the high intensity (about 10^{21} muons/yr should be filled in the storage ring). A conceptional design is shown in figure 8.17. Even if many technical challenges have to be solved, it offers a unique source for future accelerator-based neutrino physics. First experimental steps towards its realization are on thier way, among them are the HARP experiment at CERN, which determines the target for optimal production of secondaries, the study of muon scattering (MUSCAT experiment) and muon cooling (MICE experiment). For additional information see also [Nuf01, Nuf02, Hub02].

Chapter 9

Atmospheric neutrinos

In recent years the study of atmospheric neutrinos has become one of the most important fields in neutrino physics. Atmospheric neutrinos are produced in meson and muon decays, created by interactions of cosmic rays within the atmosphere. The study of these neutrinos revealed evidence for neutrino oscillations. With energies in the GeV range and baselines from about 10 km to as long as the Earth diameter ($L \approx 10^4$ km) mass differences in the order of $\Delta m^2 \gtrsim 10^{-4}$ eV^2 or equivalent values in the L/E ratio from 10–10^5 km GeV^{-1} are probed. Most measurements are based on relative quantities because absolute neutrino flux calculations are still affected by large uncertainties. The obtained results depend basically on four factors: the primary cosmic-ray flux and its modulations, the production cross sections of secondaries in atmospheric interactions, the neutrino interaction cross section in the detector and the detector acceptance and efficiency. More quantitatively the observed number of events is given by

$$\frac{\mathrm{d}N_l(\theta, p_l)}{\mathrm{d}\Omega_\theta\,\mathrm{d}p_l} = t_{\mathrm{obs}} \sum_\pm \int N_t \frac{\mathrm{d}\phi_{\nu_l}^\pm(E_\nu, \theta)}{\mathrm{d}\Omega_\theta\,\mathrm{d}E_\nu} \frac{\mathrm{d}\sigma^\pm(E_\nu, p_l)}{\mathrm{d}p_l} F(q^2)\,\mathrm{d}E_\nu \qquad (9.1)$$

where l stands for $e^\pm$ or $\mu^\pm$, p_l the lepton momentum, E_ν the neutrino energy, θ the zenith angle, t_{obs} the observation time, N_t the number of target particles, $\phi_{\nu_l}^\pm(E_\nu, \theta)$ the neutrino flux and $\sigma(E_\nu, p_l)$ the cross section. $F(q^2)$ takes into account the nuclear effects such as the Fermi momenta of target nucleons, Pauli blocking of recoil nucleons etc. The summation ($\pm$) is done for ν_l and $\bar{\nu}_l$, since current observations do not distinguish the lepton charge. For further literature see [Sok89, Ber90b, Gai90, Lon92, 94, Gri01, Jun01, Kaj01, Lea01, Lip01, Gai02]. We want to discuss the first two steps now in a little more detail.

9.1 Cosmic rays

The primary cosmic rays hitting the atmosphere consist of about 98% hadrons and 2% electrons. The hadronic component itself is dominated by protons

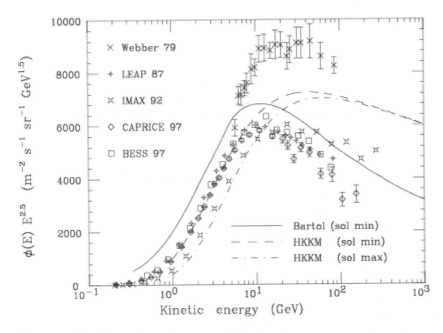

Figure 9.1. Compilation of balloon measurement of the flux of low-energy cosmic rays impinging on the Earth's atmosphere. For comparison two flux calculations by Bartol [Agr96] and HKKM [Hon95] are shown.

($\approx$87%) mixed with α-particles ($\approx$11%) and heavier nuclei ($\approx$2%). The chemical composition can be determined directly by satellite and balloon experiments in an energy range up to 1 TeV (figure 9.1). For higher energies only indirect methods like air showers can be used. Because the neutrino flux depends on the number of nucleons rather than on the number of nuclei, a significant fraction of the flux is produced by He and CNO (+ heavier nuclei). The differential energy spectrum follows a power law of the form

$$N(E)\,dE \propto E^{-\gamma}\,dE \tag{9.2}$$

with $\gamma \simeq 2.7$ for $E < 10^{15}$ eV. From this point the spectrum steepens (the 'knee') to $\gamma \simeq 3$. The exact position of the knee depends on the atomic number A as was shown recently by KASCADE, with lighter nuclei showing the knee at lower energies [Swo02]. At about 10^{18} eV the spectrum flattens again (the 'ankle') and datasets well above are still limited by statistics. This ultra-high-energy part of cosmic rays will be discussed in more detail in chapter 12. The part of the cosmic-ray spectrum dominantly responsible for the current atmospheric neutrino investigations is in the energy range below 1 TeV. The intensity of primary nucleons in that energy range can be approximated by

$$I_N(E) \approx 1.8 E^{-2.7} \text{ nucleons cm}^{-2} \text{ s}^{-1} \text{ sr}^{-1} \text{ GeV}^{-1} \qquad (9.3)$$

with E as the energy per nucleon. In the low-energy range several effects can occur. First of all, there is the modulation of the primary cosmic-ray spectrum with solar activity. A measurement of the latter is the sunspot number. The solar wind prohibits low-energy galactic cosmic rays from reaching the Earth, resulting in an 11 yr anticorrelation of cosmic-ray intensity with solar activity. This effect is most prominent for energies below 10 GeV. Such particles have, in contrast, a rather small effect on atmospheric neutrino fluxes, because the geomagnetic field prevents these low-energy particles from entering the atmosphere anyway. The geomagnetic field bends the trajectories of cosmic rays and determines the minimum rigidity called the cutoff rigidity (for an extensive discussion on this quantity see [Hil72]) for particles to arrive at the Earth [Lip00a]. The dynamics of any high energy particle in a magnetic field configuration B depends on the rigidity R given by

$$R = \frac{pc}{ze} = r_L \times B \qquad (9.4)$$

with p as the relativistic 3-momentum, z as the electric charge and r_L as the gyroradius. Particles with different masses and charge but identical R show the same dynamics in a magnetic field. The cutoff rigidity depends on the position at the Earth surface and the arrival direction of the cosmic ray. Figure 9.2 shows a contour map of the calculated cutoff rigidity at Kamioka (Japan) [Hon95], where Super-Kamiokande is located. The geomagnetic field, therefore, produces two prominent effects: the latitude (the cosmic-ray flux is larger near the geomagnetic poles) and the east–west (the cosmic-ray flux is larger for east-going particles) effect. The last one is an azimuthal effect not depending on any new physics and can be used to check the shower simulations [Lip00b]. Such a measurement was performed recently by Super-Kamiokande [Fut99]. With a statistics of 45 kt × yr and cuts on the lepton momentum ($400 < p_l < 3000$ MeV/c and zenith angle $|\cos\theta| < 0.5$) to gain sensitivity, an east–west effect is clearly visible (figure 9.3).

For higher energetic neutrinos up to 100 GeV, the primary energy is up to 1 TeV, where the details of the flux are not well measured.

9.2 Interactions within the atmosphere

The atmospheric neutrinos stem from the decay of secondaries produced in interactions of primary cosmic rays with the atmosphere. The dominant part is the decay chain

$$\pi^+ \rightarrow \mu^+ \nu_\mu \qquad \mu^+ \rightarrow e^+ \nu_e \bar{\nu}_\mu \qquad (9.5)$$

$$\pi^- \rightarrow \mu^- \bar{\nu}_\mu \qquad \mu^- \rightarrow e^- \bar{\nu}_e \nu_\mu. \qquad (9.6)$$

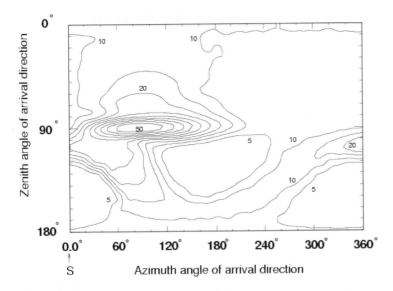

Figure 9.2. Contour map of the cutoff rigidity (in GeV) relevant for Kamioka (from [Kaj01]).

Depending on the investigated neutrino energy additional contributions come from kaon decay, especially the modes

$$K^{\pm} \rightarrow \mu^{\pm} \nu_{\mu}(\bar{\nu}_{\mu}) \tag{9.7}$$

$$K_{L} \rightarrow \pi^{\pm} e^{\pm} \nu_{e}(\bar{\nu}_{e}). \tag{9.8}$$

The latter, so called K_{e3} decay, is the dominant source for ν_e above $E_\nu \approx 1$ GeV. In the low energy range ($E_\nu \approx 1$ GeV) there is the previously mentioned contribution from muon-decay. However, for larger energies the Lorentz boost for muons is high enough in a way that they reach the Earth surface. For example, most muons are produced in the atmosphere at about 15 km. This length corresponds to the decay length of a 2.4 GeV muon, which is shortened to 8.7 km by energy loss (a vertical muon loses about 2 GeV in the atmosphere by ionization according to the Bethe–Bloch formula). Therefore, at E_ν larger than several GeV this component can be neglected. At higher energies the contribution of kaons becomes more and more important.

Several groups have performed simulations to calculate the atmospheric neutrino flux [Bar89, Per94, Hon95, Agr96, Hon01, Wen01, Tse01, Bat03]. The general consensus of all these studies is that the ratio of fluxes

$$R = \frac{\nu_e + \bar{\nu}_e}{\nu_\mu + \bar{\nu}_\mu} \tag{9.9}$$

can be predicted with an accuracy of about 5% because several uncertainties

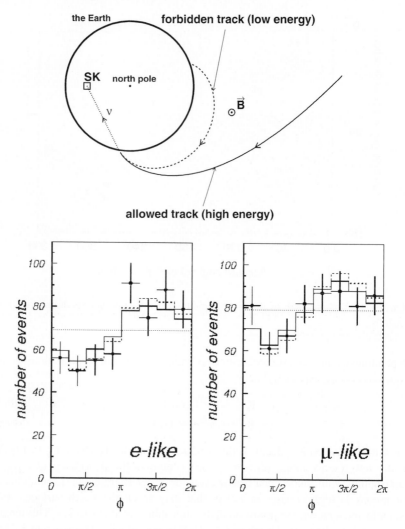

Figure 9.3. Top: Schematic explanation for the occurrence of the east–west effect. Bottom: The east–west effect as observed with Super-Kamiokande (from [Fut99]).

cancel. However, in the absolute flux predictions there is some disagreement on the level of 20–30% in the spectra and overall normalization of the neutrino flux. Let us investigate the differences in more detail. The fluxes for 'contained events' (see section 9.3) are basically produced by cosmic primaries with energies below about 20 GeV. As already described this energy range is affected by geomagnetic effects and solar activities. The next step and source of main uncertainty is the production of secondaries, especially pions and kaons in proton–air collisions. Various Monte Carlo generators are used to describe this process; however, the

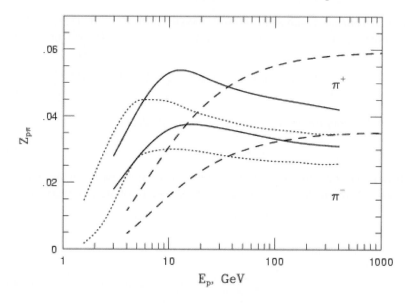

Figure 9.4. Z-factors for pions as a function of proton energy taken from two different calculations (from [Agr96]).

experimental datasets are rather poor. A useful way to compare the various interaction models used in the event generators is to evaluate the spectrum-weighted moments (Z-factors) of the inclusive cross section. The most important range of interaction energies for production of neutrinos with energies between 300 MeV and 3 GeV is a primary energy between $5 < E_N < 50$ GeV. In general, the primary energy is typically an order of magnitude higher than the corresponding neutrino energy. The Z-factors are given as (for more details see [Gai90])

$$Z_{p\pi^\pm} = \int_0^1 \mathrm{d}x\, x^{\gamma-1} \frac{\mathrm{d}n_{\pi^\pm}(x, E_N)}{\mathrm{d}x} \tag{9.10}$$

where $x = E_\pi/E_N$, E_N is the total energy of the incident nucleon in the laboratory system, E_π is the energy of the produced pion and γ as given in (9.2). Analogous factors can be derived for other secondaries like Z_{pK^+}. The Z-factors used for two simulations as a function of proton energy are shown in figure 9.4. There is a clear discrepancy between the calculations. Furthermore, past accelerator experiments have only measured pion production in pp collisions and p–Be collisions. They have to be corrected to p–air collisions. The transformation to heavier nuclei with the use of an energy-independent enhancement factor is a further source of severe uncertainty. Recently two new experimental approaches have arrived which might help to improve the situation considerably. First of all, there are measurements of muons in the atmosphere.

Strongly connected with neutrino production from meson decay is the production of muons. Assume the two-body decay $M \rightarrow m_1 + m_2$. The magnitude of the momenta of secondaries in the rest frame of M are then given by

$$p_1^* = p_2^* = p^* = \frac{M^4 - 2M^2(m_1^2 + m_2^2) + (m_1^2 - m_2^2)^2}{2M}. \tag{9.11}$$

In the laboratory frame the energy of the decay product is

$$E_i = \gamma E_i^* + \beta \gamma p^* \cos \theta^* \tag{9.12}$$

where β and γ are the velocity and Lorentz factor of the parent in the laboratory system. Therefore, the limits on the laboratory energy of the secondary i are

$$\gamma(E_i^* - \beta p^*) \leq E_i \leq \gamma(E_i^* + \beta p^*). \tag{9.13}$$

In the absence of polarization there is, in addition,

$$\frac{dn}{d\Omega^*} = \frac{dn}{2\pi \, d\cos\theta^*} \propto \frac{dn}{dE_i} = \text{constant} \tag{9.14}$$

meaning that, in such cases, a flat distribution for a product of a two-body decay between the limits of (9.13) results. For example, for process (9.7) this results in

$$\frac{dn}{dE_\nu} = \frac{dn}{dE_\mu} = \frac{0.635}{1 - (m_\mu^2/m_K^2)p_K} \tag{9.15}$$

with p_K as the laboratory momentum of the kaon and the factor 0.635 stems from the branching ratio of decay (9.7). Often we deal with decays of relativistic particles, resulting in $\beta \rightarrow 1$, which would imply for decays $M \rightarrow \mu\nu$ kinematic limits on the laboratory energies of the secondaries of

$$E \frac{m_\mu^2}{m_M^2} \leq E_\mu \leq E \tag{9.16}$$

and

$$0 \leq E_\nu \leq \left(1 - \frac{m_\mu^2}{m_M^2}\right) E \tag{9.17}$$

with E as the laboratory energy of the decay meson. Average values are:

$$\langle E_\mu \rangle / E_\pi = 0.79 \quad \text{and} \quad \langle E_\nu \rangle / E_\pi = 0.21 \quad \text{for } \pi \rightarrow \mu\nu \tag{9.18}$$
$$\langle E_\mu \rangle / E_K = 0.52 \quad \text{and} \quad \langle E_\nu \rangle / E_K = 0.48 \quad \text{for } K \rightarrow \mu\nu \tag{9.19}$$

It is a consequence of the kinematics that if one of the decay products has a mass close to the parent meson, it will carry most of the energy.

There are several recent ground-level measurements of atmospheric muon fluxes, i.e. those by CAPRICE [Boe99], AMS [Alc00] and BESS [San00],

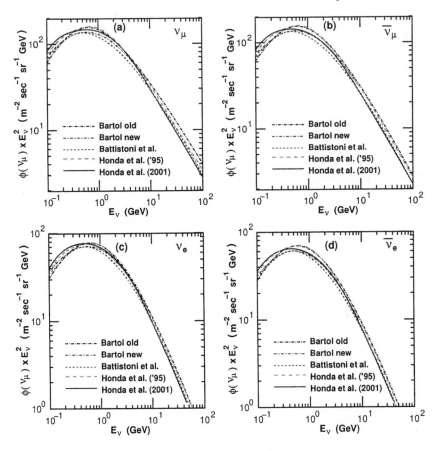

Figure 9.5. Comparison of atmospheric neutrino flux calculations for the location of Kamioka averaged over all directions (from [Gai02]).

which are in agreement with each other at a level of ±5%. Other important measurements have been obtained at high altitude (10–30 km) during the ascent of stratospheric balloons by the MASS, CAPRICE, HEAT and BESS detectors. Since low-energy muons are absorbed in the atmosphere and decay with a high probability ($c\tau_\mu \approx 6.3 p_\mu$ [GeV] km) only these high altitude measurements allow a precise measurement of muons that are most strictly associated with sub-GeV neutrino events.

A second important step is the HARP experiment at CERN [Har99]. This fixed-target experiment uses a proton beam between 2–15 GeV to investigate secondary particle production in various materials. Among them are nitrogen and oxygen targets. For the first time, pion production in proton–nitrogen and proton–oxygen collisions will be directly measured with high accuracy.

A compilation of various atmospheric neutrino flux calculations is shown in figure 9.5. As can be seen it consists basically of ν_μ and ν_e neutrinos. At very high energies ($E_\nu \gg$ TeV) neutrinos from charm production become an additional source [Thu96]. A possible atmospheric ν_τ flux is orders of magnitude less than the ν_μ flux. Now we have the flux at hand, let us discuss the experimental observation.

9.3   Experimental status

Relevant neutrino interaction cross sections for detection have already been discussed in chapter 4. The observed neutrino events can be divided by their experimental separation into contained (fully and partially), stopping, through-going and upward-going events. Basically two types of experiments have been done using either Cerenkov detection or calorimetric tracking devices. Because of its outstanding role in the field, we want to describe the Super-Kamiokande detector in a little bit more detail. For a discussion of former experiments see [Fuk94] (Kamiokande), [Bec92] (IMB), [Kaf94] (Soudan), [Ber90] (Frejus) and [Agl89] (Nusex).

9.3.1   Super-Kamiokande

Super-Kamiokande is a water Cerenkov detector containing 50 kt of ultra-pure water in a cylindrical stainless steel tank [Fuk03a] (figure 9.6). The tank is 41.4 m high and 39.3 m in diameter and separated into two regions: a primary inner volume viewed by 11 146 50 inch diameter photomultipliers (PMTs) and a veto region, surrounding the inner volume and viewed by 1885 20 inch PMTs. For analysis an inner fiducial volume of 22.5 kt is used. Neutrino interactions occurring inside the fiducial volume are called contained events. Fully-contained (FC) events are those which have no existing signature in the outer veto detector and comprise the bulk of the contained event sample. In addition, a partially-contained (PC) sample is identified in which at least one particle (typically an energetic muon) exits the inner detector. The FC sample is further divided into sub-GeV ($E_{\text{Vis}} < 1.33$ GeV) and multi-GeV ($E_{\text{Vis}} > 1.33$ GeV), where E_{Vis} is the total visible energy in the detector (figure 9.7). The events are characterized as either showering (e-like) or non-showering (μ-like) based on the observed Cerenkov light pattern. Two examples are shown in figure 9.8. Criteria have been developed to distinguish between both and were confirmed by accelerator beams.

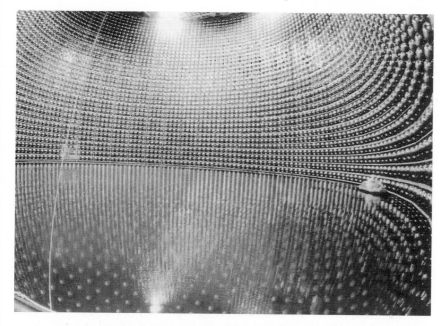

Figure 9.6. Photograph of the Super-Kamiokande detector during filling (with permission of the Kamioka Observatory, ICRR, Tokyo).

9.3.1.1   The ν_μ/ν_e ratio

Historically important for any hint of neutrino oscillation was the R-ratio defined as

$$R = \frac{[N(\mu\text{-like})/N(\text{e-like})]_{\text{obs}}}{[N(\mu\text{-like})/N(\text{e-like})]_{\text{exp}}}. \tag{9.20}$$

Here the absolute flux predictions cancel and if the observed flavour composition agrees with expectation then $R = 1$. Therefore, any deviation of R from 1 is a hint for possible oscillation, even if it cannot be judged without additional information whether ν_μ or ν_e are responsible. A compilation of R-values is given in table 9.1. As can be seen, besides Frejus and Nusex all other datasets prefer an R-value different from 1 and centre around $R = 0.6$. More detailed and convincing evidence can be found by investigating the zenith-angle dependence of the observed events separately.

9.3.1.2   Zenith-angle distributions

Neutrinos are produced everywhere in the atmosphere and can, therefore, reach a detector from all directions. Those produced directly above the detector, characterized by a zenith angle $\cos\theta = 1$, have a typical flight path of about 10 km, while those coming from the other side of the Earth ($\cos\theta = -1$)

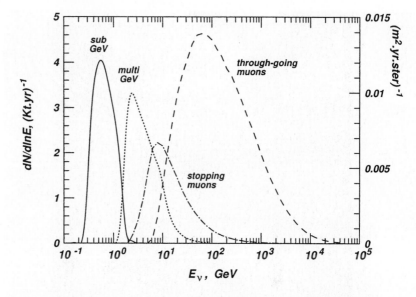

Figure 9.7. Distributions of neutrino energies that give rise to four classes of events at Super-Kamiokande. The contained events are split in sub-GeV and multi-GeV, while stopping and through-going muons refer to neutrino-induced muons produced outside the detector (from [Gai02]).

Table 9.1. Compilation of existing R measurements. The statistics are now clearly dominated by Super-Kamiokande. The no oscillation case corresponds to $R = 1$.

Experiment	R	Stat. significance (kT × y)
Super-Kamiokande (sub-GeV)	$0.638 \pm 0.017 \pm 0.050$	79
Super-Kamiokande (multi-GeV)	$0.675 \pm^{0.034}_{0.032} \pm 0.080$	79
Soudan2	$0.69 \pm 0.10 \pm 0.06$	5.9
IMB	$0.54 \pm 0.05 \pm 0.11$	7.7
Kamiokande (sub-GeV)	$0.60^{+0.06}_{-0.05} \pm 0.05$	7.7
Kamiokande (multi-GeV)	$0.57^{+0.08}_{-0.07} \pm 0.07$	7.7
Frejus	$1.00 \pm 0.15 \pm 0.08$	2.0
Nusex	$0.96^{+0.32}_{-0.28}$	0.74

have to travel more than 12 000 km before interacting. Since the production in the atmosphere is isotropic we can expect the neutrino flux to be up/down symmetric. Slight modifications at low energies are possible because of the previously mentioned geomagnetic effects. Such an analysis can be performed as long as the created charged lepton (e,μ) follows the neutrino direction, which

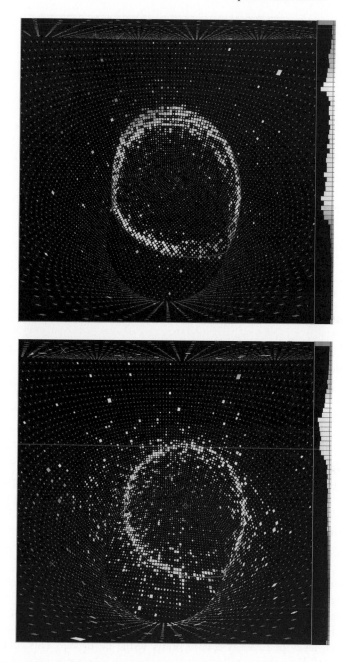

Figure 9.8. Two characteristic events as observed in Super-Kamiokande: top, sharp Cerenkov ring image produced by an muon; bottom, Diffuse Cerenkov ring image produced by an electron (with kind permission of the Super-Kamiokande collaboration).

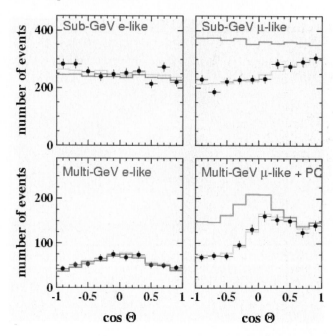

Figure 9.9. Super-Kamiokande zenith-angle distribution for e-like (left) and μ-like events (right), also divided into sub-GeV (upper row) and multi-GeV samples (lower row). A clear deficit is seen in the upward-going muons. The full curves are the Monte Carlo expectations together with an oscillation fit of $\Delta m^2 = 2.4 \times 10^{-3}$ and $\sin^2 2\theta = 1.0$.

is reasonable for momenta larger than about 400 MeV. In 1289 days of real data-taking Super-Kamiokande observed 2864 (624) e-like events and 2788 (558) μ-like in their sub-GeV (multi-GeV) data samples, which are shown in figure 9.9. It is obvious that, in contrast to e-like data which follow the Monte Carlo prediction, there is a clear deficit in the data becoming more and more profound for zenith angles smaller than horizontal, meaning less ν_μ are coming from below.

An independent check of the results from contained events can be done with upward-going muons. Upward-going events are classified as $\cos\theta < 0$. They are produced by neutrinos interacting in the rock below the detector producing muons which traverse the complete detector from below. Here about 1268 days of data-taking can be used [Fuk00]. The typical neutrino energy is about 100 GeV. Lower energetic neutrinos produce upward going stopping muons (1247 days of data-taking) and their energy is comparable to the PC events. This contains two implications. First, the overall expected suppression is larger in this case, since the L/E argument of the oscillation probability is larger. Second, even neutrinos from the horizon will experience significant oscillation. The ratio stopping/through-going events can also be used to remove the normalization because of uncertain

$\nu_\mu \cdot \nu_\tau$·**allowed region**

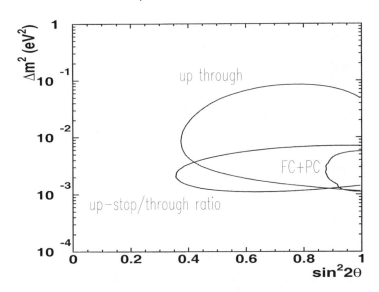

Figure 9.10. Allowed Δm^2 *versus* $\sin^2 2\theta$ regions of the various Super-Kamiokande atmospheric neutrino signals by ν_μ–ν_τ oscillations.

absolute fluxes. Upward-through-going muons have to be compared directly with absolute flux predictions. Now let us take a closer look into the oscillation analysis.

9.3.1.3 *Oscillation analysis*

All datasets—FC, PC, stopping upward muons and through-going upward muons—are divided into angular bins and their distributions are analysed. Furthermore, the FC events are also binned in energy. In the common fit the absolute normalization is allowed to vary freely and other systematic uncertainties are taken into account by additional terms, which can vary in the estimated ranges. The best-fit value obtained is $\Delta m^2 = 2.5 \times 10^3$ eV^2 and maximal mixing, having a $\chi^2 = 159.2/175$ degrees of freedom. A fit without any oscillations results in a $\chi^2 = 315/154$ degrees of freedom. The allowed regions for certain confidence levels are shown in figure 9.10 if interpreted as ν_μ–ν_τ oscillations.

A very important check of the oscillation scenario can be done by plotting the L/E ratio. The L/E ratio for atmospheric neutrinos varies over a large range from about 1–10^5 km GeV^{-1}. Plotting the event rate as a function of L/E results in a characteristic two-bump structure, corresponding to down-going and up-going particles as shown in figure 9.11. The valley in between is populated mostly

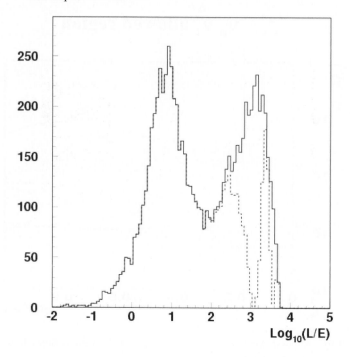

Figure 9.11. L/E double-bump structure. The bump at low values corresponds to downward-going events, the one at high L/E is due to upward-going events (from [Lip01]).

by particles with directions close to horizontal, the event rate per unit L/E is lower here because the pathlength L changes rapidly with the neutrino zenith angle θ. However, this structure is smeared out because of the imperfect energy measurement and the uncertainty in the real production point of the neutrino. According to (8.24) the probability P ($\nu_\mu \rightarrow \nu_\mu$) should show an oscillatory behaviour with minima for L/E ratios and n as an integer number of

$$L/E = n \times \frac{2\pi}{\Delta m^2} = n \times \frac{1236}{\Delta m^2_{-3}} \text{ km GeV}^{-1} \qquad (9.21)$$

with Δm^2_{-3} as the value of Δm^2 in units of 10^{-3} eV^2. Obviously, the first minimum occurs for $n = 1$.

The energy of the neutrino is determined by a correction to the final-state lepton momentum. At $p = 1$ GeV/c the lepton carries about 85% of the neutrino energy, while at 100 MeV/c it typically carries 65%. The flight distance L is determined following [Gai98] using the estimated neutrino energy and the reconstructed lepton direction and flavour. Figure 9.12 shows the data/Monte Carlo ratio for FC data as a function of L/E and momenta larger than 400 MeV/c.

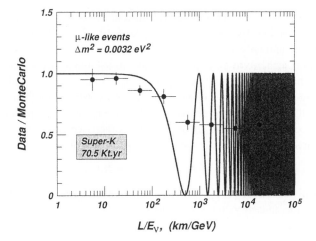

Figure 9.12. Oscillation probability as a function of L/E (from [Gai02]).

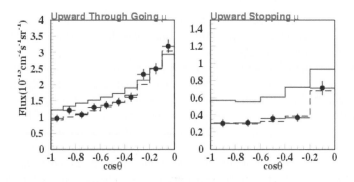

Figure 9.13. Super-Kamiokande upward-going muons flux *versus* absolute prediction. Left: Flux of through-going muons from horizontal ($\cos\theta = 0$) to vertical upward ($\cos\theta = -1$). Right: Upward-going muons which stop in the detector. Also shown are Monte Carlo expectations without oscillations and best-fit values assuming oscillations.

A clear decrease in μ-like events can be seen; however, the oscillation pattern cannot be resolved because of the previously mentioned uncertainties in energy measurements. So for large L/E ν_μ has undergone numerous oscillations and averages these out to roughly 50% of the initial rate.

There is an additional check on the oscillation scenario by looking at the zenith-angle distribution of upward-going muons and compare it with absolute flux predictions. As can be seen in figure 9.13, a deficit is also visible here and an oscillation scenario describes the data reasonably well.

Having established a ν_μ disappearance the question concerning the reason for the deficit arises. Scenarios other than oscillations such as neutrino decay [Bar99], flavour-changing neutral currents [Gon99] or violation of the equivalence principle [Hal96] have been proposed. They all show a different L/E behaviour and only neutrino decay still remains as an alternative. Assuming the oscillation scenario to be correct we have to ask ourselves which one is the oscillation channel. There are three options: $\nu_\mu \rightarrow \nu_e$, $\nu_\mu \rightarrow \nu_\tau$ and $\nu_\mu \rightarrow \nu_S$. A strong argument against ν_e comes from the non-observation of any effect in CHOOZ and Palo Verde (see chapter 8). However, subdominant contributions might still be there. How to distinguish the other two solutions? There are basically three ways. First of all, there is the NC production of π^0:

$$\nu + N \rightarrow \nu + \pi^0 + X. \tag{9.22}$$

This rate will be reduced for ν_S because it does not participate in NC interactions. The ratio of the current value of (π^0/e)-like events with respect to Monte Carlo expectation is

$$R = \frac{(\pi^0/e)_{\text{obs}}}{(\pi^0/e)_{\text{exp}}} = 1.11 \pm 0.08(\text{stat.}) \pm 0.26(\text{sys.}) \tag{9.23}$$

also containing, however, a large theoretical error because of the badly known π^0-production cross section. This might improve with K2K which should be able to measure it more precisely.

Another option would be to search directly for ν_τ appearance in Super-Kamiokande. Taking their statistics and oscillation parameters they should expect approximately 74 ν_τ events. A first analysis of this kind was performed using higher ring multiplicities and resulted in a possible 2σ effect [Ven01]. Last but not least there could be matter effects, because ν_S does not interact at all, resulting in a different effective potential from that of ν_μ as described in chapter 7. Density profiles of the Earth, relevant for the prediction, can be calculated using the Earth model. Basically, the Earth can be described as a 2-component system: the crust and the core. The crust has an average density of $\rho = 3$ g cm^{-3} and an $Y_e = 0.5$ (see chapter 8). However, for large distances $\rho = \rho(t)$ must be used. For the core the density increases up to $\rho = 13$ g cm^{-3} and we can use a step function to describe the two subsystems [Lis97, Giu98]. Furthermore, $N_n \approx N_e/2$ is valid everywhere. Thus, we can write

$$2\sqrt{2}G_F E N_e \simeq 2.3 \times 10^4 \text{ eV}^2 \left(\frac{\rho}{3 \text{ g cm}^{-3}} \right) \left(\frac{E}{\text{GeV}} \right). \tag{9.24}$$

The main effect is such, that matter effects suppress oscillation at high energy. Super-Kamiokande performed a search and excludes ν_μ–ν_S at 99%CL if $\Delta m^2 > 0$. Also for $\Delta m^2 < 0$ most regions are excluded by 99% but a small region remains which is only excluded by 90% [Sob01]. Note, that matter effects between ν_μ and ν_e could also be important.

To sum up, Super-Kamiokande has convincingly proven a ν_μ disappearance effect with a preferred explanation via $\nu_\mu \rightarrow \nu_\tau$ oscillation as shown in figure 9.10. Now we want to take a look at other experiments.

9.3.2 Soudan-2

The Soudan-2 experiment is a 963 t iron-tracking calorimeter located in the Soudan mine in Northern Minnesota at a depth of 2100 mwe [Man01]. The tracking is performed with long plastic drift tubes (0.6 cm μs^{-1} drift time) placed into steel sheets. The sheets are stacked to form a tracking lattice with a honeycomb geometry. Topologies of events include single-track events with a dE/dx compatible with a muon from ν_μ CC interactions, single-shower events as ν_e CC events (both types mostly from quasi-elastic scattering) and multi-prong events. Proton recoils larger than about 350 MeV/c can also be imaged and allow an improved neutrino energy measurement.

The dataset obtained after 5.9 kt × yr results after background subtraction in 101.9 ± 12.7 track events and 146.7 ± 12.5 shower events. Two datasets are prepared, a High Resolution (HiRes) event sample, with shower or track events with a measured recoil proton and lepton momentum larger than 150 MeV/c (if no recoiling nucleon then $p_{\text{lep}} > 600$ MeV) and multiprong events with $E_{\text{vis}} > 700$ MeV and a vector sum of $p_{\text{vis}} > 450$ MeV/c. In addition, lepton momenta larger than 250 MeV/c are required for the latter. A fit to describe the zenith angle distribution assuming that ν_e is not affected by oscillations results in a best fit value of $\Delta m^2 = 5.2 \times 10^{-3}$ eV^2 and $\sin^2 2\theta = 0.97$. The corresponding exclusion plot is shown in figure 9.16. A second sample of PC events having on average a higher energy (E_ν about 3.1 GeV, while the HiRes sample has $\langle E_\nu \rangle \approx 1.3$ GeV) has been prepared, containing 52.7 ± 7.3 μ-like and 5.0 ± 2.2 e-like events.

9.3.3 MACRO

The MACRO detector [Amb01, Amb02] was installed in the Gran Sasso Underground Laboratory (LNGS) and took data in full version from 1994 to 2000 (data were also taken in the construction phase from 1989–94). It consisted of streamer tubes and scintillators in the form of six super-modules covering a total of 76.6 m × 12 m × 9.3 m. The observed muons coming from neutrino events were characterized as upward-through-going, upward-stopping, downward internal and upward internal (figure 9.14). Upward-through-going events from neutrino interactions in the rock below the detector require muon energies of at least 1 GeV; therefore, $\langle E_\nu \rangle \approx 100$ GeV. Also the internal upward-going events can be isolated by time of flight measurements, here $\langle E_\nu \rangle \approx 4$ GeV. The other two samples cannot be distinguished because of the absence of a time of flight measurement and so these are analysed together. Upward-going muon events can be identified by time of flight Δt, resulting in a $1/\beta$ close to -1, with $\beta = c\Delta t/L$. However, the background of downward-going muons ($1/\beta$ close to $+1$) is more than a factor 10^5 higher and tails produce some background. Therefore, upward-going muons were selected with the requirement $-1.25 < 1/\beta < -0.75$. After subtracting the background, a total of 809 events in the upward-through-going

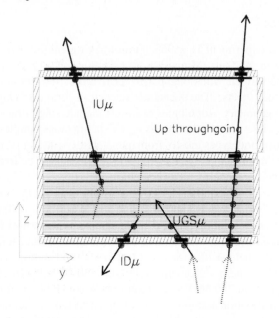

Figure 9.14. Schematic picture of the MACRO event classification.

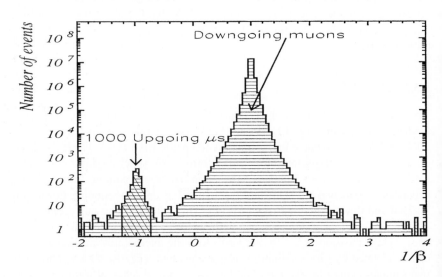

Figure 9.15. MACRO number of events as a function of $1/\beta$. Beside the background spike of down-going muons (about 33.8×10^6 events) a clear peak centred around $1/\beta = -1$ caused by upward-going muons is seen.

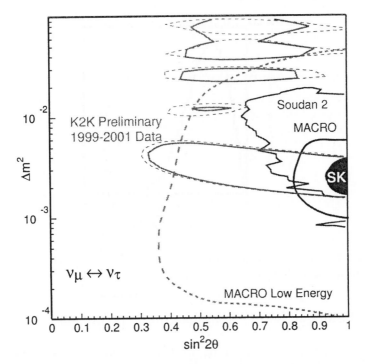

Figure 9.16. Region of combined evidence from Super-Kamiokande, Soudan2 and MACRO, together with the parameter space allowed by the K2K long-baseline experiment (from [Kea02]).

sample was observed (figure 9.15). Using the flux of the Bartol group [Agr96] and the parton distributions of GRV94 [Glu95] a total number of 1122 events were expected. The angular distribution can be best described by assuming $\nu_\mu \to \nu_\tau$ oscillations, the best fit results in $\Delta m^2 = 2.5 \times 10^{-3}$ eV^2 and maximal mixing with a $\chi^2 = 9.7/9$ degrees of freedom. The oscillation scenario into ν_S is disfavoured at a 99% CL at this best-fit value [Bar01]. The analysis for internal upward events is analogous to the one just mentioned; however, now a vertex within the detector is required. Thus, after background subtraction 135 ± 12(stat.) events are observed while expecting 247 ± 25(sys.) ± 62(theo.). Also the third data sample results in a difference between data and Monte Carlo, observing 229 events and expecting 329 ± 33(sys.) ± 92(theo.). The corresponding MACRO parameter regions are shown in figure 9.16.

As can be seen, at least three experiments seem to show a deficit in atmospheric ν_μ. To investigate the possible solutions in more detail, a large programme of new experiments has been launched to confirm this deficit by accelerator-based experiments.

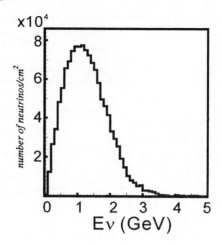

Figure 9.17. Neutrino energy spectrum of the K2K neutrino beam. Because of the relatively low beam energy no ν_τ appearance searches can be performed.

9.4 Future activities—long-baseline experiments

Two strategies are followed using accelerator neutrino beams. First of all, experiment should confirm a ν_μ disappearance and, second, perform a ν_τ appearance search. The last one has to deal with smaller statistics because of the τ production threshold of 3.5 GeV and, therefore, a reduced cross section as well as the involved efficiency for τ-detection. Three projects are on their way in Japan (KEK-Super-Kamiokande), the US (Fermilab-Soudan) and Europe (CERN–Gran Sasso). We now discuss the experiments in chronological order.

9.4.1 K2K

The first of the accelerator-based long-baseline experiments is the KEK-E362 experiment (K2K) [Oya98] in Japan sending a neutrino beam from KEK to Super-Kamiokande. It uses two detectors: one about 300 m away from the target and Super-Kamiokande at a distance of about 250 km. Super-Kamiokande is described in more detail in section 10.2.2. The neutrino beam is produced by 12 GeV protons from the KEK-PS hitting an Al-target of 2 cm diameter $\times$ 65 cm length. Using a decay tunnel of 200 m and a magnetic horn system for focusing π^+ an almost pure ν_μ-beam is produced. The contamination of ν_e from μ and K-decay is of the order 1%. The protons are extracted in a fast extraction mode allowing spills of a time width of 1.1 μs every 2.2 s. With 6×10^{12} pots (protons on target) per spill about 1×10^{20} pots can be accumulated in three years. The average neutrino beam energy is 1.4 GeV, with a peak at about 1 GeV (figure 9.17). In this energy range quasi-elastic interactions are dominant. Kinematics allows to reconstruct E_ν even if only the muon is measured via

$$E_\nu = \frac{m_N E_\mu - \frac{m_\mu^2}{2}}{m_N - E_\mu + P_\mu \cos\theta_\mu} \tag{9.25}$$

with m_N as the mass of the neutron and θ_μ as the angle of the muon with respect to the beam. The near detector consists of two parts: a 1 kt water-Cerenkov detector and a fine-grained detector. The water detector is implemented with 820 20 inch PMTs and its main goal is to allow a direct comparison with Super-Kamiokande events and to study systematic effects of this detection technique. The fine-grained detector basically consists of four parts and should provide information on the neutrino beam profile as well as the energy distribution. First of all, there are 20 layers of scintillating fibre trackers intersected with water. The position resolution of the fibre sheets is about 280 μm and allows track reconstruction of charged particles and, therefore, the determination of the kinematics in the neutrino interaction. In addition, to trigger counters there is a lead-glass counter and a muon detector. The 600 lead-glass counters are used for measuring electrons and, therefore, to determine the ν_e-beam contamination. The energy resolution is about $8\%/\sqrt{E}$. The muon chambers consist of 900 drift tubes and 12 iron plates. Muons generated in the water target via CC reactions can be reconstructed with a position resolution of 2.2 mm. The energy resolution is about 8–10%. The detection method within Super-Kamiokande is identical to that of their atmospheric neutrino detection.

Because of the low beam energy K2K is able to search for $\nu_\mu \rightarrow \nu_e$ appearance and a general ν_μ disappearance. The main background for the search in the electron channel might be quasi-elastic π^0 production in NC reactions, which can be significantly reduced by a cut on the electromagnetic energy. However, the near detector will allow a good measurement of the cross section of π^0 production in NC. The proposed sensitivity regions are given by $\Delta m^2 > 1 \times 10^{-3}$ eV^2(3×10^{-3} eV^2) and $\sin^2 2\theta > 0.1(0.4)$ for $\nu_\mu \rightarrow \nu_e$ ($\nu_\mu \rightarrow \nu_\tau$) oscillations.

In the first year of data-taking K2K accumulated 2.29×10^{19} pot (figure 9.18). K2K observes 56 events but expected $80.1^{+6.2}_{-5.4}$ from the near detector measurement, a clear deficit [Nak01, Nis03, Ahn03]. The best-fit values are $\sin^2 2\theta = 1$ and $\Delta m^2 = 2.8 \times 10^{-3}$ eV^2. This number is in good agreement with the oscillation parameters deduced from the atmospheric data. If this deficit becomes statistically more significant, this would be an outstanding result. In connection with the Japanese Hadron Facility (JHF), an upgrade of KEK is planned to a 50 GeV proton beam, which could start producing data around 2007. The energy of a possible neutrino beam could then be high enough to search for ν_τ appearance.

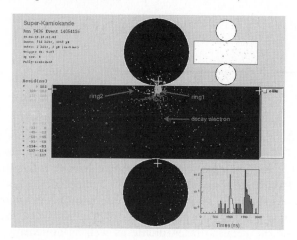

Figure 9.18. The first long-baseline event ever observed by the K2K experiment (with kind permission of the Super-Kamiokande and K2K collaboration).

9.4.2 MINOS

A neutrino program (NuMI) is also associated with the new Main Injector at Fermilab. This long-baseline project will send a neutrino beam to the Soudan mine about 730 km away from Fermilab. Here the MINOS experiment [Mic03] is installed. Using a detection principle similar to CDHS (see chapter 4), it consists of a 980 t near detector located at Fermilab about 900 m away from a graphite target and a far detector at Soudan. The far detector is made of 486 magnetized iron plates, producing an average toroidal magnetic field of 1.3 T. They have a thickness of 2.54 cm and an octagonal shape measuring 8 m across. They are interrupted by about 25 800 m^2 active detector planes in the form of 4.1 cm wide solid scintillator strips with x and y readout to get the necessary tracking informations. Muons are identified as tracks transversing at least five steel plates, with a small number of hits per plane. The total mass of the detector is 5.4 kt.

Several strategies are at hand to discriminate among the various oscillation scenarios. The proof of ν_μ–ν_τ oscillations will be the measurement of the NC/CC ratio in the far detector. The oscillated ν_τ will not contribute to the CC reactions but to the NC reactions. In the case of positive evidence a 10% measurement of the oscillation parameters can be done by comparing the rate and spectrum of CC events in the near and far detector. Also the channel ν_μ–ν_S can be explored again by looking at the NC/CC ratio, which should be compared to what is expected for a ν_τ final state. Three beam options are discussed which are shown in figure 9.19, where the low option was recently chosen. With an average neutrino energy of 3 GeV, this implies a pure ν_μ-disappearance search. A 10 kt × yr exposure will cover the full atmospheric evidence region. The MINOS project is currently under construction and data-taking should start by 2005.

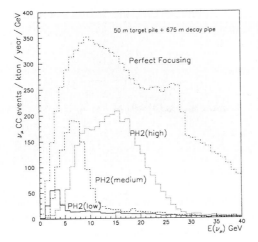

Figure 9.19. Three different options for the neutrino beam (NuMI) used by MINOS at Fermilab. The low-energy version has been chosen. Also shown is the spectrum for perfect focusing of secondary particles.

9.4.3 CERN–Gran Sasso

Another programme under preparation in Europe is the long-baseline programme using a neutrino beam (CNGS) from CERN to the Gran Sasso Laboratory [Els98]. The distance is 732 km. In contrast to K2K and MINOS, the idea here is to search directly for ν_τ appearance. The beam protons from the SPS at CERN will be extracted with energies up to 450 GeV hitting a graphite target at a distance of 830 m from the SPS. After a magnetic horn system for focusing the pions, a decay pipe of 1000 m will follow. The neutrino beam is optimized in such a way to allow the best possible ν_τ-appearance search.

Two experiments are under consideration for the Gran Sasso Laboratory to perform an oscillation search. The first one is the ICARUS experiment [Rub96]. This liquid Ar TPC with a modular design offers excellent energy and position resolution. In addition, very good imaging quality is possible, hence allowing good particle identification. A prototype of 600 t has been built and is approved for installation. An update to three or four modules is planned, which would correspond to about 3 kt. Beside a ν_μ deep inelastic scattering search by looking for a distortion in the energy spectra, an appearance search can also be done because of the good electron identification capabilities (figure 9.20). A ν_τ-appearance search can be obtained by using kinematical criteria as in NOMAD (see section 8.8.1). For ICARUS, a detailed analysis has been done for the $\tau \rightarrow e\nu\nu$ channel (figure 9.21) and is under investigation for other decay channels as well [Rub01].

The second proposal is a ν_τ-appearance search with a 2 kt lead-emulsion

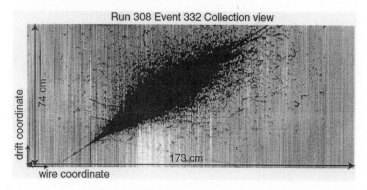

Figure 9.20. A broad electromagnetic shower as observed with the ICARUS T600 test module on the surface. This impressively shows the data quality obtainable with LAr TPCs (with kind permission of the ICARUS collaboration).

sandwich detector (OPERA) [Gul00]. The principle is to use lead as a massive target for neutrino interactions and thin (50 μm) emulsion sheets working conceptually as emulsion cloud chambers (ECC) (figure 9.22). The detector has a modular design, with a brick as the basic building block, containing 58 emulsion films. Some 3264 bricks together with electronic trackers form a module. Twenty-four modules will form a supermodule of about 652 t mass. Three supermodules interleaved with a muon spectrometer finally form the full detector. In total about 235 000 bricks have to be built. The scanning of the emulsions is done by high speed automatic CCD microscopes. The τ, produced by CC reactions in the lead, decays in the gap region, and the emulsion sheets are used to verify the kink in the decay, a principle also used in CHORUS. For decays within the lead an impact parameter analysis can be done to show that the required track does not come from the primary vertex. In addition to the $\tau \rightarrow$ e, μ, π decay modes three pion decays can also be examined. The analysis here uses an event by event basis and the experiment is, in general, considered background free. In five years of data-taking, correspoding to 2.25×10^{20} pot a total of 18.3 events are expected for $\Delta m^2 = 3.2 \times 10^{-3}$ eV^2.

Upgrades towards 600 kt–1 Mt water Cerenkov detectors are considered (Hyper-K, UNO) as well as a 80 kt LA-TPC (LANNDD) are discussed in context of the neutrino factory described in chapter 8. These are multipurpose detectors, which can also be used for atmospheric and supernova neutrino studies and proton decay. Two more experiments, AQUA-RICH and MONOLITH, can also, in principle, be used for artificial neutrino beams. However, they are mainly designed for atmospheric neutrinos.

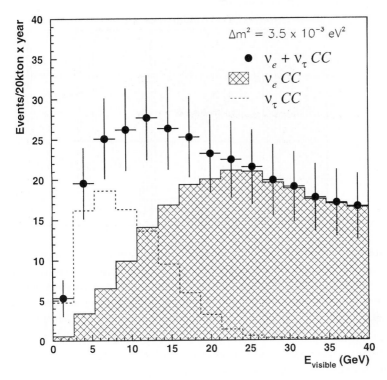

Figure 9.21. Expected ν_τ appearance signal in ICARUS using the $\tau \to$ e decay channel. The number of events are shown as a function of visible energy due to ν_e CC interactions intrinisic to the beam (hatched), the additional events due to the oscillation contribution (hatched line) as well as the sum spectrum (points).

9.4.4 MONOLITH

MONOLITH [Mon00] is a proposed 34 kt magnetized iron tracking calorimeter. It consists of two modules, each 14.5 m × 15 m × 13 m, made out of 125 horizontal iron plates each 8 cm thick, which are interleaved by active tracking devices in the form of glass resistive plate chambers (RPCs) The magnetized iron produces a field of 1.3 T. This allows the muon charge to be measured; therefore, discriminating between ν_μ and $\bar{\nu}_\mu$. This can be important in studying matter effects. Measuring the hadronic energy and the momentum of the semi-contained muons will allow a reasonably good reconstruction of the neutrino energy. The neutrino angular distribution, which determines the resolution of L, is sufficient to allow a good L/E resolution.

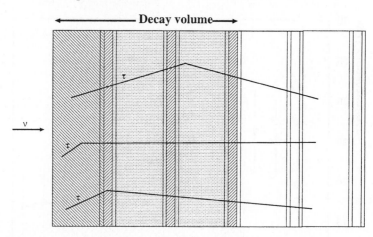

Figure 9.22. The emulsion cloud chamber (ECC) principle which will be used by OPERA. A τ-lepton produced in a charged current interaction can be detected via two mechanisms. If the τ-decay happens after the τ has traversed several emulsion sheets, there will be a kink between the various track segments (upper curve). In a early decay an impact parameter analysis can be done, because the interesting track is not pointing to the primary vertex (lower two curves). For simplicity the additional tracks from the primary vertex are not shown.

9.4.5 Very large water Cerenkov detectors

To gain statistics and rely on well-known techniques there is also the possibility to build even larger water detectors. Two such proposals are made. One is the UNO detector using 650 kt of water consisting of three cubic compartments, the second is Hyper-Kamiokande, a 1 Mt device made out of eight cubes 50 m $\times$ 50 m $\times$ 50 m^3 each.

9.4.6 AQUA-RICH

The basic principle and ideas of AQUA-RICH are summarized in [Ant99]. By using the RICH technique, particle velocities can be measured by the ring radius and direction by the ring centre. An improvement over existing Cerenkov detectors is the measurement of higher ring multiplicities and, therefore, more complicated events can be investigated. However, the main new idea is to measure momenta via multiple scattering. Multiple scattering causes a displacement and an angular change as a particle moves through a medium. The projected angular distribution $\theta_{b;c}$ of a particle with velocity β, momentum p and charge Z after traversing the path L in a medium of absorption length X_0 is Gaussian with the width

$$\sigma_{ms} = \theta_{b;c}^{rms} = \frac{k_{ms}}{\beta cp} Z \sqrt{\frac{L}{X_0}} \tag{9.26}$$

with $k_{ms} = 13.6$ MeV as the multiple scattering constant. Momentum resolution better than 10% for 10 GeV muons could be obtained in simulations, sufficient to see the oscillation pattern in atmospheric neutrinos. A 1 Mt detector is proposed.

Chapter 10

Solar neutrinos

Solar neutrinos are one of the longest standing and most interesting problems in particle astrophysics. From the astrophysical point of view, solar neutrinos are the only objects beside the study of solar oscillations (helioseismology) which allow us a direct view into the solar interior. The study of the fusion processes going on in the Sun offers a unique perspective. From the particle physics point of view, the baseline Sun–Earth with an average of 1.496×10^8 km and neutrino energies of about 1 MeV offers a chance to probe neutrino oscillation parameters of $\Delta m^2 \approx 10^{-10}$ eV^2, which is not possible by terrestrial means. The Sun is a pure source of ν_e resulting from fusion chains. During recent decades it has been established that significantly fewer solar ν_es are observed than would be expected from theoretical modelling. It is extremely important to find out to what extent this discrepancy points to 'new physics' like neutrino oscillations, rather than to an astrophysical problem such as a lack of knowledge of the structure of the Sun or of reactions in its interior or a 'terrestrial' problem of limited knowledge of capture cross sections in neutrino detectors. Nowadays the amount of data especially those from the Sudbury Neutrino Observatory experiment indeed strongly favour the neutrino oscillation hypothesis, and this is the third piece of evidence for a non-vanishing neutrino mass. In the following chapter the situation is discussed in more detail.

10.1   The standard solar model

10.1.1   Energy production processes in the Sun

According to our understanding, the Sun, like all stars, creates its energy via nuclear fusion [Gam38]. For a general discussion of the structure of stars and stellar energy generation see, e.g., [Cox68, Cla68, Sti02]. Hydrogen fusion to helium proceeds according to

$$4p \rightarrow {}^4He + 2e^+ + 2\nu_e \tag{10.1}$$

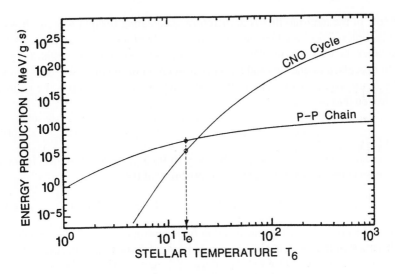

Figure 10.1. Contributions of the pp and CNO cycles for energy production in stars as a function of the central temperature. While the pp cycle is dominant in the Sun, the CNO process becomes dominant above about 20 million degrees (from [Rol88]).

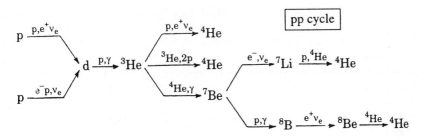

Figure 10.2. The route of proton fusion according to the pp cycle. After the synthesis of ^{3}He the process branches into three different chains. The pp cycle produces 98.4% of the solar energy.

The two positrons annihilate with two electrons resulting in an energy relevant equation

$$2e^- + 4p \rightarrow \ ^4He + 2\nu_e + 26.73 \ \text{MeV}. \tag{10.2}$$

Therefore, an energy of $Q = 2m_e + 4m_p - m_{He} = 26.73$ MeV per ^{4}He fusion is released. Using the solar constant $S = 8.5 \times 10^{11}$ cm^{-2} s^{-1} at the Earth a first guess for the total neutrino flux at the Earth can be obtained:

$$\Phi_\nu \approx \frac{S}{13 \ \text{MeV per } \nu_e} = 6.5 \times 10^{10} \ \text{cm}^{-2} \ \text{s}^{-1}. \tag{10.3}$$

Details of the neutrino flux and, therefore, its creating fusion processes however are complex. There are two ways to bring this about: one is the pp cycle [Bet38], the other the CNO cycle [Wei37, Bet39]. Figure 10.1 shows the contribution of both processes to energy production as a function of temperature. The pp cycle (see figure 10.2) is dominant in the Sun. The first reaction step is the fusion of hydrogen into deuterium:

$$p + p \rightarrow {}^2H + e^+ + \nu_e \qquad (E_\nu \leq 0.42\,\text{MeV}). \qquad (10.4)$$

The primary pp fusion proceeds this way to 99.6%. In addition, the following process occurs with a much reduced probability of 0.4%:

$$p + e^- + p \rightarrow {}^2DH + \nu_e \qquad (E_\nu = 1.44\,\text{MeV}). \qquad (10.5)$$

The neutrinos produced in this reaction (pep neutrinos) are mono-energetic. The conversion of the created deuterium to helium is identical in both cases:

$$^2DH + p \rightarrow {}^3He + \gamma + 5.49\,\text{MeV}. \qquad (10.6)$$

Neutrinos are not produced in this reaction. From that point onwards the reaction chain divides. With a probability of 85% the ^{3}He fuses directly into ^{4}He:

$$^3He + {}^3He \rightarrow {}^4He + 2p + 12.86\,\text{MeV}. \qquad (10.7)$$

In this step, also known as the pp I-process, no neutrinos are produced. However, two neutrinos are created in total, as the reaction of equation (10.4) has to occur twice, in order to produce two ^{3}He nuclei which can undergo fusion. Furthermore, ^{4}He can also be created with a probability of $2.4 \times 10^{-5}\%$ by

$$^3He + p \rightarrow {}^4He + \nu_e + e^+ + 18.77\,\text{MeV}. \qquad (10.8)$$

The neutrinos produced here are very energetic (up to 18.77 MeV) but they have a very low flux. They are called *hep* neutrinos. The alternative reaction produces ^{7}Be:

$$^3He + {}^4He \rightarrow {}^7Be + \gamma + 1.59\,\text{MeV}. \qquad (10.9)$$

Subsequent reactions again proceed via several sub-reactions. The *pp* II-process leads to the production of helium with a probability of 15% via

$$^7Be + e^- \rightarrow {}^7Li + \nu_e \qquad (E_\nu = 0.862\,\text{MeV or } E_\nu = 0.384\,\text{MeV}). \quad (10.10)$$

This reaction produces ^{7}Li in the ground state 90% of the time and leads to the emission of monoenergetic neutrinos of 862 keV. The remaining 10% decay into an excited state by emission of neutrinos with an energy of 384 keV. Thus mono-energetic neutrinos are produced in this process. In the next reaction step, helium is created via

$$^7Li + p \rightarrow 2\,{}^4He + 17.35\,\text{MeV}. \qquad (10.11)$$

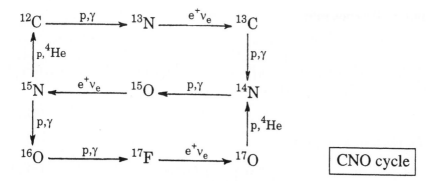

Figure 10.3. Representation of the CNO process. This also burns hydrogen to helium with C, N and O acting as catalysts and is responsible for 1.6% of the solar energy.

Assuming that ^{7}Be has already been produced, this pp II-branch has a probability of 99.98%. There is also the possibility of proceeding via ^{8}B (the pp III-chain) rather than by ^{7}Li via

$$^7\text{Be} + \text{p} \rightarrow {}^8\text{B} + \gamma + 0.14 \text{ MeV} \tag{10.12}$$

^{8}B undergoes β^+-decay via

$$^8\text{B} \rightarrow {}^8\text{Be}^* + \text{e}^+ + \nu_e \qquad (E_\nu \approx 15 \text{ MeV}). \tag{10.13}$$

The endpoint is uncertain because the final state in the daughter nucleus is very broad. The neutrinos produced here are very energetic but also very rare. Nevertheless, they play an important role. ^{8}Be undergoes α-decay into helium:

$$^8\text{Be}^* \rightarrow 2\,{}^4\text{He} + 3 \text{ MeV}. \tag{10.14}$$

The CNO cycle only accounts for about 1.6% of the energy production in the Sun, which is why it is mentioned here only briefly. The main reaction steps are:

$$^{12}\text{C} + \text{p} \qquad\qquad \rightarrow {}^{13}\text{N} + \gamma \tag{10.15}$$

$$^{13}\text{N} \rightarrow {}^{13}\text{C} + \text{e}^+ + \nu_e \qquad (E_\nu \leq 1.2 \text{ MeV}) \tag{10.16}$$

$$^{13}\text{C} + \text{p} \qquad\qquad \rightarrow {}^{14}\text{N} + \gamma \tag{10.17}$$

$$^{14}\text{N} + \text{p} \qquad\qquad \rightarrow {}^{15}\text{O} + \gamma \tag{10.18}$$

$$^{15}\text{O} \rightarrow {}^{15}\text{N} + \text{e}^+ + \nu_e \qquad (E_\nu \leq 1.73 \text{ MeV}) \tag{10.19}$$

$$^{15}\text{N} + \text{p} \qquad\qquad \rightarrow {}^{12}\text{C} + {}^4\text{He}. \tag{10.20}$$

This process and its subsidiary cycle which is not discussed here because of its negligible importance are illustrated in figure 10.3.

We have now introduced the processes relevant for neutrino production. To predict the expected neutrino spectrum we need further information—in particular about the cross sections of the reactions involved [Par94, Lan94].

10.1.2 Reaction rates

Before dealing with details of the Sun we first state some general comments on the reaction rates [Cla68, Rol88, Bah89, Raf96, Adl98a]. They play an important role in the understanding of energy production in stars. Consider a reaction of two particles T_1 and T_2 of the general form

$$T_1 + T_2 \rightarrow T_3 + T_4. \qquad (10.21)$$

The reaction rate is given by

$$R = \frac{n_1 n_2}{1 + \delta_{12}} \langle \sigma v \rangle_{12} \qquad (10.22)$$

where n_i is the particle density, σ the cross section, v the relative velocity and δ the Kronecker symbol to avoid double counting of identical particles. $\langle \sigma v \rangle$ is the temperature averaged product. At typical thermal energies of several keV inside the stars and Coulomb barriers of several MeV, it can be seen that the dominant process for charged particles is quantum mechanical tunnelling, which was used by Gamow to explain α-decay [Gam38]. It is common to write the cross section in the form

$$\sigma(E) = \frac{S(E)}{E} \exp(-2\pi\eta) \qquad (10.23)$$

where the exponential term is Gamow's tunnelling factor, the factor $1/E$ expresses the dependence of the cross section on the de Broglie wavelength and η is the so-called Sommerfeld parameter, given by $\eta = Z_1 Z_2 e^2 / \hbar v$. Nuclear physics now only enters into calculations through the so-called S-factor $S(E)$, which, as long as no resonances appear, has a relatively smooth behaviour. This assumption is critical, since we have to extrapolate from the values at several MeV, measured in the laboratory, down to the relevant energies in the keV region [Rol88]. For the averaged product $\langle \sigma v \rangle$ we also need to make an assumption on the velocity distribution of the particles. In normal main-sequence stars such as our Sun, the interior has not yet degenerated so that a Maxwell–Boltzmann distribution can be assumed. Due to the energy behaviour of the tunnelling probability and the Maxwell–Boltzmann distribution there is a most probable energy E_0 for a reaction, which is shown schematically in figure 10.4 [Bur57, Fow75]. This Gamow peak for the pp reaction, which we will discuss later, lies at about 6 keV. If we define $\tau = 3E_0/kT$ and approximate the reaction-rate dependence on temperature by a power law $R \sim T^n$, then $n = (\tau - 2)/3$. For a detailed discussion of this derivation see, e.g., [Rol88, Bah89]. Since the energy of the Gamow peak is temperature dependent, $S(E)$ is, for ease of computation, expanded in a Taylor series with respect to energy:

$$S(E) = S(0) + \dot{S}(0)E + \tfrac{1}{2}\ddot{S}(0)E^2 + \cdots \qquad (10.24)$$

where $S(0)$, $\dot{S}(0)$ etc are obtained by a fit to the experimental data. Stellar energy regions are accessible directly in laboratory experiments for the first time in the LUNA experiment, using accelerators built underground because the cross sections are small [Gre94, Fio95, Arp96]. In a first step the LUNA collaboration is operating a 50 kV accelerator at the Gran Sasso Laboratory to investigate the $^3\mathrm{He}(^3\mathrm{He}, 2p)\,^4\mathrm{He}$ reaction as the final step in the pp I chain [Arp98]. As can be seen in figure 10.5, the experimental data exceed the theoretical expectation of bare nuclei which is due to a still restricted knowledge of screening effects. In contrast to stellar plasma where atoms are fully ionized, in laboratory experiments the remaining electrons produce a shielding effect, which has to be taken into account. In addition, the reaction $d(p, \gamma)^3\mathrm{He}$ has been measured recently [Bro03]. An upgrade to a 200 kV accelerator (LUNA II) was performed and this has made additional measurements of, e.g., the $^7\mathrm{Be}(p, \gamma)\,^8\mathrm{B}$ and $^{14}\mathrm{N}(p, \gamma)\,^{15}\mathrm{O}$ cross sections possible. New measurements for the $^7\mathrm{Be}(p, \gamma)\,^8\mathrm{B}$ cross section at centre-of-mass energies between 350 and 1400 keV exist [Ham98, Jun02, Bab03, Jun03]. Earlier measurements [Kav69, Fil83] showed a 30% difference in the absolute value of $S(E)$ in this region. The new measurement seems to support the lower $S(E)$ values of [Fil83] (figure 10.5).

10.1.3 The solar neutrino spectrum

The actual prediction of the solar neutrino spectrum requires detailed model calculation [Tur88, Bah88, Bah89, Bah92, Tur93a, Tur93b, Bah95, Bah01, Cou02].

10.1.3.1 Standard solar models

The simulations which model the operation of the Sun use the basic equations of stellar evolution (see [Cla68, Rol88, Bah89]).

(i) Hydrodynamic equilibrium, i.e. the gas and radiation pressure balance the gravitational attraction:

$$\frac{\mathrm{d}p(r)}{\mathrm{d}r} = -\frac{GM(r)\rho(r)}{r^2} \tag{10.25}$$

with mass conservation

$$M(r) = \int_0^r 4\pi r^2 \rho(r)\,\mathrm{d}r.$$

(ii) Energy balance, meaning the observed luminosity L is generated by an energy generation rate ϵ:

$$\frac{\mathrm{d}L(r)}{\mathrm{d}r} = 4\pi r^2 \rho(r)\epsilon. \tag{10.26}$$

(iii) Energy transport dominantly by radiation and convection which is given in the radiation case by

$$\frac{\mathrm{d}T(r)}{\mathrm{d}r} = -\frac{3}{64\pi\sigma}\frac{\kappa\rho(r)L(r)}{r^2 T^3} \tag{10.27}$$

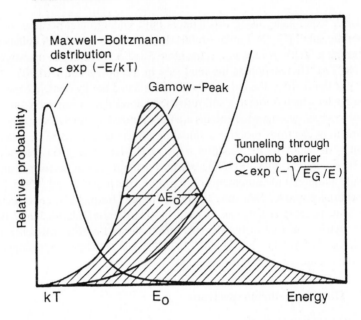

Figure 10.4. The most favourable energy region for nuclear reactions between charged particles at very low energies is determined by two effects which act in opposite directions. The first is the Maxwell–Boltzmann distribution with a maximum at kT, which implies an exponentially decreasing number of particles at high energies. The other effect is that the quantum mechanical tunnelling probability rises with growing energy. This results in the Gamow peak (not shown to true scale), at an energy E_0, which can be very much larger than kT (from [Rol88]).

with σ as the Stefan–Boltzmann constant and κ as the absorption coefficient. In the interior of the Sun up to 70% of the radius energy transport is radiation dominated, while the outer part forms the convection zone. These equations are governed by three additional equations of state for the pressure p, the absorption coefficient κ and the energy generation rate ϵ:

$$p = p(\rho, T, X) \qquad \kappa = \kappa(\rho, T, X) \qquad \epsilon = \epsilon(\rho, T, X) \qquad (10.28)$$

where X denotes the chemical composition. The Russell–Vogt theorem then ensures, that for a given M and X a unique equilibrium configuration will evolve, resulting in certain radial pressure, temperature and density profiles of the Sun. Under these assumptions, solar models can be calculated as an evolutionary sequence from an initial chemical composition. The boundary conditions are that the model has to reproduce the age, luminosity, surface temperature and mass of the present Sun. The two typical adjustable parameters are the ^{4}He abundance and the relation of the convective mixing length to the pressure scale height. Input parameters include the age of the Sun and its luminosity, as well as the equation

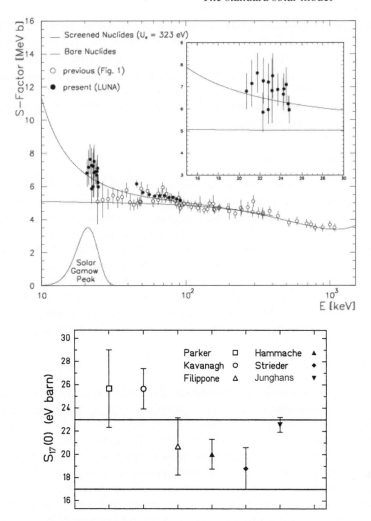

Figure 10.5. Two important cross sections in stellar astrophysics as measured with accelerators: top, compilation of ^{3}He(^{3}He, 2p) ^{4}He data (from [Arp98]); bottom, Compilation of S_{17} (0) values of the ^{7}Be(p, γ) ^{8}B reaction for various experiments (from [Jun02]).

of state, nuclear parameters, chemical abundances and opacities.

10.1.3.2 Diffusion

Evidence from several experiments strongly suggests a significant mixing and gravitational settling of He and the heavier elements in the Sun. The longstanding problem of ^{7}Li depletion in the solar photosphere can be explained if ^{7}Li is destroyed by nuclear burning processes which, however, require temperatures of about 2.6×10^6 K. Such temperatures do not exist at the base of the convection zone; therefore, ^{7}Li has to be brought to the inner regions. Also the measured sound speed profiles in the solar interior obtained by helioseismological data can be better reproduced by including diffusion processes. Therefore, these effects were included in the latest solar models.

10.1.3.3 Initial composition

The chemical abundance of the heavier elements (larger than helium) forms an important ingredient for solar modelling. Their abundance influences the radiative opacity and, therefore, the temperature profile within the Sun. Under the assumption of a homogeneous Sun, it is assumed that the element abundance in the solar photosphere still corresponds to the initial values. The relative abundances of the heavy elements are best determined in a certain type of meteorite, the type I carbonaceous chondrite, which can be linked and found to be in good agreement with the photospheric abundances [Gre93, Gre93a]. The abundance of C, N and O is taken from photospheric values, whereas the ^{4}He abundance cannot be measured and is used as an adjustable parameter.

10.1.3.4 Opacity and equation of state

The opacity is a measure of the photon absorption capacity. It depends on the chemical composition and complex atomic processes. The influence of the chemical composition on the opacity can be seen, for example, in different temperature and density profiles of the Sun. The ratio of the 'metals' Z (in astrophysics all elements heavier than helium Y are known as metals) to hydrogen X is seen to be particularly sensitive. The experimentally observable composition of the photosphere is used as the initial composition of elements heavier than carbon. In the solar core ($T > 10^7$ K) the metals do not play the central role for the opacity, which is more dependent on inverse bremsstrahlung and photon scattering on free electrons here (Thomson-scattering). The opacity or Rosseland mean absorption coefficient κ is defined as a harmonic mean integrated over all frequencies ν:

$$\frac{1}{\kappa} = \frac{\int_0^\infty \frac{1}{\kappa_\nu} \frac{\mathrm{d}B_\nu}{\mathrm{d}T} \, \mathrm{d}\nu}{\int_0^\infty \frac{\mathrm{d}B_\nu}{\mathrm{d}T} \, \mathrm{d}\nu} \tag{10.29}$$

where B_ν denotes a Planck spectrum. The implication is that more energy is transported at frequencies where the material is more transparent and at which the radiation field is more temperature dependent. The calculation of the Rosseland mean requires a knowledge of all the involved absorption and scattering cross sections of photons on atoms, ions and electrons. The calculation includes bound–bound (absorption), bound–free (photoionization), free–free (inverse bremsstrahlung) transitions and Thomson scattering. Corrections for electrostatic interactions between the ions and electrons and for stimulated emissions have to be taken into acount. The number densities n_i of the absorbers can be extracted from the Boltzmann and Saha equations. The radiative opacity per unit mass can then be expressed as (with the substitution $u \equiv h\nu/kT$)

$$\frac{1}{\kappa} = \rho \int_0^\infty \frac{15u^4 e^u/4\pi^4 (e^u - 1)^2}{(1 - e^u) \sum_i \sigma_i n_i + \sigma_s n_e} \, du \qquad (10.30)$$

where σ_s denotes the Thomson scattering cross section.

The most comprehensive compilation of opacities is given by the Livermore group (OPAL) [Ale94, Igl96]. It includes data for 19 elements over a wide range of temperature, density and composition. A detailed study of opacity effects on the solar interior can be found in [Tri97].

A further ingredient for solar model calculations is the equation of state, meaning the density as a function of p and T or, as widely used in the calculations, the pressure expressed as a function of density and temperature. Except for the solar atmosphere, the gas pressure outweighs the radiation pressure anywhere in the Sun. The gas pressure is given by the perfect gas law, where the mean molecular weight μ must be determined by the corresponding element abundances. The different degrees of ionization can be determined using the Saha equations. An equation of state in the solar interior has to consider plasma effects and the partial electron degeneracy deep in the solar core. The latest equation of state is given by [Rog96]. It is assumed here that the Sun has been a homogeneous star since joining the main sequence.

10.1.3.5 Predicted neutrino fluxes

With all these inputs it is then possible to calculate a sequence of models of the Sun which finally predict values of $T(r)$, $\rho(r)$ and the chemical composition of its current state (see table 10.1). These models are called *standard solar models* (SSM) [Tur88, Bah89, Bah92, Tur93a, Bah95, Bah01, Cou02]. They predict the location and rate of the nuclear reactions which produce neutrinos (see figure 10.6). Finally, these models give predictions for the expected neutrino spectrum and the observable fluxes on Earth (see figure 10.7 and table 12.2). It can clearly be seen that the largest part of the flux comes from the pp neutrinos (10.4). In addition to the flux, in order to predict the signal to be expected in the various detectors, it is necessary to know the capture or reaction cross sections for neutrinos.

Although the neutrino fluxes on Earth have values of the order of 10^{10} cm^{-2} s^{-1}, their detection is extremely difficult because of the small cross sections. We now turn to the experiments, results and interpretations.

Table 10.1. Properties of the Sun according to the standard solar model (SSM) of [Bah89].

	$t = 4.6 \times 10^9$ yr (today)	$t = 0$
Luminosity $L_\odot$	$\equiv 1$	0.71
Radius $R_\odot$	696 000 km	605 500 km
Surface temperature T_S	5773 K	5 665 K
Core temperature T_c	15.6×10^6 K	—
Core density	148 g cm^{-3}	—
X (H)	34.1%	71%
Y (He)	63.9%	27.1%
Z	1.96%	1.96%

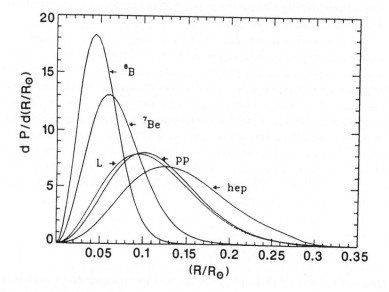

Figure 10.6. Production of neutrinos from different nuclear reactions as a function of the distance from the Sun's centre, according to the standard solar model. The luminosity produced in the optical region (denoted by L) as a function of radius is shown as a comparison. It can be seen to be very strongly coupled to the primary pp fusion (from [Bah89]).

10.2   Solar neutrino experiments

In principle there are two kinds of solar neutrino experiments: radiochemical and real-time experiments. The principle of the *radiochemical experiments* is the reaction

$$^A_N Z + \nu_e \rightarrow ^{A}_{N-1}(Z+1) + e^- \qquad (10.31)$$

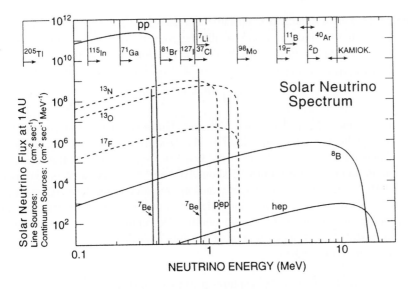

Figure 10.7. The solar neutrino spectrum at the Earth, as predicted by detailed solar model calculations. The dominant part comes from the pp neutrinos, while at high energy hep and ^{8}B neutrinos dominate. The threshold energies of different detector materials are also shown (see e.g. [Ham93]).

Table 10.2. Two examples of SSM predictions for the flux Φ_ν of solar neutrinos on the Earth (from [Bah01] and [Cou02], together with predictions for the two running geochemical experiments using Cl and Ga.

Source	Φ_ν (10^{10} cm^{-2} s^{-1})		Ga (SNU)	Cl (SNU)
	[Bah01]	[Cou02]	[Bah01]	[Bah01]
pp	5.95	5.92	69.7	0
pep	1.40×10^{-2}	1.43×10^{-2}	2.8	0.22
^{7}Be	4.77×10^{-1}	4.85×10^{-1}	34.2	1.15
^{8}B	5.05×10^{-4}	4.98×10^{-4}	12.1	5.76
^{13}N	5.48×10^{-2}	5.77×10^{-2}	3.4	0.09
^{15}O	4.80×10^{-2}	4.97×10^{-2}	5.5	0.33
^{17}F	5.63×10^{-4}	3.01×10^{-4}	0.01	0.01
$\sum(\Phi\sigma)_{Cl}$ [SNU]	$7.6^{+1.34}_{-1.1}$	7.48 ± 0.97		
$\sum(\Phi\sigma)_{Ga}$	128^{+9}_{-7}	128.1 ± 8.9		

where the daughter nucleus is unstable and decays with a 'reasonable' half-life, since it is this radioactive decay of the daughter nucleus which is used in the detection. The production rate of the daughter nucleus is given by

$$R = N \int \Phi(E)\sigma(E)\,\mathrm{d}E \qquad (10.32)$$

where Φ is the solar neutrino flux, N the number of target atoms and σ the cross section for the reaction of equation (10.31). Dealing with discrete nuclear states this implies knowledge of the involved Gamow–Teller nuclear matrix elements. Given an incident neutrino flux of about 10^{10} cm^{-2} s^{-1} and a cross section of about 10^{-45} cm^2, about 10^{30} target atoms are required to produce one event per day. We therefore require very large detectors of the order of several tons and expect the transmutation of one atom per day. Detecting this is no easy matter. It is convenient to define a new unit more suitable for such low event rates, the SNU (solar neutrino unit) where

$$1 \text{ SNU} = 10^{-36} \text{ captures per target atom per second.}$$

Any information about the time of the event, the direction and energy (with the exception of the lower limit, which is determined by the energy threshold of the detector) of the incident neutrino is lost in these experiments since only the average production rate of the unstable daughter nuclei over a certain time interval can be measured.

The situation in *real-time experiments* is different. The main detection method here is neutrino–electron scattering and neutrino reactions on deuterium, in which Cerenkov light is created by electrons, which can then be detected. In the case of scattering it is closely correlated with the direction of the incoming neutrino. However, these detectors have an energy threshold of about 5 MeV and are, therefore, only sensitive to ^{8}B neutrinos. The ^{8}B flux is about four orders of magnitude lower than the pp flux and, therefore, the target mass here has to be in the kiloton range. In discussing the existing experimental data we will follow the historic sequence.

10.2.1 The chlorine experiment

The first solar neutrino experiment, and the birth of neutrino astrophysics in general, is the chlorine experiment of Davis [Dav64, Row85a, Dav94a, Dav94b, Cle98], which has been running since 1968. The reaction used to detect the neutrinos is

$$^{37}\text{Cl} + \nu_e \rightarrow {}^{37}\text{Ar} + \text{e}^- \qquad (10.33)$$

which has an energy threshold of 814 keV. The detection method utilizes the decay

$$^{37}\text{Ar} \rightarrow {}^{37}\text{Cl} + \text{e}^- + \bar{\nu}_e \qquad (10.34)$$

Figure 10.8. The chlorine detector of Davis for the detection of solar neutrinos in the about 1400 m deep Homestake Mine in Lead, South Dakota (USA) in about 1967. The 380 000 l tank full of perchloro-ethylene is shown. Dr Davis is standing above. (Photograph from Brookhaven National Laboratory.)

which has a half-life of 35 days and results in 2.82 keV x-rays or Auger electrons from K-capture (90%). With the given threshold this experiment is not able to measure the pp neutrino flux. The contributions of the various production reactions for neutrinos to the total flux are illustrated in table 10.2 according to one of the current solar models. The solar model calculations predict values of (7.5 ± 1.0) SNU [Bah01, Cou02], where the major part comes from the ^{8}B neutrinos. All, except the ^{8}B neutrinos, lead only to the ground state of ^{37}Ar whereas ^{8}B is also populating excited states including the isotopic analogue state. The cross section for the reaction (10.33) averaged over the ^{8}B spectrum has been measured recently to be [Auf94, Bah95]

$$1.14 \pm 0.11 \times 10^{-42} \text{ cm}^2. \tag{10.35}$$

The experiment (figure 10.8) operates in the Homestake gold mine in South Dakota (USA), where a tank with 615 t perchloro-ethylene (C_2Cl_4) which serves as the target, is situated at a depth corresponding to 4100 mwe (meter water equivalent). The natural abundance of ^{37}Cl is about 24%, so that the number of target atoms is 2.2×10^{30}. The argon atoms which are produced are volatile in solution and are extracted about once every 60–70 days. The extraction efficiency is controlled by adding a small amount of isotopical pure inert ^{36}Ar or ^{38}Ar. To do this, helium is flushed through the tank taking the volatile argon out of the solution and allowing the collection of the argon in a cooled charcoal trap. The trapped argon is then purified, concentrated in several steps and finally filled into

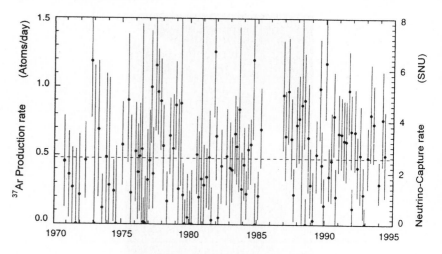

Figure 10.9. The neutrino flux measured from the Homestake ^{37}Cl detector since 1970. The average measured value (broken line) is significantly smaller than the predicted one. This discrepancy is the origin of the so-called solar neutrino problem (from [Dav96]).

special miniaturized proportional counters. These are then placed in a very low activity lead shielding and then the corresponding decay is observed. In order to further reduce the background, both the energy information of the decay and the pulse shape are used. A production rate of one argon atom per day corresponds to 5.35 SNU. The results from more than 20 yr measuring time are shown in figure 10.9. The average counting rate of 108 runs is [Cle98]

$$2.56 \pm 0.16(\text{stat.}) \pm 0.15(\text{sys.}) \text{ SNU.} \tag{10.36}$$

This is less than the value predicted by the standard solar models. This discrepancy is the primary source of the so-called *solar neutrino problem*.

10.2.2 Super-Kamiokande

A real-time experiment for solar neutrinos is being carried out with the Super-Kamiokande detector [Fuk03a], an enlarged follow-up version of the former Kamioka detector. This experiment is situated in the Kamioka mine in Japan and has a shielding depth of 2700 mwe. Super-Kamiokande started operation on 1 April 1996 and has already been described in detail in chapter 9. The fiducial mass used for solar neutrino searches is 22 kt. The detection principle is the Cerenkov light produced in neutrino-electron scattering within the water. Energy and directional information are reconstructed from the corresponding number and timing of the hit photomultipliers. The cross section for neutrino–electron scattering is given in chapter 3. From that the differential cross section with

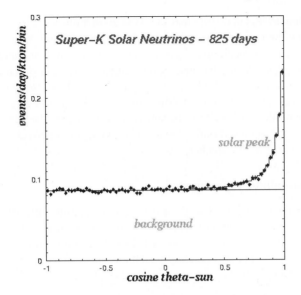

Figure 10.10. Angular distribution of the events in the Super-Kamiokande detector, relative to the direction of the Sun, after a measuring time of 1496 days (with kind permission of the SNO collaboration and Lawrence Berkeley National Laboratory).

respect to the scattering angle can be deduced as

$$\frac{\mathrm{d}\sigma}{\mathrm{d}\cos\theta} = 4\frac{m_e}{E_\nu}\frac{(1 + m_e/E_\nu)^2 \cos\theta}{[(1 + m_e/E_\nu)^2 - \cos^2\theta]^2}\frac{\mathrm{d}\sigma}{\mathrm{d}y} \qquad (10.37)$$

with

$$y = \frac{2(m_e/E_\nu)\cos^2\theta}{(1 + m_e/E_\nu)^2 - \cos^2\theta}. \qquad (10.38)$$

Therefore, for $E_\nu \gg m_e$ the electron keeps the neutrino direction with $\theta \lesssim (2m_e/E_\nu)^{1/2}$. This directional information is clearly visible as shown in figure 10.10.

Given the threshold of about 5 MeV, the detector can measure essentially only the ^{8}B neutrino flux. The experimental observation after 1496 days is shown in figure 10.10, resulting in $22\,400\pm800$ neutrino events. From the measurements a time-averaged flux of ^{8}B neutrinos of [Smy03]

$$\Phi(^8\text{B}) = 2.80 \pm 0.19 \pm 0.33 \times 10^6 \text{ cm}^{-2}\text{ s}^{-1} \qquad \text{Kamiokande (final)} \quad (10.39)$$
$$\Phi(^8\text{B}) = 2.35 \pm 0.02(\text{stat.}) \pm 0.08(\text{sys.}) \times 10^6 \text{ cm}^{-2}\text{ s}^{-1} \qquad \text{Super-K} \quad (10.40)$$

is obtained, equivalent to 46.5% of the SSM prediction of [Bah01]. A possible hep flux is smaller than

$$\Phi(\text{hep}) < 7.3 \times 10^4 \text{ cm}^{-2}\text{ s}^{-1} \qquad (10.41)$$

corresponding to less than 7.9 times the SSM prediction.

It should be noted that the observed flux at Super-Kamiokande for neutrino oscillations is a superposition of ν_e and ν_μ, ν_τ. The dominant number of events is produced by ν_e scattering because of their higher cross section (see chapter 4). The high statistics of Super-Kamiokande not only allows the total flux to be measured, they also yield more detailed information which is very important for neutrino oscillation discussions. In particular, it measures the spectral shape of the ^{8}B spectrum, annual variations in the flux and day/night effects and these will be discussed later in this chapter.

10.2.3 The gallium experiments

Both experiments described so far are unable to measure directly the pp flux, which is the reaction directly coupled to the Sun's luminosity (figure 10.6). A suitable material to detect these neutrinos is gallium [Kuz66]. There are currently two experiments operating which are sensitive to the pp neutrino flux: GALLEX/GNO and SAGE. The detection relies on the reaction

$$^{71}\text{Ga} + \nu_e \rightarrow {}^{71}\text{Ge} + e^- \qquad (10.42)$$

with a threshold energy of 233 keV. The natural abundance of ^{71}Ga is 39.9%. The detection reaction is via electron capture

$$^{71}\text{Ge} + e^- \rightarrow {}^{71}\text{Ga} + \nu_e \qquad (10.43)$$

resulting in Auger electrons and x-rays from K and L capture from the ^{71}Ge decay producing two lines at 10.37 keV and 1.2 keV. The detection of the ^{71}Ge decay (half-life 11.4 days, 100% via electron capture), is achieved using miniaturized proportional counters similar to the chlorine experiment. Both energy and pulse shape information are also used for the analysis.

10.2.3.1   GALLEX

The dominantly European group GALLEX (gallium experiment) [Ans92a, Ans95b] used 30.3 t of gallium in the form of 101 t of GaCl$_3$ solution. This experiment was carried out in the Gran Sasso underground laboratory from 1991 to 1997. The produced GeCl$_4$ is volatile in GaCl$_3$ and was extracted every three weeks by flushing nitrogen through the tank. Inactive carriers of ^{72}Ge, ^{74}Ge and ^{76}Ge were added to control the extraction efficiency, which was at 99%. The germanium was concentrated in several stages and subsequently transformed into germane (GeH$_4$), which has similar characteristics as methane (CH$_4$), which when mixed with argon, is a standard gas mixture (P10) in proportional counters. The germane is, therefore, also mixed with a noble gas (Xe) to act as a counter gas, the mixture being optimized for detection efficiency, drift velocity and energy resolution.

In addition, there was the first attempt to demonstrate the total functionality of a solar neutrino experiment using an artifical 2 MCi (!) ^{51}Cr source. This yielded mono-energetic neutrinos, of which 81% had $E_\nu = 746$ keV. This test was performed twice. The results of the two calibrations of GALLEX with the artifical chromium neutrino source resulted in the following ratios between the observed number of ^{71}Ge decays and the expectation from the source strength [Ans95a, Ham96a]:

$$R = 1.04 \pm 0.12 \quad \text{and} \quad 0.83 \pm 0.08. \tag{10.44}$$

This confirms the full functionality and sensitivity of the GALLEX experiment to solar neutrinos. At the end of the experiment ^{71}As, which also decays into ^{71}Ge, was added to study extraction with *in situ* produced ^{71}Ge. The final result of GALLEX is [Ham99]

$$77.5 \pm 6.2(\text{stat.})^{+4.3}_{-4.7}(\text{sys.}) \text{ SNU} \tag{10.45}$$

with theoretical predictions of 128 ± 8 SNU [Bah01, Cou02]. Clearly the experiment is far off.

10.2.3.2   GNO

After some maintenance and upgrades of the GALLEX equipment, the experiment was renewed in the form of a new collaboration—GNO. After 58 runs GNO reports a value of [Bel03]

$$62.9 \pm 5.4(\text{stat.}) \pm 2.5(\text{sys.}) \text{ SNU} \tag{10.46}$$

which combined with the 65 GALLEX runs averages to

$$69.3 \pm 4.1 \pm 3.6 \text{ SNU.} \tag{10.47}$$

The single run signal is shown in figure 10.11. Upgrades on the amount of Ga are under consideration.

10.2.3.3   SAGE

The Soviet–American collaboration SAGE [Gav03] uses 57 t of gallium in metallic form as the detector and has operated the experiment in the Baksan underground laboratory since 1990. The main difference with respect to GALLEX lies in the extraction of ^{71}Ge from metallic gallium.

The SAGE experiment was recently calibrated in a similar way to GALLEX [Abd99]. The current result of SAGE is [Gav03a]

$$69.1^{+4.3}_{-4.2} \text{ SNU.} \tag{10.48}$$

The data are shown in figure 10.12. Both gallium experiments are in good agreement and show fewer events than expected from the standard solar models.

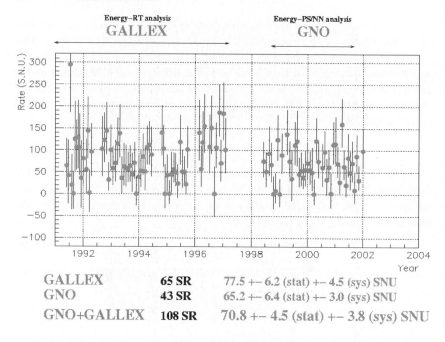

GALLEX	65 SR	77.5 +− 6.2 (stat) +− 4.5 (sys) SNU
GNO	43 SR	65.2 +− 6.4 (stat) +− 3.0 (sys) SNU
GNO+GALLEX	108 SR	70.8 +− 4.5 (stat) +− 3.8 (sys) SNU

Figure 10.11. Results from 65 GALLEX runs and 43 GNO runs (from [Kir03]).

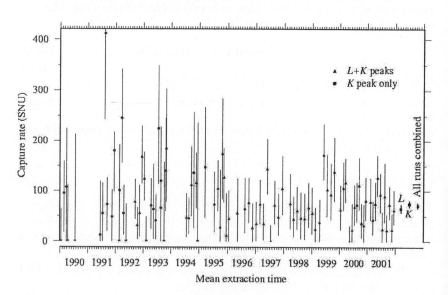

Figure 10.12. Davis plot of 11 years of data-taking with SAGE.

Figure 10.13. Construction of the Sudbury neutrino observatory (SNO) in a depth of 2070 m in the Craighton mine near Sudbury (Ontario). This Cerenkov detector uses heavy water rather than normal water. The heavy water tank is shielded by an additional 7300 t of normal water. The support structure for the photomultipliers is also shown (with kind permission of the SNO collaboration).

Both GALLEX and SAGE provide the first observation of pp neutrinos and an experimental confirmation that the Sun's energy really does come from hydrogen fusion.

10.2.4 The Sudbury Neutrino Observatory (SNO)

The most recent experiment in solar neutrino detection is the Sudbury Neutrino Observatory which has run since 1999. Being a real-time Cerenkov detector like Super-Kamiokande, this experiment uses 1000 t of heavy water (D_2O) instead of H_2O. Placed in a transparent acrylic tank, it is surrounded by 9700 photomultipliers and several kilotons of H_2O as shielding [Bog00] (figure 10.13). The threshold of SNO is about 5 MeV. Heavy water allows several reactions to be studied. The first one is charged weak currents only sensitive to ν_e:

$$\nu_e + d \rightarrow e^- + p + p \quad \text{(CC)} \quad (10.49)$$

with a threshold of 1.442 MeV. In addition to this absorption reaction on deuterium (10.49), a second process for detecting all types of neutrinos the elastic scattering

$$\nu + e^- \rightarrow \nu + e^- \quad \text{(ES)} \quad (10.50)$$

which is dominated by ν_e scattering can be used as in Super-Kamiokande. The remarkable aspect, however, is the additional determination of the *total* neutrino flux, independent of any oscillations, due to the flavour-independent reaction via neutral weak currents

$$\nu + d \rightarrow \nu + p + n \quad \text{(NC)} \quad (10.51)$$

which has a threshold of 2.225 MeV. Cross-section calculations can be found in [Nak02]. The produced neutrons are detected via 6.3 MeV gamma-rays produced in the reaction

$$n + d \rightarrow {}^3\text{H} + \gamma. \quad (10.52)$$

To enhance the detection efficiency of neutrons and therefore improve on the NC flux measurement, two strategies are envisaged: Cl in the form of 2 t of NaCl has been added to the heavy water to use the gamma-rays up to 8.6 MeV produced in the ${}^{35}\text{Cl}(n, \gamma)\ {}^{36}\text{Cl}$ process and a set of He-filled proportional counters will be deployed to perform neutron detection via

$$^3\text{He} + n \rightarrow {}^3\text{H} + p \quad (10.53)$$

which will allow NC and CC reactions to be discriminated on an event-by-event basis. By comparing the rate of the two processes (equations (10.49) and (10.51)) it is directly possible to test the oscillation hypothesis. However, a comparison of the CC absorption process with the scattering one also provides very important information. As mentioned, ν_μ and ν_τ also contribute to scattering but with a lower cross section (see chapter 4) and the ratio is given by

$$\frac{CC}{ES} = \frac{\nu_e}{\nu_e + 0.14(\nu_\mu + \nu_\tau)}. \quad (10.54)$$

The first measurement by SNO of the CC reaction resulted in [Ahm01]

$$\Phi(^8\text{B}) = 1.75 \pm 0.07(\text{stat.})^{+0.12}_{-0.11}(\text{sys.}) \pm 0.05(\text{theor.}) \times 10^6 \text{ cm}^{-2} \text{ s}^{-1} \quad (10.55)$$

significantly less than that of Super-Kamiokande. This is already a hint that additional active neutrino flavours come from the Sun and participate in the scattering process. The real breakthrough came with the measurement of the first NC data [Ahm02]. This flavour-blind reaction indeed measured a total solar neutrino flux of

$$\Phi_{NC} = 5.09^{+0.44}_{-0.43}(\text{stat.})^{+0.46}_{-0.43}(\text{Sys.}) \times 10^6 \text{ cm}^{-2} \text{ s}^{-1} \qquad (10.56)$$

in excellent agreement with the standard model (figure 10.14). The flux in non-ν_e neutrinos is

$$\Phi(\nu_\mu, \nu_\tau) = 3.41 \pm 0.45 \pm 0.43 \times 10^6 \text{ cm}^{-2} \text{ s}^{-1} \qquad (10.57)$$

on a 5.3σ level different from zero. This result is in excellent agreement with the recent measurement using the data set including salt and therefore having an enhanced NC sensitivity. Here, a total NC flux assuming no constraint on the spectral shape of [Ahm03]

$$\Phi_{NC}^{\text{Salt}} = 5.21 \pm 0.27(\text{stat.}) \pm 0.38(\text{sys.}) \times 10^6 \text{ cm}^{-1} \text{ s}^{-1}$$

and assuming a ^{8}B spectral shape of

$$\Phi_{NC}^{\text{Salt}} = 4.90 \pm 0.24^{+0.29}_{-0.27} \times 10^6 \text{ cm}^{-1} \text{ s}^{-1}$$

have been obtained. The CC/NC ratio is given by $0.036 \pm 0.026(\text{stat.}) \pm 0.024(\text{sys.})$ and maximal mixing is rejected at the 5.40 level. A global fit determines as best-fit point values of $\Delta m^2 = 7.1^{+2.1}_{-0.6} \times 10^{-5}$ ev^2 and $\theta = 32.5^{+2.4}_{-2.3}$ degrees. After several decades it now seems unavoidable to blame neutrino properties for the solar neutrino problem. It is no longer a problem of missing neutrinos but a fact that the bulk of solar neutrinos arrive at the Earth in the wrong flavour.

10.3 Attempts at theoretical explanation

All experiments so far indicate a deficit of solar neutrinos compared to the theoretically predicted flux (see table 10.2). If we accept that there is a real discrepancy between experiment and theory, there are two main solutions to the problem. One is that our model of the Sun's structure may not be correct or our knowledge of the neutrino capture cross sections may be insufficient, the other is the possibility that the neutrino has as yet unknown properties. The problem is basically solved by the new SNO results, because there is no way to produce ν_μ or ν_τ in the Sun and the total observed flux agrees well with the SSM predictions (figure 10.15). Let us discuss two possible sources for the flavour conversion—neutrino oscillations and a neutrino magnetic moment.

10.3.1 Neutrino oscillations as a solution to the solar neutrino problem

Two kinds of solutions are provided by oscillations. Either the ν_e oscillates in vacuum on its way to Earth or it has already been converted within the Sun by

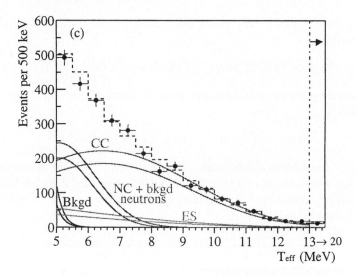

Figure 10.14. Kinetic energy T_{eff} of events with a vertex within a fiducial volume of 550 cm radius and $T_{\text{eff}} > 5$ MeV. Also shown are the Monte Carlo predictions for neutral currents (NC) and background neutrons, charged currents (CC) and elastic scattering (ES), scaled to the fit results. The broken lines represent the summed components and the bands show $\pm 1\sigma$ uncertainties (from [Ahm02]).

matter effects (see chapter 8). In the case of vacuum solutions the baseline is about 1.5×10^8 km and E_ν about 10 MeV, resulting in Δm^2 regions of around 10^{-10} eV^2. The other attractive solution is oscillation in matter and conversion via the Mikheyev, Smirnov and Wolfenstein (MSW) effect [Wol78, Mik86].

10.3.2 Neutrino oscillations in matter and the MSW effect

Matter influences the propagation of neutrinos by elastic, coherent forward scattering. The basic idea of this effect is the differing interactions of different neutrino flavours within matter. While interactions with the electrons of matter via neutral weak currents are possible for all kinds of neutrinos, only the ν_e can interact via charged weak currents (see figure 10.16). The CC for the interaction with the electrons of matter leads to a contribution to the interaction Hamiltonian of

$$H_{WW} = \frac{G_F}{\sqrt{2}} [\bar{e}\gamma^\mu (1 - \gamma_5)\nu_e][\bar{\nu}_e\gamma_\mu (1 - \gamma_5)e]. \qquad (10.58)$$

By a Fierz transformation this term can be brought to the form

$$H_{WW} = \frac{G_F}{\sqrt{2}} [\bar{\nu}_e\gamma^\mu (1 - \gamma_5)\nu_e][\bar{e}\gamma^\mu (1 - \gamma_5)e]. \qquad (10.59)$$

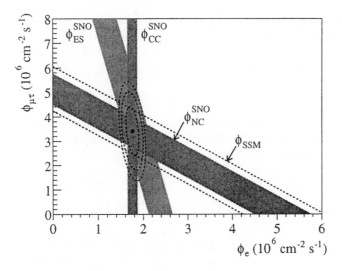

Figure 10.15. Flux of ^{8}B solar neutrinos which are μ or τ flavour *versus* flux of ν_e deduced from the three neutrino reactions in SNO. The diagonal bands show the total ^{8}B flux as predicted by the SSM (broken lines) and that measured with the NC reaction at SNO (full line). The intercepts of these bands with the axes represent the $\pm 1\sigma$ errors (from [Ahm02]).

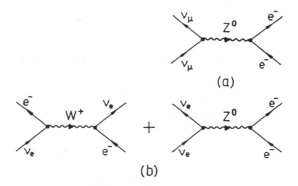

Figure 10.16. Origin of the Mikheyev–Smirnov–Wolfenstein effect. Whereas weak NC interactions are possible for all neutrino flavours, only the ν_e also has the possibility of interacting via charged weak currents.

Calculating the four-current density of the electrons in the rest frame of the Sun, we obtain

$$\langle e|\bar{e}\gamma^i (1 - \gamma_5)e|e\rangle = 0 \tag{10.60}$$
$$\langle e|\bar{e}\gamma^0 (1 - \gamma_5)e|e\rangle = N_e. \tag{10.61}$$

The spatial components of the current must disappear (no permanent current density throughout the Sun) and the zeroth component can be interpreted as the electron density of the Sun. For left-handed neutrinos we can replace $(1 - \gamma_5)$ by a factor of 2, so that equation (10.59) can be written as

$$H_{WW} = \sqrt{2}G_F N_e \bar{\nu}_e \gamma_0 \nu_e. \tag{10.62}$$

Thus the electrons contribute an additional potential for the electron neutrino $V = \sqrt{2}G_F N_e$. With this additional term, the free energy–momentum relation becomes (see chapter 7)

$$p^2 + m^2 = (E - V)^2 \simeq E^2 - 2EV \qquad \text{(for } V \ll E). \tag{10.63}$$

In more practical units this can be written as [Bet86]

$$2EV = 2\sqrt{2}\left(\frac{G_F Y_e}{m_N}\right)\rho E = A \tag{10.64}$$

where ρ is the density of the Sun, Y_e is the number of electrons per nucleon and m_N is the nucleon mass. In analogy to the free energy–momentum relation an effective mass $m_{\text{eff}}^2 = m^2 + A$ can be introduced which depends on the density of the solar interior. In the case of two neutrinos ν_e and ν_μ in matter, the matrix of the squares of the masses of ν_e and ν_μ in matter has the following eigenvalues for the two neutrinos $m_{1m,2m}$:

$$m_{1m,2m}^2 = \tfrac{1}{2}(m_1^2 + m_2^2 + A) \pm [(\Delta m^2 \cos 2\theta - A)^2 + \Delta m^2 \sin^2 2\theta]^{1/2} \tag{10.65}$$

where $\Delta m^2 = m_2^2 - m_1^2$. The two states are closest together for

$$A = \Delta m^2 \cos 2\theta \tag{10.66}$$

which corresponds to an electron density of

$$N_e = \frac{\Delta m^2 \cos 2\theta}{2\sqrt{2}G_F E} \tag{10.67}$$

(see also [Bet86, Gre86a, Sch97]). At this point (the resonance region) the oscillation amplitude is maximal, meaning ν_e and ν_μ oscillation between the extremes 0 and 1, independent of the vacuum mixing angle. Also the oscillation length has a maximum of

$$L_{mR} = \frac{L_0}{\sin 2\theta} = \frac{1.64 \times 10^7 m}{\tan 2\theta \times Y_e \rho / \text{g cm}^{-3}} \tag{10.68}$$

with $L_0 = L_C \cos 2\theta$ at the resonance. L_C, the scattering length of coherent forward scattering, is given as

$$L_C = \frac{4\pi E}{A} = \sqrt{2}\pi G_F N_e = \frac{\sqrt{2}\pi m_N}{G_F Y_e \rho} \tag{10.69}$$

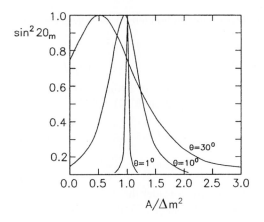

Figure 10.17. Dependence of the oscillation amplitude as a function of $A/\Delta m^2$ for different values of the vacuum mixing angle θ ($\theta = 1°, 10°, 30°$).

which can be written numerically as

$$L_C = 1.64 \times 10^7 \frac{N_A}{N_e/\text{cm}^{-3}} \text{ m} = \frac{1.64 \times 10^7}{Y_e \rho/\text{g cm}^{-3}} \text{ m}. \tag{10.70}$$

The oscillation length is stretched by a factor $\sin^{-1} 2\theta$ with respect to vacuum. The width (FWHM) of the resonance

$$\Gamma = 2\Delta m^2 \sin 2\theta \tag{10.71}$$

becomes broader for larger mixing angles θ. It is worthwhile mentioning that such a resonance cannot occur for antineutrinos ($A \rightarrow -A$) because now in the denominator of (8.63) $A/\Delta m^2 + \cos 2\theta$ cannot vanish ($0 < \theta < \pi/4$ for $\Delta m^2 > 0$).

10.3.2.1 Constant density of electrons

The energy difference of the two neutrino eigenstates in matter is modified compared to that in the vacuum by the effect discussed in the previous section to become

$$(E_1 - E_2)_m = C \cdot (E_1 - E_2)_V \tag{10.72}$$

where C is given by

$$C = \left[1 - 2 \left(\frac{L_V}{L_e} \right) \cos 2\theta_V + \left(\frac{L_V}{L_e} \right)^2 \right]^{\frac{1}{2}} \tag{10.73}$$

and the neutrino–electron interaction length L_e is given by

$$L_e = \frac{\sqrt{2}\pi\hbar c}{G_F N_e} = 1.64 \times 10^5 \left(\frac{100 \text{ g cm}^{-3}}{\mu_e \rho}\right) \qquad \text{[m]}. \qquad (10.74)$$

The equations describing the chance of finding another flavour eigenstate after a time t correspond exactly to equation (8.24) with the additional replacements

$$L_m = \frac{L_V}{C} \qquad (10.75)$$

$$\sin 2\theta_m = \frac{\sin 2\theta_V}{C}. \qquad (10.76)$$

In order to illustrate this, we consider the case of two flavours in three limiting cases. Using equations (8.24) and (10.73) the oscillation of ν_e into a flavour ν_x is given by

$$|\langle \nu_x | \nu_e \rangle|^2 = \begin{cases} \sin^2 2\theta_V \sin^2(\pi R/L_V) & \text{for } L_V/L_e \ll 1 \\ (L_e/L_V)^2 \sin^2 2\theta_V \sin^2(\pi R/L_e) & \text{for } L_V/L_e \gg 1 \\ \sin^2(\pi R \sin 2\theta_V/L_V) & \text{for } L_V/L_e = \cos 2\theta_V. \end{cases} \qquad (10.77)$$

The last case corresponds exactly to the resonance condition mentioned earlier. In the first case, corresponding to very small electron densities, the matter oscillations reduce themselves to vacuum oscillations. In the case of very high electron densities, the mixture is suppressed by a factor $(L_e/L_V)^2$. The third case, the resonance case, contains an energy-dependent oscillatory function, whose energy average results typically in a value of 0.5. This corresponds to maximal mixing. In a medium with constant electron density N_e the quantity A is constant for a fixed E and, in general, does not fulfil the resonance condition. Therefore, the effect described here does not show up. However, in the Sun we have varying density which implies that there are certain resonance regions where this flavour conversion can happen.

10.3.2.2 Variable electron density

A variable density causes a dependence of the mass eigenstates $m_{1m,2m}$ on A (N_e) which is shown in figure 10.18. Assume the case $m_1^2 \approx 0$ and $m_2^2 > 0$ which implies $\Delta m^2 \approx m_2^2$. For $\theta = 0$, resulting in $\theta_m = 0$ as well for all A, this results in

$$\begin{aligned} \nu_{1m} = \nu_1 = \nu_e & \quad \text{with} \quad m_{1m}^2 = A \\ \nu_{2m} = \nu_2 = \nu_\mu & \quad \text{with} \quad m_{2m}^2 = m_2^2. \end{aligned} \qquad (10.78)$$

The picture changes for small $\theta > 0$. Now for $A = 0$ the angle $\theta_m = \theta$ which is small and implies

$$\begin{aligned} \nu_{1m} = \nu_1 \approx \nu_e & \quad \text{with} \quad m_{1m}^2 = 0 \\ \nu_{2m} = \nu_2 \approx \nu_\mu & \quad \text{with} \quad m_{2m}^2 = m_2^2. \end{aligned} \qquad (10.79)$$

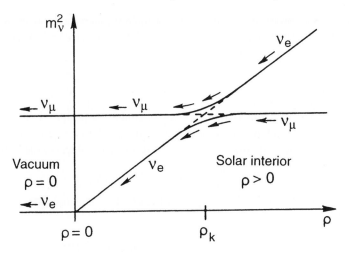

Figure 10.18. The MSW effect. The heavy mass eigenstate is almost identical to ν_e inside the Sun: in the vacuum, however, it is almost identical to ν_μ. If a significant jump at the resonance location can be avoided, the produced electron neutrino remains on the upper curve and therefore escapes detection in radiochemical experiments (after [Bet86]).

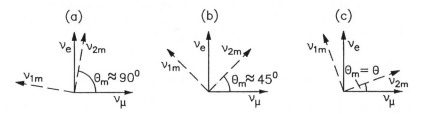

Figure 10.19. Representation of the three phases of the MSW effect in the (ν_e, ν_μ) plane: (a) $\theta_m \approx 90°$ in the solar interior; (b) $\theta_m \approx 45°$ in the resonance layer; and (c) $\theta_m \approx 0°$ at the surface of the Sun.

For large A there is $\theta_m \approx 90°$ and the states are given as

$$\nu_{1m} \approx -\nu_\mu \quad \text{with} \quad m^2_{1m} \approx m^2_2$$
$$\nu_{2m} \approx \nu_e \quad \text{with} \quad m^2_{2m} \approx A \tag{10.80}$$

opposite to the $\theta = 0$ case (figure 10.19). This implies an inversion of the neutrino flavour. While ν_{1m} in vacuum is more or less ν_e, at high electron density it corresponds to ν_μ, the opposite is valid for ν_{2m}. This flavour flip is produced by the resonance where maximal mixing is possible.

Solar neutrinos are produced in the interior of the Sun, where the density is rather high $\rho \approx 150$ g cm^{-3}. Therefore, assuming $A/\Delta m^2 \gg 1$ equivalent to $\theta_m \approx 90°$, the produced ν_e are basically identical to ν_{2m}, the heavier mass eigenstate. A ν_e produced in the interior of the Sun, therefore, moves along the upper curve and passes a layer of matter where the resonance condition is fulfilled. Here maximal mixing occurs, $\theta_m \approx 45°$, and

$$\nu_{2m} = \frac{1}{\sqrt{2}}(\nu_e + \nu_\mu). \tag{10.81}$$

Passing the resonance from right to left and remaining on the upper curve, the state ν_{2m} at the edge of the Sun is now associated with ν_μ. The average probability that a ν_e produced in the solar interior, passes the resonance and leaves the Sun still as ν_e is given by

$$P(\nu_e \rightarrow \nu_e) = \tfrac{1}{2}(1 + \cos 2\theta_m \cos 2\theta) \tag{10.82}$$

with θ_m as the mixing angle at the place of neutrino production. The conversion is, therefore,

$$P(\nu_e \rightarrow \nu_\mu) = \tfrac{1}{2}(1 - \cos 2\theta_m \cos 2\theta) \approx \cos^2 \theta. \tag{10.83}$$

The smaller the vacuum mixing angle is, the larger the flavour transition probability becomes. To make use of the MSW mechanism two further conditions have to be fulfilled.

Existence of an MSW resonance. First of all, the point of neutrino production (N_e) has to be above the resonance, otherwise the neutrino will never experience such a resonance. Expressed in neutrino energies, this requires

$$E > E_R = \frac{\Delta m^2 \cos 2\theta}{2\sqrt{2}G_F N_e} = 6.6 \times 10^6 \frac{\Delta m^2/\text{eV}^2}{Y_e \rho/\text{g cm}^{-3}} \cos 2\theta \ \text{MeV}. \tag{10.84}$$

Using for the solar interior $\rho = 150 \ \text{g cm}^{-3}$ and $Y_e \approx 0.7$ and a small mixing angle $\cos \theta \approx 1$, this corresponds to

$$E_R = 6.3 \times 10^4 \cos 2\theta \frac{\Delta m^2}{\text{eV}^2} \ \text{MeV}. \tag{10.85}$$

As a result solar neutrinos of $E_\nu = 10$ MeV (0.1 MeV) pass through a resonance and experience the MSW mechanism if $\Delta m^2 < 1.6 \times 10^{-4} \ \text{eV}^2$ ($<1.6 \times 10^{-6} \ \text{eV}^2$).

Adiabaticity. Another requirement for a produced state ν_{2m} to remain in its eigenstate is adiabaticity, especially important in the resonance region. It requires a slow variation of the electron density $N_e(r)$ along the neutrino path, in such a way that it can be considered as constant over several oscillation lengths L_m. Quantitatively, this condition can be discussed with the help of an adiabaticity parameter γ, defined as

$$\gamma = \frac{\Delta m^2}{2 E x_R} \frac{\sin^2 2\theta}{\cos 2\theta} = 2.53 \frac{\Delta m^2/\text{eV}^2}{E/\text{MeV} x_R/\text{m}^{-1}} \frac{\sin^2 2\theta}{\cos 2\theta} \tag{10.86}$$

with

$$x_R = \left| \frac{1}{N_e} \frac{dN_e}{dr} \right|_R = \left| \frac{d \ln N_e}{dr} \right|_R \qquad (10.87)$$

x_R is the relative change in density per unit length within the resonance layer R where it is especially important. Assuming a density profile for the Sun of the form

$$N_e(r) \propto \exp\left(-\frac{10.5r}{R_\odot}\right) \rightarrow x_R = \frac{10.5}{R_\odot} = 1.50 \times 10^{-8} \text{ m}^{-1} \qquad (10.88)$$

the adiabaticity parameter γ is given in the case of the Sun as

$$\gamma = 1.69 \times 10^8 \frac{\Delta m^2/\text{eV}^2}{E/\text{MeV}} \frac{\sin^2 2\theta}{\cos 2\theta}. \qquad (10.89)$$

The condition for adiabaticity is, therefore, $\gamma \gg 1$, meaning

$$\gamma \gg 1 \qquad \text{i.e. } E \ll \frac{\Delta m^2/\text{eV}^2}{2x_R} \frac{\sin^2 2\theta}{\cos 2\theta}. \qquad (10.90)$$

The non-adiabatic region is defined as $\gamma \lesssim 1$. The transition from adiabatic to non-adiabatic behaviour happens in the range of a critical energy E_c, defined in the case of the Sun as

$$\frac{\pi}{2}\gamma_c = 1 \qquad \text{i.e. } E_c = \frac{\pi \Delta m^2/\text{eV}^2}{4x_R} \frac{\sin^2 2\theta}{\cos 2\theta} = 2.65 \times 10^8 \frac{\Delta m^2}{\text{eV}^2} \frac{\sin^2 2\theta}{\cos 2\theta}. \qquad (10.91)$$

The energy range for the adiabatic MSW mechanism in the Sun is, therefore, restricted from below by E_0 and from above by E_c. The existence of an adiabatic region at all requires $E_0 < E_c$ resulting in a minimal mixing angle, $\sin^2 2\theta \gtrsim 2.4 \times 10^{-4}$. For an exact description the neutrino state has to be tracked along its path through the Sun. Its behaviour on x_R in the resonance region is especially important. To study the behaviour there, the density $\rho(r)$ is replaced by simpler integrable density profiles, which reproduce the true solar density and its derivative in the resonance region. Behaviours like $\rho(r) = a + br$ or $\rho(r) = a \exp(-br)$ which can be integrated exactly are convenient for this. For the linear change in electron density the transition probability from ν_{2m} into ν_{1m} in the resonance region is then given by

$$P_c = \exp\left(-\frac{\pi}{2}\gamma\right). \qquad (10.92)$$

This is called the Landau–Zener probability and is well known from atomic physics. The general survival probability of ν_e in the Sun is given by (Parke formula)

$$P(\nu_e \rightarrow \nu_e) = \tfrac{1}{2}(1 + (1 - 2P_c \cos 2\theta_m \cos 2\theta)) \qquad \text{with } \cos 2\theta_m \approx 1$$
$$\approx \sin^2 \theta + P_c \cos 2\theta. \qquad (10.93)$$

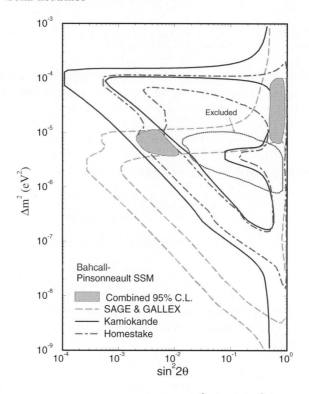

Figure 10.20. Contour ('Iso-SNU') plot in the Δm^2 *versus* $\sin^2 2\theta$ plane. For each experiment the total rate defines a triangular-shaped region as an explanation of the results. The different energy thresholds cause a shift of the curves and only the overlap regions describe all data. Additional information like day–night effects constrain the regions further.

The adiabatic limit is obtained for $\gamma \gg 1$, $P_c = 0$. The extreme non-adiabatic limit ($\gamma \ll 1$, $P_c = 1$) is

$$P(\nu_e \rightarrow \nu_e) = \tfrac{1}{2}(1 - \cos 2\theta_m \cos 2\theta) \approx \cos^2 \theta \approx 1 \qquad (10.94)$$

meaning flavour conversion no longer occurs.

10.3.3   Experimental signatures and results

Having extensively discussed the MSW effect, we now want to examine which parameter space is consistent with the experimental results. Every experiment measures the probability P ($\nu_e \rightarrow \nu_e$) which manifests itself in a triangular-shaped Iso-SNU band in the Δm^2–$\sin^2 2\theta$ plot (figure 10.20). Because of the different energy intervals investigated by the experiments, the bands are shifted against each other, but further information is available. An energy-dependent suppression could be visible in two observables: a distortion in the ^{8}B β-spectrum and a day–night effect. The MSW effect could occur on Earth during the night, if the neutrinos have to travel through matter, just as it does on the Sun. The density in the mantle is 3–5.5 g cm^{-3} and that of the core 10–13 g cm^{-3}. This density change is not big enough to allow for the full MSW mechanism but for fixed E_ν there is a region in Δm^2 where the resonance condition is fulfilled. Taking $E_\nu = 10$ MeV, $\rho \approx 5$ g cm^{-3} and $Y_e \approx 0.5$, a value of $\Delta m^2 \approx 4 \times 10^{-6}$ eV^2 for $\cos 2\theta \approx 1$ results. Because of the now strong oscillations a reconversion of ν_μ or ν_τ to ν_e can result (ν_e regeneration). Therefore, the measured ν_e flux should be higher at night than by day (day–night effect). For the same reason there should be an annual modulation between summer and winter, because neutrinos have to travel shorter distances through the Earth in summer than in winter time. However, this effect is smaller than the day–night effect. One additional requirement exists for the day–night effect to occur, namely the resonance oscillation length L_{mR} should be smaller than the Earth's diameter; therefore, $\sin^2 2\theta$ should not be too small. Taking these values for ρ and Y_e, it follows that, at resonance, half of the oscillation length is smaller than the Earth's diameter results if $\sin^2 2\theta \gtrsim 0.07$.

Super-Kamiokande and SNO did not observe any of these effects. The day–night effect using 1496 live days is given by Super-Kamiokande as [Fuk02, Smy03a]

$$A = 2\frac{N-D}{N+D} = -0.018 \pm 0.016(\text{stat.})^{+0.013}_{-0.012}(\text{sys.}) \qquad (10.95)$$

and in SNO at flux ratios >5 MeV [Ahm02a]

$$A = +14 \pm 6.3(\text{stat.})^{+1.5}_{-1.4}(\text{sys.})\% \qquad \text{CC alone} \qquad (10.96)$$

$$A = +7.0 \pm 4.9(\text{stat.})^{+1.3}_{-1.2}(\text{sys.})\% \qquad \text{NC} \qquad (10.97)$$

where, in the latter case, no asymmetry in the total flux was assumed. Also no spectral deformation of the ^{8}B spectrum is visible (figure 10.21).

Taking all the data together and performing combined fits for the MSW mechanism (figure 10.22) two solutions are favoured:

- the large mixing angle (LMA) solution with $\Delta m^2 \approx 7 \times 10^{-5}$ eV^2 and $\sin^2 2\theta \approx 1$; and
- the long oscillation wavelength (LOW) solution with $\Delta m^2 \approx 10^{-7}$ eV^2 and $\sin^2 2\theta \approx 1$.

Using the KamLAND result as well fixes the LMA as the correct solution: the nice agreement between these two completely independent measurements is shown in figure 10.23.

10.3.4 The magnetic moment of the neutrino

Another possible source for flavour conversion of solar neutrinos could be that as a result of a magnetic moment μ_ν of the neutrino (see chapter 4), the magnetic

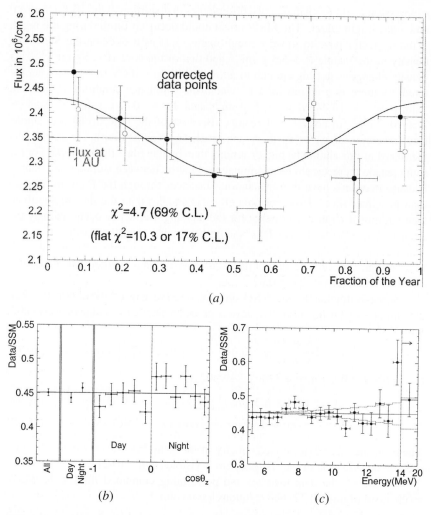

Figure 10.21. The high statistics of Super-Kamiokande allows a search for various effects on the ^{8}B spectrum. (*a*) Seasonal effects: Only the annual modulation of the flux due to the eccentricity of the Earth orbit could be observed (full line). Vacuum oscillations would have produced an additional effect. (*b*) Day–night effect: The solar zenith angle (θ_e) dependence of the neutrino flux normalized to the SSM prediction. The width of the night-time bins was chosen to separate solar neutrinos that pass through the Earth core ($\cos\theta_e > 0.84$) from those that pass through the mantle. (*c*) Spectral distortions: The measured ^{8}B and hep spectrum relative to the SSM predictions using the spectral shape from [Ort00]. The data from 14–20 MeV are combined in a single bin. The horizontal full line shows the measured total flux, while the dotted band around this line indicates the energy correlated uncertainty.

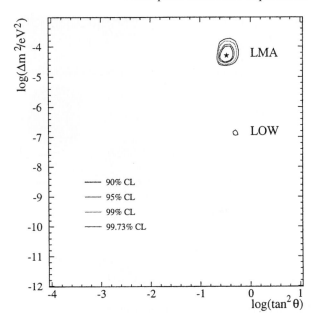

Figure 10.22. Regions in the Δm^2–$\tan^2 2\theta$ describing all obtained solar neutrino observation. As can be seen the LMA and LOW solutions are the only ones remaining with a strong preference of the LMA solution. The use of $\tan^2 2\theta$ instead of $\sin^2 2\theta$ stems from the following fact: The transition probability for vacuum oscillations is symmetric under $\Delta m^2 \rightarrow -\Delta m^2$ or $\theta \rightarrow \theta + \pi/4$. However, the MSW transition is only symmetric under simultaneous transformations $(\Delta m^2, \theta) \rightarrow (-\Delta m^2, \theta \pm \pi/4)$ (see [Fog99, deG00]). For $\Delta m^2 > 0$ resonance is only possible for $\theta < \pi/4$ and thus traditionally MSW solutions were plotted in $(\Delta m^2, \sin^2 2\theta)$. In principle, solutions are possible for $\theta > \pi/4$. To account for that, $\tan^2 2\theta$ is used now (from [Ahm02a]).

field of the Sun leads to a spin precession which transforms left-handed neutrinos into *sterile* right-handed ones (sterile because the right-handed state does not take part in the weak interaction). The stronger the solar activity, expressing itself, e.g., in the number of Sunspots, is the stronger the magnetic field and the more effective the transformation of neutrinos would be; hence, they would not show up in the experiments on Earth, producing a smaller rate in the radiochemical detectors. The transformation probability is given by

$$P(\nu_{e_L} \rightarrow \nu_{e_R}) = \sin^2 \int_0^R \mu B(r') \, dr'. \tag{10.98}$$

Assuming the thickness of the convection zone x is about 2×10^{10} cm, a magnetic moment of about 10^{-10}–$10^{-11} \mu_B$ [Moh91] would be adequate to solve the solar neutrino problem via spin precession in a typical magnetic field of 10^3 G.

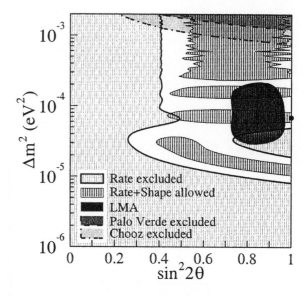

Figure 10.23. Regions in the Δm^2–$\sin^2 2\theta$ plot describing the observed deficit of KamLAND. As can be seen, there is an overlap of allowed parameters with the LMA solution of solar neutrino results (from [Egu03]).

However, the pure spin–flavour precession $\nu_{eL} \rightarrow \nu_{eR}$, which is sterile with respect to weak interactions, in solar magnetic field cannot explain the data. The reason is that the SNO NC data require a dominantly active neutrino flavour and in addition only an energy-independent suppression is obtained.

In the general case flavour transformations are also possible, for example transitions of the form $\nu_e \rightarrow \bar{\nu}_\mu$ or $\bar{\nu}_\tau$ for Majorana neutrinos or $\nu_{eL} \rightarrow \nu_{\mu R}$ or $\nu_{\tau R}$ for Dirac neutrinos [Lim88]. This is caused by the so-called 'transition magnetic moments' which are non-diagonal, and are also possible for Majorana neutrinos (see chapter 6). By allowing spin–flavour precession (SFP) like $\nu_{eL} \rightarrow \bar{\nu}_{\mu R}$, it has been shown that a resonance behaviour in matter can occur (resonant spin flavour precession, RSFP) (see [Akh88, Lim88]). The transition probability can be written for an uniform magnetic field and matter of constant density as [Akh97]

$$P(\nu_{eL} \rightarrow \bar{\nu}_{\mu R}; r) = \frac{(2\mu B_\perp)^2}{(\Delta m^2/2E - \sqrt{2}G_F N_{\text{eff}})^2 + (2\mu B_\perp)^2} \sin^2\left(\frac{1}{2}\sqrt{D}r\right)$$

(10.99)

where D is the denominator of the pre-sine factor and N_{eff} is given by $N_e - N_n/2$ (Dirac) and $N_e - N_n$ (Majorana) respectively. The resonance condition here is

$$\sqrt{2}G_F(N_e - N_n) = \frac{\Delta m^2}{2E}$$

(10.100)

quite similar to the MSW condition (10.67). Typical values are $\Delta m^2 \approx 10^{-8}$ –

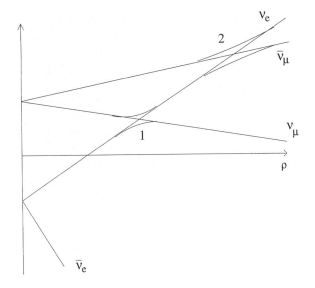

Figure 10.24. Schematic drawing of the two resonances which can occur for neutrinos coming from the solar interior. First, they experience the RSFP resonance and afterwards, if the conversion is not complete, they go through the MSW resonance.

10^{-7} eV^2 and $B = 10$–50 kG for $\mu_\nu = 10^{-11}\mu_B$. The most general case is the occurrence of both resonances, RSFP and MSW (figure 10.24).

By transversing the Sun, neutrinos first undergo SFP and, in the case of an incomplete conversion, then the MSW resonance. Depending on the involved Δm^2 and E, the predicted conversion probability can be quite complicated and detailed predictions for the experiments depend on the chosen parameters. It is interesting to note that in the case of adiabacity in the RSFP scenario an MSW resonance will never occur. A detailed discussion can be found in [Akh97]. Support for this scenario could come from the detection of solar $\bar{\nu}_e$ which can be produced via $\nu_{eL} \rightarrow \bar{\nu}_{\mu R} \rightarrow \bar{\nu}_{eR}$. An analysis using Super-Kamiokande data results in a limit on a possible $\bar{\nu}_e$ flux of less than 0.8% of the ^{8}B ν_e flux [Gan03]. SNO will be able to measure this flux via the reaction

$$\bar{\nu}_e + D \rightarrow e^+ + n + n \qquad (10.101)$$

and a threshold of 4.03 MeV and KamLAND by using inverse β-decay as for their reactor neutrino detection (see chapter 8). Taking the LMA as the solution, it turns out that the required density for RSFP is beyond the central density of the Sun and cannot occur. Any expected antineutrino flux is then of the order of 0.5% of the SSM flux and below [Akh03]. A further point often discussed in this context is whether the production rate in the chlorine experiment shows an anticorrelation with sun-spot activity over a period of 11 years. As sun-spots are a phenomenon which is connected to the magnetic field of the convection zone, while solar neutrinos come from the inner regions, a connection seems to be comprehensible only if a magnetic moment for neutrinos is allowed. Statements concerning an anticorrelation between the number of detected solar neutrinos and Sun-spot activity are so far inconclusive [Dav94a, Dav94b, Dav96]. In addition to this 11-year variation there should also be a half-yearly modulation [Vol86], which is connected to the fact that the rotation axis of the Sun is tilted by about 7° to the vertical of the Earth's orbit, and that we therefore cross the equatorial plane twice a year. The magnetic field in the solar interior is nearly zero in the equatorial plane, which then leads to a considerably smaller precession.

A fit to all experimental data show that the LMA solution is realized in nature and the MSW effect is at work [Bah03a, Ahm02a]. However, from the typical 80 fit parameters, all but three, namely the integral rates of the chlorine and gallium experiments, rely on only about 2% of the total solar neutrino flux. Therefore, it is desirable to obtain the same amount of information for low-energy neutrinos. This requires real-time measurements in the region below 1 MeV.

10.4   Future experiments

Great efforts are being made to improve measurements of the solar neutrino spectrum due to its importance to both astrophysics and elementary particle physics. The discrimination power among the various solutions is shown by the survival probability as a function of E_ν in figure 10.25. It is clearly visible that the strongest dependence shows up in the low-energy range ($E_\nu \lesssim 1$ MeV). Therefore, a real-time measurement in this region is crucial. In addition, the measurement of the ratio

$$R = \frac{\langle {}^3\text{He} + {}^4\text{He}\rangle}{\langle {}^3\text{He} + {}^3\text{He}\rangle} = \frac{2\phi({}^7\text{Be})}{\phi(\text{pp}) - \phi({}^7\text{Be} + {}^8\text{B})} = 0.174 \quad \text{(SSM)} \quad (10.102)$$

is a very important test of SSM predictions and stellar astrophysics. Its value reflects the competition between the two primary ways of terminating the pp chain [Bah03]. Motivated by this, the use of detectors with various threshold energies in radiochemical experiments and through direct measurements of the energy spectrum of solar neutrinos in real time and at low energies are explored. Not all will be discussed here in detail (see [Kir99] for more details). The success of using electron scattering and charged and neutral current measurements to disentangle the solution of the solar neutrino problem suggests an application of this at lower energies as well. A division of proposed experiments into two groups according to the detection reactions is reasonable and a compilation shown in table 10.3. As reaction processes to detect them, scattering and inverse β-decay are usually considered.

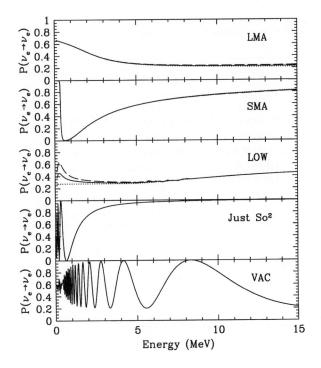

Figure 10.25. Survival probability of electrons coming from the Sun due to the various solutions described in the text. The largest effects can be expected in the low energy region ($E_\nu < 1$ MeV).

10.4.1 Measuring ^{7}Be neutrinos with the Borexino experiment

Borexino is designed to measure the important (862 keV) ^{7}Be line in real time [Bor91]. It is currently installed in the Gran Sasso underground laboratory. The Borexino experiment uses 300 t of a liquid scintillator, of which 100 t can be used as a fiducial volume (i.e. in order to reduce background only this reduced volume is used for data-taking). The detection reaction is

$$\nu_e + e \rightarrow \nu_e + e \qquad (10.103)$$

but in contrast to existing real-time experiments scintillation instead of the Cerenkov effect is used for detection, allowing a much lower threshold. The mono-energetic ^{7}Be line produces a recoil spectrum of electrons which has a maximum energy of 665 keV ('Compton edge'). The signal should hence be visible as a plateau in a region between 250–650 keV and, according to the SSM of [Bah01], should amount to about 50 events per day. This rate will be suppressed in a characteristical way by the LMA solutions.

Sufficient suppression of the radioactive background will be decisive for

Table 10.3. Real-time experiments under study for sub-MeV neutrinos using electron scattering (ES) or charged current reactions (CC) for detection.

Experiment	Idea	Principle
CLEAN	Liquid He, Ne	ES
ZEPLIN	Liquid Xe	ES
XMASS	Liquid Xe	ES
HERON	Liquid He	ES
SUPER MUNU	TPC (He)	ES
GENIUS	Ge semicond.	ES
LOW-C14	Borexino CTF	ES
MOON	^{100}Mo	CC
LENS	^{115}In Liquid Scint.	CC

the success of this experiment. Extremely low purity levels are necessary in the scintillators (less than 10^{-16} g (U, Th)/g) and in the materials used. The ability to achieve such low background levels could be demonstrated in a smaller pilot experiment (counting test facility, CTF). The current contamination of ^{14}C in the organic scintillator is preventing a real-time pp measurement.

10.4.2 Real-time measurement of pp neutrinos using coincidence techniques

The final goal of solar neutrino spectroscopy will be a real-time measurement of pp neutrinos. A proposal for doing this with nuclear coincidence techniques was made recently [Rag97]. The detection principle using coincidences relies on the following two reactions:

$$\nu_e + (A, Z) \rightarrow (A, Z+1)_{gs} + e^- \rightarrow (A, Z+2) + e^- + \bar{\nu}_e \quad (10.104)$$
$$\nu_e + (A, Z) \rightarrow (A, Z+1)^* + e^- \rightarrow (A, Z+1)_{gs} + \gamma. \quad (10.105)$$

Therefore, either coincidence between two electrons for the ground-state transitions or the coincidence of an electron with the corresponding de-excitation photon(s) is required. To prevent the mother isotope from decaying by single β-decay, the idea is to use double β-decay candidates. Three candidates are found which would allow pp neutrino measurements using excited-state transitions, namely ^{82}Se, ^{160}Gd or ^{176}Yb. By using different excited states, it is even possible to compare different contributions of the solar neutrino flux. All these ideas require several tons of material. An example for a proposal to use ground state transitions is ^{100}Mo (MOON) [Eji00a]. Here about 3.4 t of the double beta emitter ^{100}Mo are considered, having an energy threshold of 158 keV for solar neutrinos. The produced daughter will decay with a lifetime of 15 s via β-decay.

An alternative to double beta isotopes might be the use of ^{115}In as a fourfold forbidden unique β-decay isotope under study in LENS. Its low threshold of $E_\nu = 128$ keV has meant that it has been considered for some time but only recently has scintillator technology made it feasible. Additional isotopes were proposed recently [Zub03]. Here also a possible antineutrino tag with a threshold of 713 keV is proposed by using the $\beta^+\beta^+$ emitter ^{106}Cd. Solar pp antineutrinos are, in principle, unobservable by the nuclear coincidence method, because one always has to account for at least the positron mass.

Besides this coincidence technique there always remains the possibility of using neutrino–electron scattering as a real-time pp-neutrino reaction. For recent reviews on ongoing activities see [Kir99, Sch03]. In summary, after several decades of missing solar neutrinos the problem has finally been solved by SNO. Their result shows clearly that the full solar neutrino flux is arriving on Earth, but the dominant part is not ν_e. Also the various solutions discussed reduced to a single one with the help of new data from all solar neutrino experiments and KamLAND. Matter effects are responsible for the flavour conversion and the LMA solution is the correct one with $sm^2 \approx 7.10^{-5}$ eV^2 and $\tan^2\theta \approx 0.4$ implying non-maximal mixing.

Having dealt in great detail with solar neutrinos, we now discuss another astrophysical source of neutrinos, which has caused a lot of excitement and discussion over the past few years.

Chapter 11

Neutrinos from supernovae

Among the most spectacular events in astrophysics are phenomena from the late phase of stellar evolution, namely the explosion of massive stars. Such events are called supernovae and some of them are extremly luminous neutrino sources. Neutrinos are emitted in a period of about 10 s and roughly equal in number those emitted by the Sun during its life. The physics of supernova explosions is rather complex; however, additional information can be found in [Sha83, Woo86a, Arn89, Pet90, Whe90, Woo92, Woo94, Bur95, Mül95c, Raf96, Bur97, Raf99, Ful01, Ham03].

11.1 Supernovae

Supernovae arise from two different final stages of stars. Either they are caused by thermonuclear explosions of a white dwarf within a binary system or they are explosions caused by the core collapse of massive stars ($M > 8M_\odot$). In the first case a compact star accretes matter from its main sequence companion until it is above a critical mass called the Chandrasekhar mass. The second mechanism is due to the fact that no further energy can be produced by nuclear fusion of iron-like nuclei created in the interiors of massive stars. Therefore, such stars become unstable with respect to gravity and are called core collapse supernovae.

Supernovae are classified spectroscopically by the appearance of H-lines in their spectra. Those with no H-lines are called type I supernovae and those with H-lines correspond to supernovae of type II. Type I supernovae are further subdivided due to other spectral features (see [Whe90, Ham03]). In addition to type II, type Ib and Ic are considered as core collapse supernovae as well. The whole supernova phenomenon is probably more complex than the simple classification system suggests. This is supported by SN 1993J, which originally looked like a type II supernova but later in its evolution showed more the characteristics of a type Ib [Lew94]. Since no neutrinos are produced in association with type Ia supernovae only core collapse supernovae will be considered here.

Table 11.1. Hydrodynamic burning phases during stellar evolution (from [Gro90]).

Fuel	T (10^9 K)	Main product	Burning time for $25M_\odot$	Main cooling process
^{1}H	0.02	^{4}He, ^{14}N	7×10^6 a	Photons, neutrinos
^{4}He	0.2	^{12}C, ^{16}O, ^{22}Ne	5×10^5 a	Photons
^{12}C	0.8	^{20}Ne, ^{23}Na, ^{24}Mg	600 a	Neutrinos
^{20}Ne	1.5	^{16}O, ^{24}Mg, ^{28}Si	1 a	Neutrinos
^{16}O	2.0	^{28}Si, ^{32}S	180 d	Neutrinos
^{28}Si	3.5	^{54}Fe, ^{56}Ni, ^{52}Cr	1 d	Neutrinos

11.1.1 The evolution of massive stars

The longest phase in the life of a star is hydrogen burning (see chapter 10). After the hydrogen is burned off the energy production in the interior is no longer sufficient to withstand gravitation. The star does not compensate for this by a reduced luminosity but by contraction. According to the virial theorem, only half of the energy released in this process produces an internal rise in pressure, while the other half is released. From the equation of state for a non-degenerate ideal gas is $p \sim \rho \cdot T$, it follows that a pressure increase is connected with a corresponding temperature increase. If a sufficiently high temperature has been reached, the burning of helium via the triple α-process to ^{12}C ignites (helium flash). This causes the outer shells to inflate and a red giant develops. After a considerably shorter time than that used for the hydrogen burning phase, the helium in the core has been fused, mainly into carbon, oxygen and nitrogen and the same cycle, i.e. contraction with an associated temperature increase, now leads to a successive burning of these elements. The lower temperature burning phases (He, H) move towards the stellar surface in this process. From the burning of carbon onwards, neutrinos become the dominant energy loss mechanism, which leads to a further reduction of the burning time scales. Additional burning phases follow as shown in table 11.1. The last possible reaction is the burning of silicon to nickel and iron. The silicon burning corresponds less to fusion than to the photo-disintegration of ^{28}Si with simultaneous building up of a kind of chemical equilibrium between the reactions of the strong (n, p, α, ... induced reactions) and the electromagnetic interactions. The equilibrium occurs from about $T \approx 3.5 \times 10^9$ K and the distribution of the elements created is, at that time, dominated mainly by ^{56}Ni, and later, at higher temperatures, by ^{54}Fe. This last burning process lasts only for the order of a day. This ends the hydrostatic burning phases of the star. A further energy gain from fusion of elements is no longer possible, as the maximum binding energy per nucleon, of about 8 MeV/nucleon, has been reached in the region of iron.

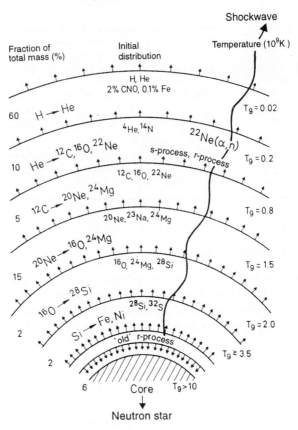

Figure 11.1. Schematic representation of the structure, composition and development of a heavy star (about $25M_\odot$). In the hydrostatic burning phases of the shells, elements of higher atomic number, up to a maximum of Fe and Ni, are built up from the initial composition (the major components of which are labelled). The gravitational collapse of the core leads to the formation of a neutron star and the ejection of $\approx95\%$ of the mass of the star (supernova explosion). The ejected outer layers are traversed by the detonation shock wave which initiates explosive burning (from [Gro90]).

The number of burning phases a star goes through depends on its mass. Stars with more than about $12M_\odot$ carry out burning up to iron. Such stars then have a small, dense, iron core and an extended envelope. The burning regions form shells on top of each other give the interior of stars an onion-like structure (see figure 11.1). The stability of the iron core is mainly guaranteed by the pressure of the degenerate electrons. The origin of the degenerate electron gas can be understood from figure 11.2. It shows a phase diagram which characterizes the state of the matter inside stars. For very large densities the Pauli principle has to

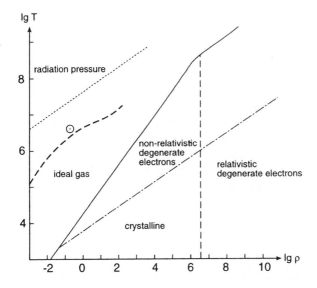

Figure 11.2. Schematic ρ–T phase diagram for the characterization of matter inside stars. The areas shown are those in which the equation of state is dominated either by radiation pressure (above the dotted line) or by a degenerate electron gas (below the full line). The latter can be relativistic or non-relativistic (to the right or left of the vertical dotted line). The dot-dashed line characterizes a temperature below which the ions prefer a crystalline state. The heavy dotted line shows the standard solar model and the present Sun (from [Kip90]).

be taken into account. This implies that each cell in phase space of size h^3 can contain a maximum of two electrons. The entire phase space volume is then given by

$$V_{Ph} = \tfrac{4}{3}\pi R^3 \tfrac{4}{3}\pi p^3. \tag{11.1}$$

Higher densities at constant radius produce an increased degeneracy. The pressure is then no longer determined by the kinetic energy but by the Fermi energy or Fermi momentum p_F. This implies that $p_e \sim p_F^5$ (non-relativistic) and $p_e \sim p_F^4$ (relativistic). In addition, the pressure no longer depends on the temperature but exclusively on the electron density n_e, according to [Sha83]

$$p = \frac{1}{m_e}\frac{1}{5}(3\pi^2)^{\frac{2}{3}}n_e^{\frac{5}{3}} \qquad \text{non-relativistic} \tag{11.2}$$

$$p = \tfrac{1}{4}(3\pi^2)^{\frac{1}{3}}n_e^{\frac{4}{3}} \qquad \text{relativistic.} \tag{11.3}$$

This has the fatal consequence of gravitational collapse. The previous cycle of pressure increase $\rightarrow$ temperature increase $\rightarrow$ ignition $\rightarrow$ expansion $\rightarrow$ temperature drop now no longer functions. Released energy leads only to a temperature increase and thus to unstable processes but no longer to pressure increase.

The stability condition of a star in hydrostatic equilibrium is given by [Lan75, Sha83]

$$E = \frac{3\gamma - 4}{5\gamma - 6} \frac{GM^2}{R} \qquad (11.4)$$

$\gamma = \partial \ln p / \partial \ln \rho$ is the adiabatic index, p the pressure and ρ the density. The adiabatic index for a non-relativistic degenerate electron gas is, for example, 5/3, while it is 4/3 in the relativistic case. The appearance of a critical mass, the *Chandrasekhar limit*, reflects the violation of the stability condition ($\gamma > 4/3$) for hydrostatic equilibrium, because of the pressure dependence $p \propto \rho^{4/3} \propto n_e^{4/3}$ of a non-relativistic degenerate electron gas. The Chandrasekhar mass is given by ([Cha39, 67], also see [Hil88])

$$M_{Ch} = 5.72 Y_e^2 M_\odot \qquad (11.5)$$

where Y_e is the number of electrons per nucleon. This instability causing core collapse in massive stars is the pre-condition for a supernova explosion.

11.1.2   The actual collapse phase

The typical parameter values for a $15 M_\odot$ star with a core mass of $M_{Ch} \approx 1.5 M_\odot$ are a central temperature of approximately 8×10^9 K, a central density of 3.7×10^9 g cm^{-3} and a Y_e of 0.42. The Fermi energy of the degenerate electrons is roughly 4–8 MeV. These are typical values at the start of the collapse of a star. The cause for the core collapse is the photo-disintegration of nuclei of the iron group, such as via the reaction

$$^{56}\text{Fe} \rightarrow 13\,^4\text{He} + 4\text{n} - 124.4\,\text{MeV} \qquad (11.6)$$

and electron capture by free protons and heavy nuclei

$$\text{e}^- + \text{p} \rightarrow \text{n} + \nu_e \qquad \text{e}^- + {}^Z\text{A} \rightarrow {}^{Z-1}\text{A} + \nu_e. \qquad (11.7)$$

The latter process becomes possible because of the high Fermi energy of the electrons. The number of electrons is strongly reduced by this process (11.7). Mainly neutron-rich, unstable nuclei are produced. Since it was the degenerate electrons which balanced the gravitational force, the core collapses quickly. The lowering of the electron concentration can also be expressed by an adiabatic index $\gamma < 4/3$. The inner part ($\approx 0.6 M_\odot$) keeps $\gamma = 4/3$ and collapses homologously ($v/r \approx 400$–700 s^{-1}), while the outer part collapses at supersonic speed. Homologous means that the density profile is kept during the collapse. The behaviour of the infall velocities within the core is shown in figure 11.3 at a time of 2 ms before the total collapse. The matter outside the sonic point defined by $v_{coll} = v_{sound}$ collapses with a velocity characteristic for free fall.

The outer layers of the star do not notice the collapse of the iron core, due to the low speed of sound. More and more neutron-rich nuclei are produced in the core, which is reflected in a further decrease of Y_e taken away by the produced neutrinos. The emitted neutrinos can initially leave the core zone unhindered. Neutrinos are trapped starting at density regions around 10^{12} g cm^{-3} in which typical nuclei have masses of between 80 and 100 and about 50 neutrons. Here, matter becomes opaque for neutrinos as the outward diffusion speed becomes considerably smaller than the inward collapse speed of a few milliseconds. The dominant process for the neutrino opacity is coherent neutrino–nucleus scattering via neutral weak currents with a cross section of (see, e.g., [Fre74, Hil88])

$$\sigma \simeq 10^{-44} \text{ cm}^2 N^2 \left(\frac{E_\nu}{\text{MeV}}\right)^2 \tag{11.8}$$

where N is the neutron number in the nucleus. The reason for the behaviour on the neutron number relies on the fact that 10 MeV neutrinos interact with the nucleus coherently and the NC cross section on protons is reduced by $1-4\sin^2\theta_W$ (see chapter 4). Hence, neutrinos are trapped and move with the collapsing material (*neutrino trapping*). The transition between the 'neutrino optical' opaque and the free-streaming region defines a *neutrino sphere* (see (11.10)). The increase in neutrino trapping by neutrons acts as an inverse process to that of (11.7) and consequently stabilizes electron loss and leads to an equilibrium with respect to weak interactions. Therefore, no further neutronization occurs and the lepton number per baryon $Y_{\text{lepton}} = Y_e + Y_\nu$ is conserved at the value of Y_e at the beginning of neutrino-trapping $Y_L \approx 0.35$ [Bet86a].

Henceforward the collapse progresses adiabatically. This is equivalent to a constant entropy, as now neither significant energy transport nor an essential change in composition takes place. Figure 11.4 shows the mean mass numbers and nuclear charge number of the nuclei formed during neutronization, together with the mass fractions of neutrons, protons, as well as the number of electrons per nucleon. We thus have a gas of electrons, neutrons, neutrinos and nuclei whose pressure is determined by the relativistic degenerate electrons. Thus the 'neutron star' begins as a hot lepton-rich quasi-static object, which develops into its final state via neutrino emission, i.e. it starts off as a *quasi-neutrino star* [Arn77].

The collapsing core finally reaches densities that normally appear in atomic nuclei ($\rho > 3 \times 10^{14}$ g cm^{-3}). For higher densities, however, the nuclear force becomes strongly repulsive, matter becomes incompressible and bounces back (equivalent to $\gamma > 4/3$) (see e.g. [Lan75, Bet79, Kah86]). Exactly how much energy this shock wave contains depends, among other factors, on the equation of state of the very strongly compressed nuclear matter [Kah86, Bet88], whether the bounce back of the core is hard or soft. A soft bounce back provides the shock with less initial energy. Unfortunately, the equation of state is not very well known, since extrapolation into areas of supernuclear density is required. At bounce a strong sound wave is produced, which propagates outwards and steepens into a shock wave near $M \approx 0.6 M_\odot$ after about 1 ms after bounce. This is

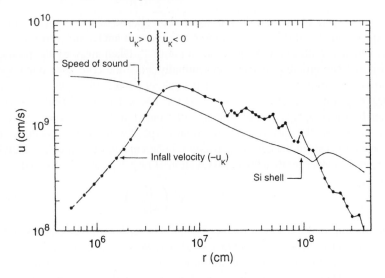

Figure 11.3. Infall velocity of the material in the core of a supernova about 2 ms before the complete collapse of the star. Within the homologous inner core ($r < 40$ km) the velocity is smaller than the local velocity of sound. In the region $r > 40$ km (outer core), the material collapses with supersonic speed (from [Arn77]).

illustrated in figure 11.5. This shock formations corresponds to a radius of about 100 km, where the density is $\rho \approx 10^{14}$ g cm^{-3}.

The outgoing shock dissociates the incoming iron nuclei into protons and neutrons. This has several consequences. The shock wave loses energy by this mechanism and if the mass of the iron core is sufficiently large, the shock wave does not penetrate the core and a supernova explosion does not take place. It becomes an accretion shock at a radius between 100–200 km and $\rho \approx \times 10^{10}$ g cm^{-3}. It was assumed that the stalled shock could be revived by neutrino heating. This should finally result in an explosion and has been called *delayed explosion* mechanism [Bet85]. However, new simulations seem to indicate that still no explosion is happening [Tho02]. The nuclear binding energy of $0.1 M_\odot$ iron is about 1.7×10^{51} erg and thus comparable to the explosion energy. However, the dissociation into nucleons leads to an enormous pressure increase, which results in a reversal of the direction of motion of the incoming matter in the shock region. This transforms a collapse into an explosion. As the shock moves outwards, it still dissociates heavy nuclei, which are mainly responsible for neutrino trapping. Moreover, the produced free protons allow quick neutronization via $e^- + p \rightarrow n + \nu_e$ if passing the neutrino sphere in which the density is below 10^{11} g cm^{-3}. The produced neutrinos are released immediately and are often called 'prompt ν_e burst' or 'deleptonization burst'. If the shock wave leaves the iron core without getting stalled, the outer layers

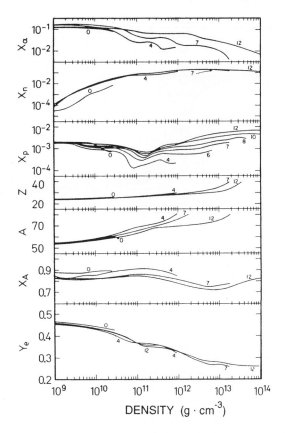

Figure 11.4. Change in the core composition during the gravitational collapse (the numbers correspond to various stages of the collapse). X_n, X_p, X_α, X_A denote the *mass* fraction (not the *number* densities) of the neutrons, protons, α-particles and nuclei. Y_e denotes the electrons per nucleon (from [Bru85]).

represent practically no obstacle and are blown away, which results in an optical supernova in the sky. Such a mechanism is known as a *prompt explosion* [Coo84, Bar85]. The structure of a star at the end of its hydrostatic burning phases can only be understood with the help of complex numerical computer simulations [Bet82, Sha83, Woo86a, Woo92, Woo95, Bur95, Mez98]. Current one-dimensional spherical computer simulations suggest that core bounce and shock formation are not sufficient to cause a supernova explosion [Bur95, Jan95, Jan96, Mez98, Bur00, Ram00, Mez01, Bur02, Tho02], only one group reports successful explosions [Tot98]. In more recent work, a further, decisive boost is given to the shock by the formation of neutrino-heated hot bubbles [Bet85, Col90, Col90a], which furthermore produce a considerable mixing of the emitted material. This

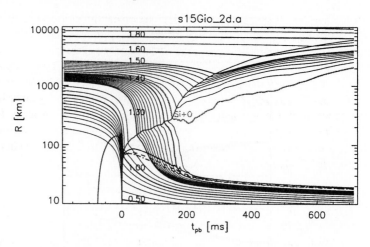

Figure 11.5. The development of a supernova explosion of type II, according to Wilson and Mayle. As the nuclear matter is over-compressed in the collapse, a rebound occurs and produces a shock wave. However, this is weakened by thermal decomposition of the incoming matter during the collapse (from [Kei03]).

strong mixing has been observed in supernova 1987A (see section 11.4.1.2) and has been confirmed by two-dimensional computer simulations [Mül95c, Bur95], which allow these effects to be described for the first time. The simulation also showed the importance of large-scale convection in the process. The problem during the last ten years has been that the shock is stalled about 100–150 km from the centre of the star and, in general, the inclusion of neutrino absorption and the implied energy deposition permitted only a moderate explosion. The newest supercomputers allowed the simulation of dying stars in *two* or *three* dimensions, following both the radial and lateral directions. Convection brings hot material from near the neutrino sphere quickly up in regions behind the shock and cooler material down to the neutrino sphere where it helps absorbing energy from the neutrino flow. This helps in revitalizing the stalled (for about 100 ms) shock. The simulations also lead to the general belief that newly-born neutron stars are convective [Her94, Bur95, Jan95, Jan96, Mez98]. A further aspect of convection is that typical explosions show asymmetrically ejected matter, resulting in a recoil of the remaining core which gives it a speed of hundreds of kilometres per second (rocket effect) [Bur95, Bur95a, Jan95a]. Such asymmetries might also point to the importance of rotation and magnetic fields in supernova explosions [Kho99]. Nevertheless computer simulations, in general, fail to produce explosions and further progress has to be made to understand better the details of the underlying physical processes.

Meanwhile the object below the shock has become a protoneutron star. It consists basically of two parts: an inner settled core within the radius where the shock wave was first formed, consisting out of neutrons, protons, electrons and neutrinos ($Y_{\text{lepton}} \approx 0.35$); the second part is the bloated outer part, which lost most of its lepton number during the ν_e burst at shock breakout. This part settles within 0.5–1 s of core bounce, emitting most of its energy in neutrinos. After about 1 s the protoneutron star is basically an object by itself. It has a radius of about 30 km which slowly contracts further and cools by emission of (anti)neutrinos of all flavours and deleptonizes by the loss of ν_e. After 5–10 s it has lost most of its lepton number and energy, a period called Kelvin–Helmholtz cooling. The energy released in a supernova corresponds to the binding energy of the neutron star produced:

$$E_B \approx \frac{3}{5} \frac{G M_{\text{neutron star}}}{R} = 5.2 \times 10^{53} \text{ erg} \left(\frac{10 \text{ km}}{R_{\text{neutron star}}} \right) \left(\frac{M_{\text{neutron star}}}{1.4 M_\odot} \right)^2. \tag{11.9}$$

This is the basic picture of an exploding, massive star. For a detailed account see [Arn77, Sha83, Woo86a, Pet90, Col90, Bur95].

11.2   Neutrino emission in supernova explosions

We now discuss the neutrinos which could be observed from supernova explosions. The observable spectrum originates from two processes. First the deleptonization burst as the outgoing shock passes the neutrino sphere, resulting in the emission of ν_e with a duration of a few ms. The radius R_ν of this sphere might be defined via the optical depth τ_ν and be approximated by

$$\tau_\nu(R_\nu, E_\nu) = \int_{R_\nu}^{\infty} \kappa_\nu(E_\nu, r) \rho(r) \, \mathrm{d}r = \tfrac{2}{3} \tag{11.10}$$

where $\kappa_\nu(E_\nu, r)$ is the opacity. $\tau_\nu < 2/3$ characterizes the free streaming of neutrinos. The second part comes form the Kelvin–Helmholtz cooling phase of the protoneutron star resulting in an emission of all flavours (ν_e, $\bar{\nu}_e$, ν_μ, $\bar{\nu}_\mu$, ν_τ, $\bar{\nu}_\tau$; in the following ν_μ is used for the last four because their spectra are quite similar). The emission lasts typically for about 10 s (figure 11.6). It is reasonable to assume that the energy is equipartitioned among the different flavours.

Neutrinos carry away about 99% of the energy released in a supernova explosion. This global property of emitting 0.5×10^{53} erg in each neutrino flavour with typical energies of 10 MeV during about 10 s remains valid; however, it is worthwhile taking a closer look at the details of the individual spectra. Their differences offer the opportunity to observe neutrino flavour oscillations.

From considerations concerning the relevant opacity sources for the different neutrino flavours, a certain energy hierarchy might be expected. For ν_e the dominant source is $\nu_e n \to e p$ and for $\bar{\nu}_e$ it is $\bar{\nu}_e p \to e^+ n$. The spectrum of the $\bar{\nu}_e$ corresponds initially to that of the ν_e but, as more and more protons vanish, the $\bar{\nu}_e$s react basically only via neutral currents also resulting in a lower opacity

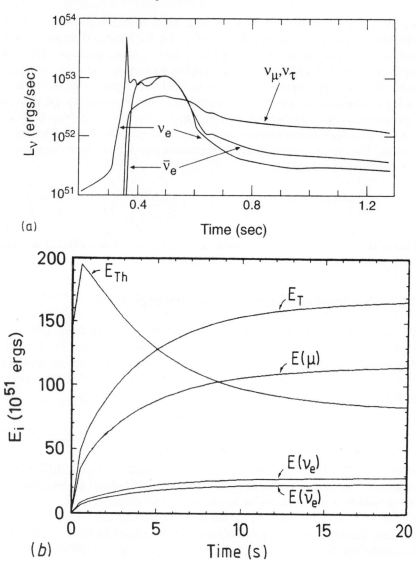

Figure 11.6. (*a*) Calculated luminosity of a $2M_\odot$ 'Fe' core of an $\approx 25M_\odot$ main sequence star as a function of the time from the start of the collapse for the various neutrino flavours (from [Bru87]). (*b*) Cooling of a hot proto-neutron star of $1.4M_\odot$ in the first 20 s after gravitational collapse. E_{Th} denotes the integrated internal energy, E_T is the total energy released and E_{ν_e} and $E_{\bar\nu_e}$ are the total energies emitted as ν_e and $\bar\nu_e$, respectively. E_μ is the energy emitted as ν_μ, $\bar\nu_\mu$, ν_τ and $\bar\nu_\tau$. All energies are in units of 10^{51} erg (from [Bur86]).

and, therefore, higher average energy (about 14–17 MeV). The typical average energy for ν_e is in the region 10–12 MeV. As $\nu_\mu(\bar{\nu}_\mu)$ and $\nu_\tau(\bar{\nu}_\tau)$ can only interact via neutral currents (energies are not high enough to produce real muons and tau-leptons), they have smaller opacities and their neutrino spheres are further inside, resulting in a higher average energy (about 24–27 MeV). The post bounce energy hierarchy expected is $\langle E_{\nu_e}\rangle < \langle E_{\bar{\nu}_e}\rangle < \langle E_{\nu_\mu}\rangle$. Obviously this implies for the radii of the neutrino spheres that $\langle R_{\nu_\mu}\rangle < \langle R_{\bar{\nu}_e}\rangle < \langle R_{\nu_e}\rangle$. However, recent more sophisticated simulations including additional neutrino processes, like NN $\rightarrow$ NN$\nu_\mu\bar{\nu}_\mu$ (nucleon-nucleon bremsstrahlung), $e^+e^- \rightarrow \nu_\mu\bar{\nu}_\mu$ and $\nu_e\bar{\nu}_e \rightarrow \nu_\mu\bar{\nu}_\mu$ (annihilation) and scattering on electrons $\nu_\mu e \rightarrow \nu_\mu e$, have revealed that the difference in $\langle E_{\nu_\mu}\rangle$ and $\langle E_{\bar{\nu}_e}\rangle$ is much smaller than previously assumed [Han98, Raf01, Kei03]. It is only on the level of 0–20% even if the fluxes itself may differ by a factor of two.

The spectra from the cooling phase are not exactly thermal. Since neutrino interactions increase with energy, the effective neutrino sphere increases with energy. Thus, even if neutrinos in their respective neutrino sphere are in thermal equilibrium with matter, the spectrum becomes 'pinched', i.e. depleted at higher and lower energies in comparison with a thermal spectrum. This is a consequence of the facts that the temperature decreases with increasing radius and the density decreases faster than $1/r$.

11.3 Detection methods for supernova neutrinos

As is clear from the discussion in the last section, basically all solar neutrino detectors can be used for supernova detection [Bur92]. Running real-time experiments can be divided into (heavy) water Cerenkov detectors like Super-Kamiokande and SNO as well as liquid scintillators like KamLAND and the Large Volume Detector (LVD), a 1.8 kt detector which has been running in the Gran Sasso Laboratory since 1992 [Ful99]. For the latter the most important detection reaction is (see also chapter 8)

$$\bar{\nu}_e + p \rightarrow n + e^+ \tag{11.11}$$

making them mainly sensitive to the $\bar{\nu}_e$ component. Additional processes with smaller cross sections are:

$$\nu_e + {}^{16}O \rightarrow {}^{16}F + e^- \quad \text{(CC)} \qquad E_{\text{Thr}} = 15.4\,\text{MeV} \tag{11.12}$$
$$\bar{\nu}_e + {}^{16}O \rightarrow {}^{16}N + e^+ \quad \text{(CC)} \qquad E_{\text{Thr}} = 11.4\,\text{MeV} \tag{11.13}$$
$$\bar{\nu} + {}^{16}O \rightarrow {}^{16}O^* + \bar{\nu} \quad \text{(NC)} \tag{11.14}$$
$$\nu_i(\bar{\nu}_i) + e^- \rightarrow \nu_i(\bar{\nu}_i) + e^- \quad (i \equiv e, \mu, \tau). \tag{11.15}$$

The CC reaction on ^{16}O has a threshold of 13 MeV and rises very fast, making it the dominant ν_e detection mode at higher energies. Liquid organic scintillators lack the reactions on oxygen but can rely on CC and NC reactions on ^{12}C:

$$\nu_e + {}^{12}C \rightarrow {}^{12}C^* + \nu_e \qquad (NC) \qquad (11.16)$$

$$\nu_e + {}^{12}C \rightarrow {}^{12}N + e^- \qquad (CC). \qquad (11.17)$$

The superallowed NC reaction might be detected by the associated 15 MeV de-excitation gamma (see chapter 4). The behaviour of the relevant cross sections with E_ν is shown in figure 11.7. The CC reaction has a threshold of $E_\nu \gtrsim 30$ MeV. Two more modifications of current detection schemes using water Cerenkov detectors became possible recently, namely SNO (see chapter 10) and huge underwater or ice detectors, described in more detail in chapter 12. The photomultipliers used in the latter are too sparse to allow an event-by-event reconstruction but a large burst of supernova neutrinos producing enough Cerenkov photons could cause a coincident increase in single photomultiplier count rates for many tubes. An important issue to measure is the energy and temperature of the emitted ν_μ and ν_τ neutrinos but they only interact via neutral currents. SNO profits again from measuring both NC and CC reactions on deuterium ((10.49) and (10.51)); in particular the NC reaction is extremely important for determining global neutrino properties and fluxes, while the CC reaction is very important to investigate the deleptonization burst. In addition, a $\bar{\nu}_e$ component can be measured via (10.102). It was recently proposed to measure NC quantities by using elastic scattering on free protons $\nu + p \rightarrow \nu + p$ where the proton recoil produces scintillation light in detectors like KamLAND [Bea02]. Furthermore, the planned ICARUS experiment (see chapter 9) will allow the ν_e spectrum to be measured via the reaction

$$\nu_e + {}^{40}Ar \rightarrow e^- + {}^{40}K^* \qquad (11.18)$$

with a threshold of 5.9 MeV and a 4.3 MeV gamma ray from the ^{40}K* de-excitation. Following the discussion of cross sections, it can be concluded that mainly $\bar{\nu}_e$ were detected from the supernova 1987A which is discussed in the next section.

11.4   Supernova 1987A

One of the most important astronomical events of the last century must have been the supernova (SN) 1987A [Arn89, Che92, Kos92, Woo97] (see figure 11.8) (the numbering scheme for supernovae contains the year of their discovery and another letter which indicates the order of occurrence). This was the brightest supernova since Kepler's supernova in 1604 and provided astrophysicists with a mass of new data and insights, as it was for the first time possible to observe it at all wavelengths. Moreover, for the first time neutrinos could be observed from this spectacular event. This first detection of neutrinos which do not originate from the Sun for many scientists marked the birth of neutrino astrophysics. Further details can be found, e.g., in [Arn89, Bah89, Che92, Kos92, McG93].

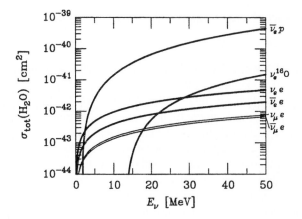

Figure 11.7. Some relevant total neutrino cross sections for supernova detection using water Cerenkov detectors. The curves refer to the total cross section per water molecule so that a factor of two for protons and 10 for electrons is already included (from [Raf96]).

(a)   (b)

Figure 11.8. The supernova 1987A. (a) The Large Magellanic Cloud before the supernova on 9 December 1977. The precursor star Sanduleak $-69°$ 202 is shown. (b) The same field of view as (a) on 26 February 1987 at $1^h\ 25^{min}$ where the $4^m.4$ brightness supernova 1987A can be seen. The length of the horizontal scale is 1 arcmin (with kind permission of ESO).

11.4.1 Characteristics of supernova 1987A

11.4.1.1 Properties of the progenitor star and the event

Supernova 1987A was discovered on 23 February 1987 at a distance of 150 000 light years (corresponding to 50 kpc) in the Large Magellanic Cloud (LMC), a companion galaxy of our own Milky Way [McN87]. The evidence that the supernova was of type II, an exploding star, was confirmed by the detection of hydrogen lines in the spectrum. However, the identification of the progenitor star Sanduleak −69° 202 was a surprise since it was a blue B3I supergiant with a mass of about $20M_\odot$. Until then it was assumed that only red giants could explode. The explosion of a massive blue supergiant could be explained by the smaller 'metal' abundance in the Large Magellanic Cloud, which is only one-third of that found in the Sun, together with a greater mass loss, which leads to a change from a red giant to a blue giant. The oxygen abundance plays a particularly important role. On one hand, the oxygen is relevant for the opacity of a star and, on the other hand, less oxygen results in less efficient catalysis of the CNO process, which causes a lower energy production rate in this cycle. Indeed it is possible to show with computer simulations that blue stars can also explode in this way [Arn91,Lan91].

A comparison of the bolometric brightness of the supernova in February 1992 of $L = 1 \times 10^{37}$ erg s^{-1} shows that this star really did explode. This is more than one order of magnitude less than the value of $L \approx 4 \times 10^{38}$ erg s^{-1}, which was measured *before* 1987, i.e. the original star has vanished. SN 1987A went through a red giant phase but developed back into a blue giant about 20 000 years ago. The large mass ejection in this process was discovered by the Hubble Space Telescope, as a ring around the supernova. Moreover, an asymmetric explosion seems now to be established [Wan02]. The course of evolution is shown in figure 11.9. The total explosive energy amounted to $(1.4\pm0.6)\times10^{51}$ erg [Che92].

11.4.1.2 γ-radiation

γ-line emission could also be observed for the first time. It seems that the double magic nucleus ^{56}Ni is mostly produced in the explosion. It has the following decay chain:

$$^{56}\text{Ni} \xrightarrow{\beta^+} {}^{56}\text{Co} \xrightarrow{\beta^+} {}^{56}\text{Fe}^* \rightarrow {}^{56}\text{Fe}. \tag{11.19}$$

^{56}Ni decays with a half-life of 6.1 days. ^{56}Co decays with a half-life of 77.1 days, which is very compatible with the decrease in the light curve as shown in figure 11.10 [Che92]. Two gamma lines at 847 and 1238 keV, which are characteristic lines of the ^{56}Co decay, were detected by the Solar Maximum Mission satellite (SMM) at the end of August 1987 [Mat88]. From the intensity of the lines the amount of ^{56}Fe produced in the explosion can be estimated as about $0.075M_\odot$.

In general, photometric measurements of light curves provide important information about supernovae [Lei03]. They mainly depend on the size and

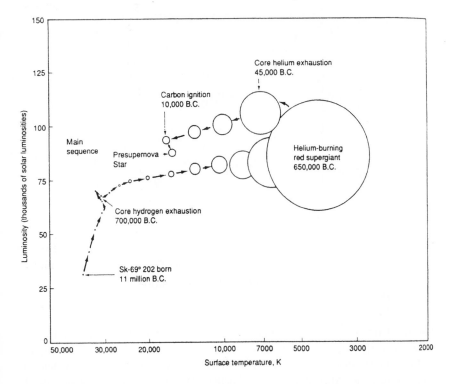

Figure 11.9. According to the theoretical model of S Woosley *et al* [Woo88], Sanduleak −62° 202 was probably born some 11 million years ago, with a mass about 18 times that of the Sun. Its initial size predetermined its future life, which is mapped in this diagram showing the luminosity against surface temperature at various stages, until the moment immediately before the supernova explosion. Once the star had burned all the hydrogen at its centre, its outer layers expanded and cooled until it became a red supergiant, on the right-hand side of the diagram. At that stage helium started burning in the core to form carbon, and by the time the supply of helium at the centre was exhausted, the envelope contracted and the star became smaller and hotter, turning into a blue supergiant. (With kind permission of T Weaver and S Woosley.)

mass of the progenitor star and the strength of the explosion. However, various additional energy inputs exist which results in modulations of the emerging radiation. By far the longest observed light curve is SN 1987A [Sun92]. The early development of the bolometric light curve together with the V-band ($\lambda = 540$ nm) light curve for more than ten years are shown in figure 11.10.

A direct search for a pulsar at the centre of SN 1987A with the Hubble Space Telescope has still been unsuccessful [Per95a]. The evidence for a pulsar in SN 1987A by powering the light curve would be very interesting in so far as it has never been possible to observe a pulsar and supernova directly from the same

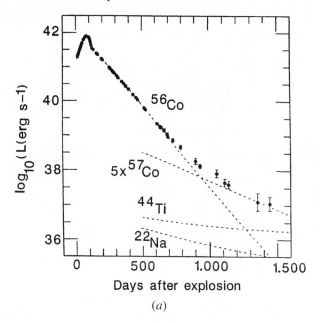

(a)

Figure 11.10. Behaviour of the light curve of supernova 1987A. (*a*) The early phase (from [Sun92]). (*b*) V-band light curve. The early phase of an expanding photosphere is driven by the shock breakout, resulting in an early peak lasting from a few hours to a couple of days. After a rapid, initial cooling the supernova enters a phase when its temperature and luminosity remains nearly constant. In this 'plateau phase' it is powered by the recombination of the previously ionized atoms in the supernave shock. Once the photosphere has receded deep enough, additional heating due to radioactive decays of ^{56}Ni and ^{56}Co dominates. Afterwards the light curve is powered solely by radioactive decay in the remaining nebula, the light curve enters the 'radioactive tail', typically after about 100 days. SN 1987A suffered from dust forming within the ejecta (after $\approx$450 days), which resulted in an increase in the decline rate in the optical as light was shifted to the infrared. After about 800 days the light curve started to flatten again due to energy released of ionized matter ('freeze out'). Later, the flattening is caused by long-lived isotopes, especially ^{57}Co and ^{44}Ti. At very late times, the emission is dominated by the circumstellar inner ring, which was ionized by the shock breakout. After about 2000 days the emission of the ring is stronger than from the supernova ejecta itself (from [Lei03]).

event and no hint of a pulsar contribution to a supernova light curve has been established.

11.4.1.3 *Distance*

The determination of the distance of the supernova has some interesting aspects. The Hubble telescope discovered a ring of diameter (1.66 ± 0.03) arcsec around

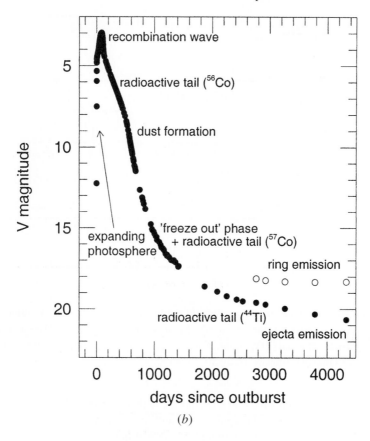

Figure 11.10. (Continued.)

SN 1987A in the UV region in a forbidden line of doubly ionized oxygen [Pan91]. Using the permanent observations of UV lines from the International Ultraviolet Explorer (IUE), in which these lines also appeared, the ring could be established as the origin of these UV lines. The diameter was determined to be $(1.27 \pm 0.07) \times 10^{18}$ cm, from which the distance to SN 1987A can be established as $d = (51.2 \pm 3.1)$ kpc [Pan91]. Correcting to the centre of mass of the LMC leads to a value of $d = (50.1 \pm 3.1)$ kpc [Pan91]. This value is not only in good agreement with those of other methods but has a relatively small error, which makes the use of this method for distance measurements at similar events in the future very attractive.

11.4.1.4 Summary

In the previous sections we have discussed only a small part of the observations and details of SN 1987A. Many more have been obtained, mainly related to the

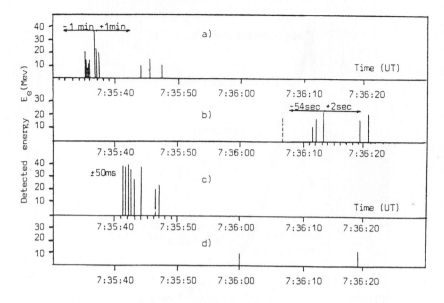

Figure 11.11. Time and energy spectrum of the four detectors which saw neutrinos from SN1987A as mentioned in the text: (*a*) the Kamiokande detector, (*b*) the Baksan detector, (*c*) the IMB detector and (*d*) the Mont Blanc detector (although it had no events at the time seen by the other experiments) (see text) (from [Ale87a, Ale88]).

increasing interaction of the ejected layers with the interstellar medium. The detection of the neutron star (pulsar) created in the supernova, or even a black hole, are eagerly awaited. Even though SN 1987A has provided a huge amount of new information and observations in all regions of the spectrum, the most exciting event was, however, the first detection of neutrinos from the star's collapse.

11.4.2 Neutrinos from SN 1987A

A total of four detectors claim to have seen neutrinos from SN 1987A [Agl87, Ale87, Ale88, Bio87, Hir87, Hir88, Bra88a]. Two of these are water Cerenkov detectors (KamiokandeII and Irvine–Michigan–Brookhaven (IMB) detector) and two are liquid scintillator detectors (Baksan Scintillator Telescope (BST) and Mont Blanc). The Cerenkov detectors had a far larger amount of target material, the fiducial volumes used are 2140 t (Kamiokande), 6800 t (IMB) and 200 t (BST). Important for detection is the trigger efficiency for $e^{\pm}$ reaching about 90% (80%) for KII (BST) at 10–20 MeV and being much smaller for IMB at these energies. In addition, IMB reports a dead time of 13% during the neutrino burst [Bra88]. The observed events are listed in table 11.2 and were obtained by reactions described in section 11.3. Within a certain timing uncertainty, three of the experiments agree on the arrival time of the neutrino pulse, while the Mont

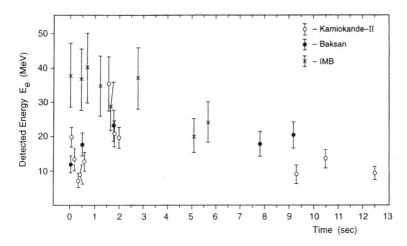

Figure 11.12. Energies of all neutrino events detected at 7.35 UT on 23 February 1987 *versus* time: The time of the first event in each detector has been set to $t = 0.0$ (from [Ale88, Ale88a]).

Blanc experiment detected them about 4.5 hr before the other detectors. Since its five events all lie very close to the trigger threshold of 5 MeV and as the larger Cerenkov detectors saw nothing at that time, it is generally assumed that these events are a statistical fluctuation and are not related to the supernova signal. The other three experiments also detected the neutrinos before the optical signal arrived, as expected. The relatively short time of a few hours between neutrino detection and optical discovery points to a compact progenitor star. The time structure and energy distribution of the neutrinos is shown in figure 11.12. If *assuming* the first neutrino events have been seen by each detector at the same time, i.e. setting the arrival times of the first event to $t = 0$ for each detector, then within 12 s, 24 events were observed (KamiokandeII + IMB + Baksan). The overall important results tested experimentally with these observations can be summarized as follows:

(i) All observed events are due to $\bar{\nu}_e$ interactions (maybe the first event of KII could be from ν_e). Fitting a Fermi–Dirac distribution an average temperature of $\langle T_\nu \rangle = (4.0 \pm 1.0)$ MeV and $\langle E_\nu \rangle = (12.5 \pm 3.0)$ MeV can be obtained ($\langle E_\nu \rangle = 3.15 k \langle T_\nu \rangle$).

(ii) The number of observed events estimates the time integrated $\bar{\nu}_e$ flux to be about $\Phi = (5 \pm 2.5) \times 10^9 \bar{\nu}_e$ cm^{-2}. The total number of neutrinos emitted from SN 1987A is then given (assuming six flavours and a distance SN–Earth of $L = 1.5 \times 10^{18}$ km) by

$$N_{tot} = 6\Phi 4\pi L^2 \approx 8 \times 10^{57}. \qquad (11.20)$$

This results in a total radiated energy corresponding to the binding energy of the neutron star of $E_{tot} = N_{tot} \times \langle E_\nu \rangle \approx (2\pm1) \times 10^{53}$ erg, in good agreement with expectation. This observation also for the first time experimentally verified that indeed more than 90% of the total energy released is carried away by neutrinos and that the visible signal corresponds to only a minute fraction of the energy.

(iii) The duration of the neutrino pulse was of the order 10 s.

The number of more detailed and specific analyses, however, far exceeds the number of observed events. The most recent systematic and comprehensive study based on a solid statistical treatment is given in [Lor02]. Several SN models were explored, dominantly from [Woo86a], half being prompt explosions and half delayed explosions. Their conclusion is in strong favour of the data of the delayed explosion mechanism and a typical radius of the resulting neutron star in the order of 10 km. For further discussion of neutrinos from SN 1987A, we refer to [Kos92, Raf96].

11.4.2.1 *Possible anomalies*

Three in some way unexplained facts remain from the observation. The least worrysome is a discrepancy in $\langle E_\nu \rangle$ between KII and IMB, implying a harder spectrum for IMB. However, IMB has a high energy threshold and relies on the high energy tail of the neutrino spectrum, which depends strongly on the assumed parameters. Therefore, both might still be in good agreement. A second point is the 7.3 s gap between the first nine and the following three events in Kamiokande. But given the small number of events this is possible within statistics and additionally the gap is filled with events from IMB and BST. Probably the most disturbing is the deviation from the expected isotropy for the $\bar{\nu}_e + p \rightarrow e^+ + n$ reaction (even a small backward bias is expected), especially showing up at higher energies. Various explanations have been proposed among them ν_e forward scattering. However, the cross section for these reaction is too small to account for the number of observed events and it would be too forward peaked as well. Unless some new imaginative idea is born, the most common explanation relies again on a possible statistical fluctuation.

11.4.3 Neutrino properties from SN 1987A

Several interesting results on neutrino properties can be drawn from the fact that practically all neutrinos were detected within about 12 s and the observed flux is in agreement with expectations. For additional bounds on exotic particles see [Raf96, Raf99].

11.4.3.1 *Lifetime*

The first point relates to the lifetime of neutrinos. As the expected flux of antineutrinos has been measured on Earth, no significant number could have

Table 11.2. Table of the neutrino events registered by the four neutrino detectors KamiokandeII [Hir87], IMB [Bio87], Mont Blanc [Agl87] and Baksan [Ale87a, Ale88]. T gives the time of the event, E gives the visible energy of the electron (positron). The absolute uncertainties in the given times are: for Kamiokande ± 1 min, for IMB 50 ms and for Baksan -54 s, $+2$ s.

Detector	Event number	T (UT)	E (MeV)
Kamioka	1	7 : 35 : 35.000	20 ± 2.9
	2	7 : 35 : 35.107	13.5 ± 3.2
	3	7 : 35 : 35.303	7.5 ± 2.0
	4	7 : 35 : 35.324	9.2 ± 2.7
	5	7 : 35 : 35.507	12.8 ± 2.9
	(6)	7 : 35 : 35.686	6.3 ± 1.7
	7	7 : 35 : 36.541	35.4 ± 8.0
	8	7 : 35 : 36.728	21.0 ± 4.2
	9	7 : 35 : 36.915	19.8 ± 3.2
	10	7 : 35 : 44.219	8.6 ± 2.7
	11	7 : 35 : 45.433	13.0 ± 2.6
	12	7 : 35 : 47.439	8.9 ± 1.9
IMB	1	7 : 35 : 41.37	38 ± 9.5
	2	7 : 35 : 41.79	37 ± 9.3
	3	7 : 35 : 42.02	40 ± 10
	4	7 : 35 : 42.52	35 ± 8.8
	5	7 : 35 : 42.94	29 ± 7.3
	6	7 : 35 : 44.06	37 ± 9.3
	7	7 : 35 : 46.38	20 ± 5.0
	8	7 : 35 : 46.96	24 ± 6.0
Baksan	1	7 : 36 : 11.818	12 ± 2.4
	2	7 : 36 : 12.253	18 ± 3.6
	3	7 : 36 : 13.528	23.3 ± 4.7
	4	7 : 36 : 19.505	17 ± 3.4
	5	7 : 36 : 20.917	20.1 ± 4.0
Mt Blanc	1	2 : 52 : 36.79	7 ± 1.4
	2	2 : 52 : 40.65	8 ± 1.6
	3	2 : 52 : 41.01	11 ± 2.2
	4	2 : 52 : 42.70	7 ± 1.4
	5	2 : 52 : 43.80	9 ± 1.8

decayed in transit, which leads to a lower limit on the lifetime for $\bar{\nu}_e$ of [Moh91]

$$\left(\frac{E_\nu}{m_\nu}\right) \tau_{\bar{\nu}_e} \geq 5 \times 10^{12} \text{ s} \approx 1.5 \times 10^5 \text{ yr.} \tag{11.21}$$

In particular, the radiative decay channel for a heavy neutrino ν_H

$$\nu_H \rightarrow \nu_L + \gamma \qquad (11.22)$$

can be limited independently. No enhancement in γ-rays coming from the direction of SN 1987A was observed in the Gamma Ray Spectrometer (GRS) on the Solar Maximum Satellite (SMM) [Chu89]. The photons emanating from neutrino decay would arrive with a certain delay, as the parent heavy neutrinos do not travel at the speed of light. The delay is given by

$$\Delta t \simeq \frac{1}{2} D \frac{m_\nu^2}{E_\nu^2}. \qquad (11.23)$$

For neutrinos with a mean energy of 12 MeV and a mass smaller than 20 eV the delay is about 10 s, which should be reflected in the arrival time of any photons from the decay. The study by [Blu92] using SMM data resulted in

$$\frac{\tau_H}{B_\gamma} \geq 2.8 \times 10^{15} \frac{m_H}{\text{eV}} \text{ s} \qquad m_H < 50 \text{ eV} \qquad (11.24)$$

$$\frac{\tau_H}{B_\gamma} \geq 1.4 \times 10^{17} \text{ s} \qquad 50 \text{ eV} < m_H < 250 \text{ eV} \qquad (11.25)$$

$$\frac{\tau_H}{B_\gamma} \geq 6.0 \times 10^{18} \frac{\text{eV}}{m_H} \text{ s} \qquad m_H > 250 \text{ eV}. \qquad (11.26)$$

Here B_γ is the branching ratio of a heavy neutrino into the radiative decay channel. Thus, there is no hint of a neutrino decay.

11.4.3.2 Mass

Direct information about the mass is obtained from the observed spread in arrival time. The time of flight t_F of a neutrino with mass m_ν and energy E_ν ($m_\nu \gg E_\nu$) from the source (emission time t_0) to the detector (arrival time t) in a distance L is given by

$$t_F = t - t_0 = \frac{L}{v} = \frac{L}{c} \frac{E_\nu}{p_\nu c} = \frac{L}{c} \frac{E_\nu}{\sqrt{E_\nu^2 - m_\nu^2 c^4}} \approx \frac{L}{c} \left(1 + \frac{m_\nu^2 c^4}{2 E^2} \right). \qquad (11.27)$$

If $m_\nu > 0$, the time of flight is getting shorter if E_ν increases. For two neutrinos with E_1 and E_2 ($E_2 > E_1$) emitted at times t_{01} and t_{02} ($\Delta t_0 = t_{02} - t_{01}$) the time difference on Earth is

$$\Delta t = t_2 - t_1 = \Delta t_0 + \frac{L m_\nu^2}{2c} \left(\frac{1}{E_2^2} - \frac{1}{E_1^2} \right). \qquad (11.28)$$

Here Δt, L, E_1, E_2 are known, Δt_0 and m_ν are unknown. Depending on which events from table 11.2 are combined and assuming simultaneously a reasonable

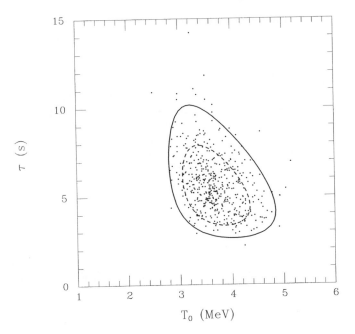

Figure 11.13. Loredo plot of parameter estimation of neutrino mass (from [Lor02]).

emission interval, mass limits lower than 30 eV (or even somewhat smaller at the price of model dependence) are obtained [Arn80, Kol87]. The new analysis by [Lor02] sharpens this bound even more and concludes that (see figure 11.13)

$$m_{\bar{\nu}_e} < 5.7 \, \text{eV} \qquad (95\% \, \text{CL}) \tag{11.29}$$

not too far away from the current β-decay results (see chapter 6). What can be done in the future with Super-Kamiokande and SNO, assuming a galactic supernova at a distance of 10 kpc, was studied in Monte Carlo simulations [Bea98] and is shown in figure 11.14.

11.4.3.3 Magnetic moment and electric charge

If neutrinos have a magnetic moment, precession could convert left-handed neutrinos into right-handed ones due to the strong magnetic field ($\sim 10^{12}$ G) of the neutron star. Such right-handed neutrinos would then be sterile and would immediately escape, thereby forming an additional energy loss mechanism and shortening the cooling time. Especially for high energy neutrinos ($E_\nu >$ 30 MeV), this is important. During the long journey through the galactic magnetic field some sterile neutrinos might be rotated back to ν_L. No shortening of the neutrino pulse is seen and the agreement of the observed number of neutrinos

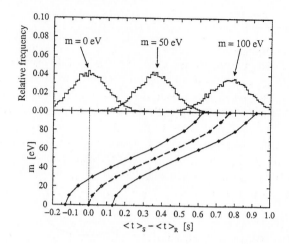

Figure 11.14. The result of a study of 10 000 simulated 10 kpc neutrino bursts in Super-Kamiokande and SNO. The top frame shows distributions of relative average time delays between NC events tagged in SNO and $\bar{\nu}_e$ events tagged in Super-Kamiokande, for three representative ν_μ, ν_τ masses. The bottom frame shows the 90% CL band on neutrino mass that one would obtain for a given measured delay. The vertical line at zero delay shows that one would obtain a 30 eV mass limit for ν_μ, ν_τ if the NC events arrive in time with the $\bar{\nu}_e$ events (from [Bea98]).

with the expected one implies an upper limit of [Lat88, Bar88]

$$\mu_\nu < 10^{-12}\mu_B. \tag{11.30}$$

However, many assumptions have been made in arriving at this conclusion, such as those of isotropy and equally strong emission of different flavours, as well as the magnetic field of the possibly created pulsar, so that the exact limit has to be treated with some caution.

A further limit exists on the electric charge of the neutrino. This is based on the fact that neutrinos with a charge e_ν would be bent along their path in the galactic field leading to a time delay of [Raf96]

$$\frac{\Delta t}{t} = \frac{e_\nu^2(B_T d_B)^2}{6E_\nu^2} \tag{11.31}$$

with B_T as the transverse magnetic field and d_B the path length in the field. The fact that all neutrinos arrived within about 10 s results in [Bar87]

$$\frac{e_\nu}{e} < 3 \times 10^{-17} \left(\frac{1\,\mu\mathrm{G}}{B_T}\right) \left(\frac{1\,\mathrm{kpc}}{d_B}\right). \tag{11.32}$$

More bounds like a test of relativity can be found in [Raf96].

11.4.3.4 Conclusion

Our knowledge of supernova explosions has grown enormously in recent years because of SN 1987A, not least from the first confirmation that supernovae type II really are phenomena from the late phase in the evolution of massive stars and that the energy released corresponds to the expectations. Also, the first detection of neutrinos has to be rated as a particularly remarkable event. As to how far the observed data are specific to SN 1987A, and to what extent they have general validity, only further supernovae of this kind can show. SN 1987A initiated a great deal of experimental activity in the field of detectors for supernova neutrinos, so that we now turn to the prospects for future experiments. The experiments themselves as well as the likely occurrence of supernovae are important.

11.5   Supernova rates and future experiments

How often do type II SN occur in our galaxy and allow us additional observations of neutrinos? Two ways are generally considered [Raf02]. First of all, historical records of supernovae in our galaxy can be explored, by counting supernova remnants or historic observations. Another way is to study supernova rates in other galaxies, which depends on the morphological structure of the galaxy. This can then be converted into a proposed rate for our Milky Way. Because of the small statistics involved and further systematic uncertainties a rate of between one and six supernovae per century seems to be realistic.

Under these assumptions let us discuss future experiments for the detection of supernova neutrinos besides the ones already discussed [Bur92, Cli92]. The main new idea available is to use high-Z detectors (larger cross section) in combination with neutron detectors. These detectors are primarily sensitive to high energy neutrinos and, therefore, offer a chance to measure ν_μ and ν_τ fluxes [Cli94]. Here also NC and CC processes are considered:

$$\nu_x + (A, Z) \rightarrow (A - 1, Z) + n + \nu_x \qquad \text{(NC)} \qquad (11.33)$$

$$\nu_e + (A, Z) \rightarrow (A - 1, Z + 1) + n + e^- \qquad \text{(CC)} \qquad (11.34)$$

$$\bar{\nu}_e + (A, Z) \rightarrow (A - 1, Z - 1) + n + e^+ \qquad \text{(CC)}. \qquad (11.35)$$

By comparing the NC/CC rates, sensitive neutrino oscillation tests can be performed: ^{208}Pb seems to be particularly promising [Ful98]. A proposal in the form of OMNIS [Smi97,Smi01,Zac02] exists. OMNIS plans to use 12 kt of Pb/Fe (4 kt of lead, 8 kt of iron) in the form of slabs interleaved with scintillators loaded with 0.1% of Gd for neutron capture.

The proposed ICARUS experiment (see chapter 8) also offers good detection capabilities especially for ν_e due to the reaction

$$\nu_e + {}^{40}\text{Ar} \rightarrow {}^{40}\text{K}^* + e^- \qquad (11.36)$$

where the coincidence of the electron with the de-excitation gamma rays can be used. This is important in detecting the breakout signal. A much larger version in the form of a 80 kt LAv TPC (LANNDD) has been suggested.

The total number of available experiments is impressive and will, at the time of the next nearby supernova, provide a lot of new information on neutrino properties [Bea99]. A global network of available detectors called SuperNova Early Warning System (SNEWS) has been established [Sch01].

A completely different way of detecting supernovae utilizes the vibrations of spacetime due to the powerful explosion. This appears in the form of gravitational waves. Up to now only indirect evidence of their existence exists: the slowdown in the period of the pulsar PSR1913+16 can be described exactly by assuming the emission of gravitational waves [Tay94]. For direct detection several detectors in the form of massive resonance masses in the form of bars (e.g. EXPLORER, Auriga, ALLEGRO) and spheres (MiniGrail) or in the form of laser interferometers (e.g. VIRGO, LIGO, GEO and others exist. The search for the small vibrations of spacetime of which supernovae are a good source has started recently and exciting results can be expected in the near future. The ultimate goal would be a laser interferometer in space. Such a project (LISA) is proposed and may become a cornerstone project of the European Space Agency.

11.5.1 Cosmic supernova relic neutrino background

Beside the supernova rate in our galaxy, for direct detection it might be worthwhile asking whether there is a supernova relic neutrino background (SRN) accumulated over cosmological times. However, this is not an easy task to perform [Tot95, Tot96, Mal97, Kap00]. The main uncertainty is in determining the supernova rate as a function of galactic evolution. Additional uncertainties arise from cosmological parameters. The SRN flux depends approximately quadratic on the Hubble constant and weakly on the density parameter (Ω_0) and the cosmological constant (Λ) (see chapter 13). A calculated flux [Tot95] is shown in figure 11.15. As can be seen, solar and terrestrial neutrino sources are overwhelming below 15 MeV, starting at about 50 MeV atmospheric neutrinos dominate. Therefore, only a slight window between 15–40 MeV exists for possible detection. Converting this flux into an event rate prediction for Super-Kamiokande results in 1.2 events per year with an uncertainty of a factor three [Tot95]. Recently, Super-Kamiokande measured an upper bound of $1.2\bar{\nu}_e$ cm^{-2} s^{-1} for the SRN flux in the energy range $E_\nu > 19$ MeV [Mal03], which is in the order of some theoretical predictions.

11.6   Neutrino oscillations and supernova signals

Finally we want to discuss in general terms the interplay between supernovae and neutrinos. Having described what has been learned about the properties of the neutrino from SN 1987A, we now concentrate on effects which neutrino

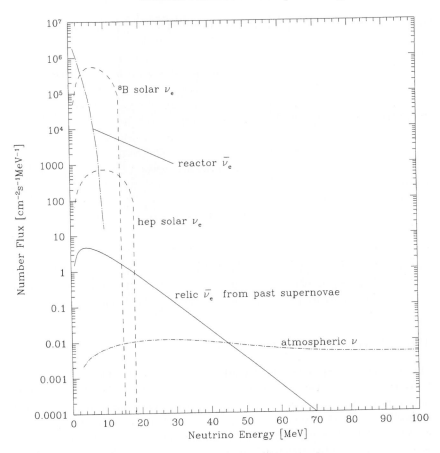

Figure 11.15. Supernova relic neutrino background compared with other sources (from [Tot95]).

oscillations might have on supernova explosions and *vice versa*. A supernova as a complex object has a variety of densities and density gradients to enable the MSW and RSFP effects to work (see chapter 10), and some of these effects will be discussed now. MSW effects have to be expected because of the LMA solution of the solar neutrino problem and must be taken into account. Possible observable effects are the disappearance of some ν_e from the neutronization burst, distortions of the ν_e energy spectrum, interchange of the original spectra, modifications to the $\bar{\nu}_e$ spectrum and Earth matter effects.

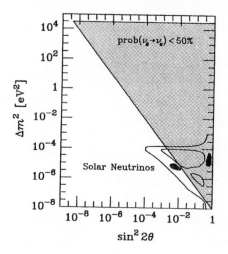

Figure 11.16. MSW triangle for the prompt ν_e burst from a stellar collapse. In the shaded area the conversion probability exceeds 50% for $E_\nu = 20$ MeV. The solar LMA and SMA solutions as well as the old Kamiokande allowed range are also shown (from [Raf96]).

11.6.1 Effects on the prompt ν_e burst

One obvious consequence which could be envisaged is that the deleptonization burst of ν_e could be much harder to detect because of ν_e–$\nu_{\mu,\tau}$ oscillations and the correspondingly smaller interaction cross section. The solar ν_e flux is depleted by the MSW mechanism so the same should occur in supernovae (see also section 10.3.2). The main difference is that instead of the possible exponential electron density profile used for solar neutrinos the electron density is now better approximated by a power law

$$n_e \approx 10^{34}\ \text{cm}^{-3} \left(\frac{10^7\ \text{cm}}{r}\right)^3 \tag{11.37}$$

where 10^7 cm is the approximate radius where the ν_e are created. [Noe87] found a conversion probability of more than 50% if

$$\Delta m^2 \sin^3 2\theta > 4 \times 10^{-9}\ \text{eV}^2 \frac{E_\nu}{10\ \text{MeV}} \tag{11.38}$$

assuming that $\Delta m^2 \lesssim 3 \times 10^4$ eV$^2 E_\nu/10$ MeV to ensure that the resonance is outside the neutrino sphere. The region with larger than 50% conversion probability region for $E_\nu = 20$ MeV is shown in figure 11.16. As can be seen, the LMA solution implies a significant conversion, making a direct observation of this component more difficult.

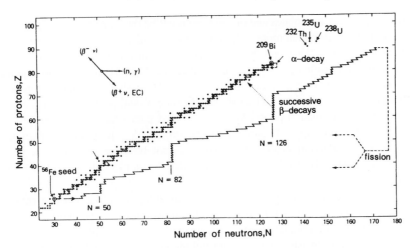

Figure 11.17. Illustration of the s- and r-process paths. Both processes are determined by (n, γ) reactions and β-decay. In the r-process the neutron-rich nuclei decay back to the peninsula of stable elements by β-decay after the neutron density has fallen. Even the uranium island is still reached in this process. The s-process runs along the stability valley (from [Rol88]).

11.6.2 Cooling phase neutrinos

Neutrino oscillations could cause a partial swap $\nu_e \leftrightarrow \nu_{\mu,\nu_\tau}$ and $\bar{\nu}_e \leftrightarrow \bar{\nu}_\mu, \bar{\nu}_\tau$, so that the measured flux at Earth could be a mixture of original $\bar{\nu}_e$, $\bar{\nu}_\mu$ and $\bar{\nu}_\tau$ source spectra. However, note that in a normal mass hierarchy no level crossing occurs for antineutrinos. The LMA mixing angles are large and imply significant spectral swapping. Applying this swapping to SN 1987A data leads to somewhat contradictory results [Smi94, Lun01, Min01, Kac01].

11.6.3 Production of r-process isotopes

Supernovae are cauldrons for the production of heavy elements. As already discussed, lighter elements are converted by fusion up to iron-group elements, where no further energy can be obtained by fusion and thus no heavier elements can be created. These heavier elements could also not have been created in sufficient amounts via charged particle reactions, due to the increased Coulomb barriers—other mechanisms must have been at work. As proposed by Burbidge, Burbidge, Fowler and Hoyle (B^{2}FH) [Bur57] heavier isotopes are produced by the competing reactions of neutron capture (n, γ-reactions) and nuclear β-decay. Depending now on the β-decay lifetimes and the lifetime against neutron capture two principal ways can be followed in the nuclide chart (figure 11.17). For rather low neutron fluxes and, therefore, a *slow* production of isotopes via neutron capture, β-decay dominates and the process is called the s-process (slow process). Element production then more or less follows the 'valley of stability' up to ^{209}Bi, where strong α-decay stops this line. However, in an environment with very high neutron densities ($n_n \approx 10^{20}$ cm^{-3}) there could be several neutron captures on one isotope before β-decay lifetimes are getting short enough to compete. This process is called the r-process (rapid process) and pushes elements in the sometimes experimental poorly known region of nuclei far beyond stability. In this way elements up to U and Th can be produced. These are the dominant processes for element production. There might be other processes at work, i.e. the p-process for producing neutron-depleted isotopes but they are of minor importance.

The point of importance here is the still unknown place where r-process production actually occurs in nature. Supernovae are a very good candidate because enough iron nuclei are available as seeds and sufficient neutron densities can be achieved. In particular, the hot bubble between the settled protoneutron star and the escaping shock wave at a few seconds after core bounce might be an ideal high-entropy environment for this process. Naturally, the r-process can only occur in a neutron-rich medium ($Y_e < 0.5$). The p/n ratio in this region is governed by neutrino spectra and fluxes because of the much higher number density with respect to the ambient e^+e^- population. The system is driven to a neutron-rich phase because normally the $\bar{\nu}_e$ are more energetic than ν_e therefore preferring β-reactions of the type $\bar{\nu}_e p \leftrightarrow n e^+$ with respect to $\nu_e n \leftrightarrow p e^-$.

If $\nu_e \leftrightarrow \nu_{\mu,\nu_\tau}$ oscillations happen outside the neutrino sphere, a subsequent flux of ν_e can be produced which is more energetic than $\bar{\nu}_e$ because the original ν_{μ,ν_τ} are more energetic (see section 11.2). It was shown that a partial swap of only 10% is sufficient to drive the medium into a proton-rich state ($Y_e > 0.5$), to be compared with the standard values of $Y_e = 0.35$–0.46. If such oscillations occur, this would prohibit r-process nucleosynthesis in supernovae [Qia93, Qia95, Qia01]. The Δm^2 parameter range for which this effect occurs is determined by the density profile of the supernova core a few seconds after bounce. The parameters lie in the cosmological interesting region, reaching down to neutrino masses of about 3 eV. For smaller masses, the oscillations occur at radii too large to have an impact on nucleosynthesis. A detailed analysis results in parameter region presented in figure 11.18 which would prevent r-process nucleosynthesis. On average $\sin^2 2\theta$ should be smaller than 10^{-5}. The effect only becomes of importance if the LSND evidence (see chapter 8) is correct. In addition, the energetic neutrinos might produce new nuclei by the neutrino process [Dom78, Woo88, Woo90, Heg03]. Their abundance might serve as 'thermometer' and allows information on the neutrino spectrum and oscillations.

11.6.4 Neutrino mass hierarchies from supernove signals

Recently, detailed studies have been performed to disentangle information about the neutrino mass hierarchies from a high statistics supernovae observation

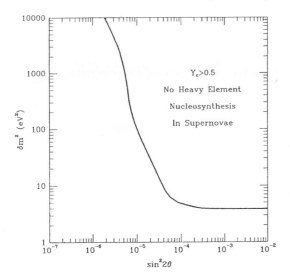

Figure 11.18. Matter-enhanced mixing between ν_μ (ν_τ) and ν_e with parameters in the labelled region drives the material in the neutrino-driven wind proton rich (corresponding to an electron fraction $Y_e > 0.5$). Hence this region is incompatible with supernova r-process nucleosynthesis (from [Qia01]).

[Dig00, Tak02, Lun03]. Assume only the solar and atmospheric evidence and that $\Delta m_{32}^2 = \Delta m_{atm}^2$ and $\Delta m_{21}^2 = \Delta m_\odot^2$. The two key features for the discussion are the neutrino mass hierarchy $|\Delta m_{32}^2| \approx |\Delta m_{31}^2| \gg |\Delta m_{21}^2|$ and the upper bound $|\Delta m_{ij}^2| < 10^{-2}$ eV^2. It should be noted that $\Delta m_{32}^2 > 0$ implies $m_3 > m_2, m_1$ and $\Delta m_{32}^2 < 0$ means $m_3 < m_2, m_1$ (see chapter 5). The major difference is that, in the first case, the small U_{e3} admixture to ν_e is the heaviest state while, in the inverted case, it is the lightest one. Since the ν_μ and ν_τ spectra are indistinguishable only the U_{ei} elements are accessible. Unitarity ($\sum U_{ei} = 1$) further implies that a discussion of two elements is sufficient, for example U_{e2} and U_{e3}, where the latter is known to be small. In that case U_{e2} can be obtained from the solar evidence:

$$4|U_{e2}|^2|U_{e1}|^2 \approx 4|U_{e2}|^2(1 - |U_{e2}|^2) = \sin^2 2\theta_\odot. \qquad (11.39)$$

The system is then determined by two pairs of parameters

$$(\Delta m_L^2, \sin^2 2\theta_L) \simeq (\Delta m_\odot^2, \sin^2 2\theta_\odot)$$
$$(\Delta m_H^2, \sin^2 2\theta_H) \simeq (\Delta m_{atm}^2, 4|U_{e3}|^2). \qquad (11.40)$$

The resonance density for the MSW effect is given by

$$\rho_{\text{res}} \approx \frac{1}{2\sqrt{2}G_F} \frac{\Delta m^2}{E} \frac{m_N}{Y_e} \cos 2\theta$$

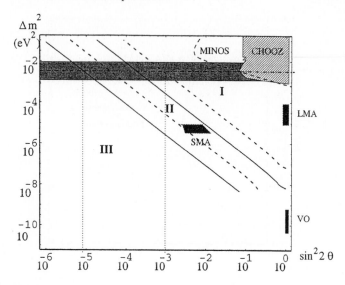

Figure 11.19. The contours of equal flip probability. The full lines denote the contours of flip probability for a 5 MeV neutrino, with the left-hand one stands for a 90% flip (highly non-adiabatic) and the right-hand one for 10% (adiabatic). The broken lines represents the corresponding curves for 50 MeV. The two vertical lines indicate the values of 4 $|U_{e3}|^2 = \sin^2 2\theta$ lying on the borders of adiabatic (I), non-adiabatic (III) and transition (II) regions for Δm^2 corresponding to the best-fit value of the atmospheric solution (from [Dig00].

$$\approx 1.4 \times 10^6 \text{ g cm}^{-3} \frac{\Delta m^2}{1 \text{ eV}^2} \frac{10 \text{ MeV}}{E} \frac{0.5}{Y_e} \cos 2\theta. \qquad (11.41)$$

This implies, for the two-parameter sets, a resonance at $\rho = 10^3$–10^4 g cm^{-3} and $\rho = 10$–30 g cm^{-3}, the latter if the L parameters lie within the LMA region. Both resonance regions are outside the supernova core, more in the outer layers of the mantle. This has some immediate consequences: the resonances do not influence the dynamics of the collapse and the cooling. In addition the possible r-process nucleosynthesis does not occur. The produced shock wave has no influence on the MSW conversion and the density profile assumed for resonant conversion can be almost static and to be the same as that of the progenitor star. Furthermore, in regions with $\rho > 1$ g cm^{-3}, Y_e is almost constant and

$$\rho Y_e \approx 2 \times 10^4 \left(\frac{r}{10^9 \text{ cm}} \right)^{-3} \qquad (11.42)$$

is a good approximation. However, the exact shape depends on details of the composition of the star.

The transition regions can be divided into three parts: a fully adiabatic part, a transition region and a section with strong violation of adiabaticity. This is

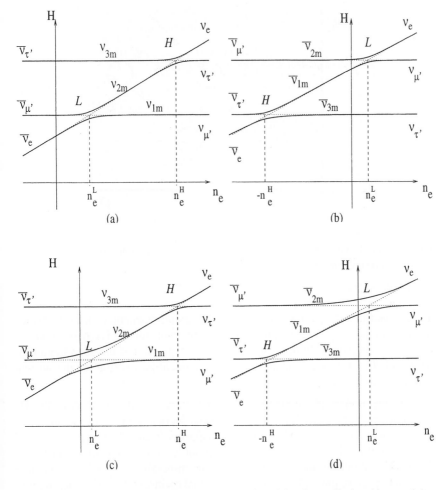

Figure 11.20. Level crossing diagrams for small solar mixing (upper line) and large mixing (lower line) as is realized by LMA. The left-hand column corresponds to normal and the right-hand to inverted mass hierarchies. The part of the plot with $N_e < 0$ is the antineutrino channel.

shown in figure 11.19. As can be seen, the H-resonance is adiabatic if $\sin^2 2\theta_{e3} = 4|U_{e3}|^2 > 10^{-3}$ and the transition region corresponds to $\sin^2 2\theta_{e3} \approx 10^{-5}$–$10^{-3}$. The features of the final neutrino spectra strongly depend on the position of the resonance. The smallness of $|U_{e3}|^2$ allows the two resonances to be discussed independently. The location of the resonances in the different mass schemes are shown in figure 11.20. Note that for antineutrinos the potential is $V = -\sqrt{2}G_F N_e$ (see chapter 10). They can be drawn in the same diagram and be envisaged as neutrinos moving through matter with an effective $-N_e$. It

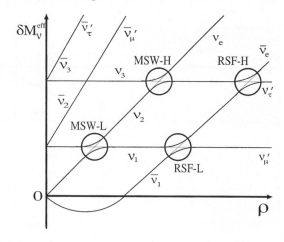

Figure 11.21. Level crossing diagrams for the combined appearance of RSFP and MSW resonances (from [And03]).

is important that the starting points are at the extremities and that the neutrinos move towards $N_e \rightarrow 0$. As can be seen in the plot, for the LMA solution the L-resonance is always in the neutrino sector independent of the mass hierarchy, while the H-resonance is in the neutrino (antineutrino) sector for the normal (inverted) hierarchy.

As mentioned, the value of $|U_{e3}|^2$ and, therefore, the position of the H-resonance is important. Consider the normal hierarchy first. In the adiabatic region, the neutronization burst almost disappears from ν_e and appears as ν_{ν_μ}. The $\bar{\nu}_e$ spectrum is a composite of the original $\bar{\nu}_e$ and ν_{ν_μ} spectrum and the ν_e spectrum is similar to the ν_{ν_μ} spectrum. The observable ν_{ν_μ} spectrum contains components of all three ($\bar{\nu}_e$, ν_e and ν_{ν_μ}) original spectra. In the transition region, the neutronization burst is a mixture of ν_e and ν_{ν_μ}, the ν_e and $\bar{\nu}_e$ spectra are composite and the ν_{ν_μ} is as in the adiabatic case. The signal in the non-adiabatic case is similar to that in the transition region but now there can be significant Earth matter effects for ν_e and $\bar{\nu}_e$.

In the inverted mass scheme, the adiabatic region shows a composite neutronization burst, the ν_e and ν_{ν_μ} spectra are composite and $\bar{\nu}_e$ is practically all ν_{ν_μ}. A strong Earth matter effect can be expected for ν_e. In the transition region, all three spectra are a composite of the original ones, the neutronization burst is a composite and Earth matter effects show up for ν_e and $\bar{\nu}_e$.

Therefore, for a future nearby supernova, an investigation of the neutronization burst and its possible disappearance, the composition and hardness of the various spectra and the observation of the Earth matter effect might allow conclusions on the mass hierarchy to be drawn.

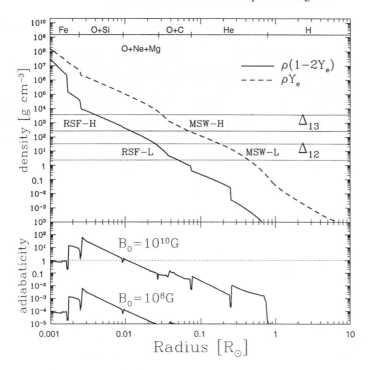

Figure 11.22. Position of the various resonances within the supernova. The two horizontal bands correspond to $\Delta_{12} = \Delta m_{12}^2 \cos\theta_{12}/2\sqrt{2}G_F E_\nu$ and $\Delta_{13} = \Delta m_{13}^2 \cos\theta_{13}/2\sqrt{2}G_F E_\nu$. The band width comes form the energy range 5–70 MeV. At intersections between the full, broken curves and the horizontal bands, the RSF and MSW conversions occur (from [And03]).

11.6.5 Resonant spin flavour precession in supernovae

The large magnetic field associated with the proto-neutron star also makes resonant spin flavour precession (RSFP) discussed in chapter 10 important if neutrinos have a corresponding magnetic moment. Various analyses have been performed mostly in a two-flavour scenario using $\bar\nu_e \to \nu_\mu$. This occurs if Y_e is smaller than 0.5 otherwise $\nu_e \to \bar\nu_\mu$ will occur. However, a full discussion will involve three flavours and matter oscillations. In this full description five resonance points occur: two MSW (ν_e–ν_μ and ν_e–ν_τ) and three RSFP ($\bar\nu_e$–ν_μ, $\bar\nu_e$–ν_τ and $\bar\nu_\mu$–ν_τ) resonances, as shown in figure 11.21. A recent analysis has been performed by [And03], using a progenitor model developed by [Woo95]. The location of the resonance is shown in figure 11.22. The density profile stays static for about 0.5 s, a period in which 50% of all neutrinos are emitted. After that, because of rapid changes in density and the corresponding effects on the resonance points, detailed predictions of the emitted neutrino spectrum are difficult. As long as this cannot be improved, it will be difficult to tell which mechanisms are at work in a possible observation.

This has only been a short glimpse of the strong interplay between neutrino and supernova physics. Many uncertainties and caveats exist, so the dominant part of the presented numbers should be handled with some caution and hopefully future observations will help to sharpen our view. Having discussed solar and supernova neutrinos, where experimental observations exist, we now come to other astrophysical neutrino sources, where the first neutrino discoveries might be possible in the near future.

Chapter 12

Ultra-high energetic cosmic neutrinos

Having discussed neutrinos from stars with energies $E_\nu < 100$ MeV which have already been observed, we now want to discuss additional astrophysical neutrinos produced with much higher energy. The observation of cosmic rays with energies beyond 10^{20} eV, even though the details are under intense debate, by the Fly's Eye and AGASA air shower arrays [Bir91, Aga01] supports the possibility of also observing neutrinos up to this energy range. Neutrinos mostly originate from the decay of secondaries like pions resulting from 'beam dump' interactions of protons with other protons or photons as in accelerator experiments. Like photons, neutrinos are not affected by a magnetic field. Even absorption which might affect photon detection is not an issue and this allows a search for hidden sources, which cannot be seen otherwise. Therefore, they are an excellent candidate for finding point sources in the sky and might help to identify the sources of cosmic rays. In addition, our view of the universe in photons is limited for energies beyond 1 TeV. The reason is the interaction of such photons with background photons $\gamma + \gamma_{BG} \rightarrow e^+e^-$. The reaction has a threshold of $4E_\gamma E_{\gamma_{BG}} \approx (2m_e)^2$. In this way TeV photons are damped due to reactions on the IR background and PeV photons by the cosmic microwave background (see chapter 13). For additional literature see [Sok89, Gai90, Ber91, Lon92, 94, Gai95, Lea00, Sch00, Gri01, Hal02].

12.1 Sources of high-energy cosmic neutrinos

The search for high-energy neutrinos might be split into two lines. One is the obvious search for point sources, in the hope that a signal will shed light on the question of what the sources of cosmic rays are. The second one would be a diffuse neutrino flux like the one observed in gamma rays. It is created by pion decay, produced in cosmic-ray interactions within the galactic disc. Instead of this more-observational-motivated division, the production mechanism itself can be separated roughly in two categories, i.e. acceleration processes and annihilation in combination with the decay of heavy particles. The acceleration process can be subdivided further into those of galactic and extragalactic origin.

12.1.1   Neutrinos produced in acceleration processes

The observation of TeV γ-sources together with the detection of a high-energy
($E_\gamma > 100$ MeV) diffuse galactic photon flux by the EGRET experiment on
the Compton Gamma Ray Observatory opened a new window into high-energy
astrophysics. Such highly energetic photons might be produced by electron
acceleration due to synchrotron radiation and inverse Compton scattering. In
addition, it is known that cosmic rays with energies up to 10^{20} eV exist, implying
the acceleration of protons in some astrophysical sources ('cosmic accelerators').
Observations of neutrinos would prove proton acceleration because of charged
pion production

$$p + p, p + \gamma \rightarrow \pi^0, \pi^\pm, K^\pm + X. \tag{12.1}$$

This process is similar to the production of artificial neutrino beams as described
in chapter 4. Associated with charged pion-production is π^0-production creating
highly-energetic photons. It would offer another source for TeV photons via
π^0-decay. Two types of sources exist in that way: diffusive production within
the galaxy by interactions of protons with the interstellar medium; and point-
like sources, where the accelerated protons interact directly in the surrounding
of the source. The latter production mechanism corresponds to an astrophysical
beam-dump experiment also creating neutrinos (figure 12.1). The dump must be
partially transparent for protons, otherwise it cannot be the source of cosmic rays.
These are guaranteed sources of neutrinos because it is known that the beam and
the target both exist (table 12.1). High energy photons are effected by interactions
with the cosmic microwave (see section 13.3) and infrared backgrounds, mostly
through $\gamma\gamma \rightarrow \ell^+\ell^-$ reactions, as they traverse intergalactic distances. This
limits the range for the search of cosmic sources, a boundary not existing for
neutrinos. The current TeV (the units are GeV/TeV/PeV/EeV/ZeV in ascending
factors of 10^3) γ-observations have failed to find proton accelerators. A positive
observation of neutrinos from point sources would be strong evidence for proton
acceleration. In addition, in proton acceleration, the neutrino and photon fluxes
are related [Hal97, Wax99, Bah01a, Man01a]. Three of the possible galactic point-
source candidates where acceleration can happen are:

- Young supernova remnants. Two mechanisms for neutrino generation by
  accelerated protons are considered. First, the inner acceleration, where
  protons in the expanding supernova shell are accelerated, e.g. by the strongly
  rotating magnetic field of the neutron star or black hole. The external
  acceleration is done by two shock fronts running towards each other.
- Binary systems. Here matter is transformed from a expanded star like a red
  giant towards a compact object like a neutron star or black hole. This matter
  forms an accretion disc which acts as a dynamo in the strong magnetic field
  of the compact object and also as a target for beam-dump scenarios.
- Interaction of protons with the interstellar medium like molecular clouds.

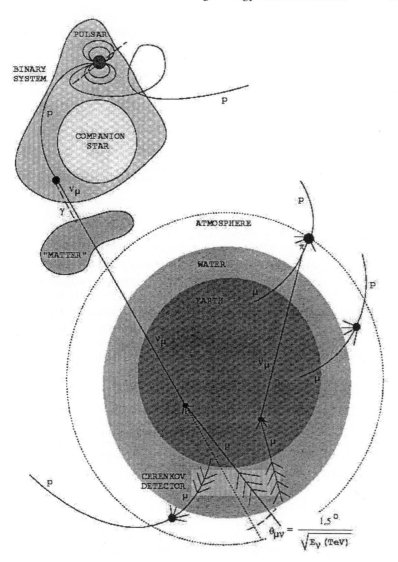

Figure 12.1. Schematic drawing of a 'cosmic beam dump' experiment at a cosmic 'accelerator'. Protons are accelerated towards high energy in a binary system and hitting matter in the accretion disc. The result is hadroproduction especially pions. While neutral pions decay into photons, charged pions decay into ν_{μ}. This shows a strong correlation between high-energy gamma rays and neutrinos. While photons might get absorbed or downscattered to lower energies neutrinos will find their way to the Earth undisturbed, allowing also a search for 'hidden' sources. On Earth they might be detected by their CC interactions, resulting in upward-going muons in a detector, because otherwise the atmospheric background is too large.

Table 12.1. Cosmic beam dumps. The first part consist of calculable source. The second part has uncertainties in the flux determination and an observation is not guaranteed.

Beam	Target
Cosmic rays	Atmosphere
Cosmic rays	Galactic disc
Cosmic rays	CMB
AGN jets	Ambient light, UV
Shocked protons	GRB photons

The Sun can also act as a source of high-energy neutrinos due to cosmic-ray interactions within the solar atmosphere [Sec91, Ing96]. As in basically all astrophysical beam dumps the target (here the atmosphere) is rather thin and most pions decay instead of interact. They can be observed at energies larger than about 10 TeV, where the atmospheric background is sufficiently low. However, the expected event rate for $E_\nu > 100$ GeV is only about 17 per km^3 yr. The galactic disc and galactic centre are good sources with a calculable flux most likely to be observable. Here the predicted flux can be directly related to the observed γ-emission [Dom93, Ing96a]. The galactic centre might be observable above 250 TeV and about 160 events km^{-2} yr^{-1} in a 5° aperture can be expected.

   These stellar sources are considered to accelerate particles up to about 10^{15} eV, where a well-known structure called 'the knee' is observed in the cosmic-ray energy spectrum (figure 12.2). The spectrum steepens from a power law behaviour of $dN/dE \propto E^{-\gamma}$ from $\gamma = 2.7$ to $\gamma = 3$. To be efficient, the size R of an accelerator should be at least the gyroradius R_g of the particle in an electromagnetic field:

$$R > R_g = \frac{E}{B} \rightarrow E_{max} = \gamma B R \qquad (12.2)$$

where, for the maximal obtainable energy E_{max}, the relativistic γ factor was included, because we may not be at rest in the frame of the cosmic accelerator. Using reasonable numbers for supernova remnants E_{max} values are obtained close to the knee position. For higher energies, stronger and probably extragalactic sources have to be considered, possible sources are shown in figure 12.3. The best candidates are active galactic nuclei (AGN), which in their most extreme form are also called quasars. A schematic picture of the quasar phenomenon is shown in figure 12.4. They are among the brightest sources in the universe and measurable to high redshifts. Moreover, they must be extremly compact, because variations in the luminosity are observed on the time scales of days. The appearance of two jets perpendicular to the accretion disc probably is an efficient place for particle acceleration. AGNs where the jet is in the line of sight to the Earth are

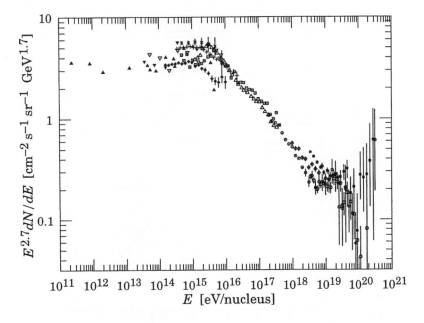

Figure 12.2. The energy spectrum of cosmic rays (from [PDG02]).

called blazars. For additional literature on high-energy phenomena see [Lon92, 94, Sch00]. In the current picture quasars correspond to the core of a young active galaxy, whose central 'engine' is a supermassive black hole ($M \approx 10^8 M_\odot$). The Schwarzschild radius given by

$$R_S = \frac{2GM}{c^2} = 2.95 \frac{M}{M_\odot} \text{ km} \tag{12.3}$$

of such black holes is about 3×10^8 km. They are surrounded by thick accretion discs. From there matter spirals towards the hole, strongly accelerated and transformed into a hot, electrically conducting plasma producing strong magnetic fields. Part of this infalling matter is absorbed into the black hole, part is redirected by the magnetic field, which then forms two plasma jets leaving to opposite sides and perpendicular to the disc. In such jets or their substructures (blobs) protons can be accelerated to very high energies.

Another extragalactic neutrino source becoming more and more prominent during the last years is gamma-ray bursters (GRBs) [Wax97, Vie98, Bot98, Mes02, Wax03]. The phenomenon has been discovered only in the gamma ray region with bursts lasting from 6 ms up to 2000 s. After being a big mystery for more than 20 yr, they are now observed with a rate of three per day and it is known that they are of cosmological origin. Recently a link between GRBs and supernova explosions has been established [Hjo03, Sta03, Del03]. Even though there is no detailed understanding of the internal mechanism of GRBs, the relativistic fireball

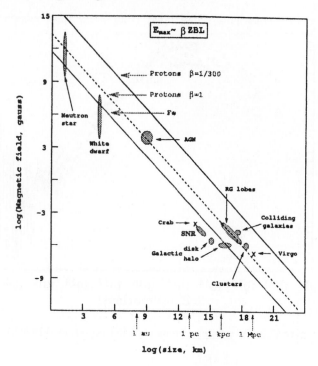

Figure 12.3. Hillas plot of sources.

model is phenomenologically successful. Expected neutrino energies cluster around 1–100 TeV for GRBs and 100 PeV for AGN jets assuming no beaming effects.

12.1.2   Neutrinos produced in annihilation or decay of heavy particles

Three kinds of such sources are typically considered:

- evaporating black holes,
- topological defects and
- annihilation or decay of (super-)heavy particles.

We concentrate here qualitatively only on the last possibility, namely neutralinos as relics of the Big Bang and candidates for cold dark matter. Neutralinos χ (discussed in chapter 5) are one of the preferred candidates for weakly interacting massive particles (WIMPs) to act as dark matter in the universe (see chapter 13). They can be accumulated in the centre of objects like the Sun or the Earth [Ber98]. The reason is that by coherent scattering on nuclei they lose energy and if they fall below the escape velocity they get trapped and, finally, by additional scattering

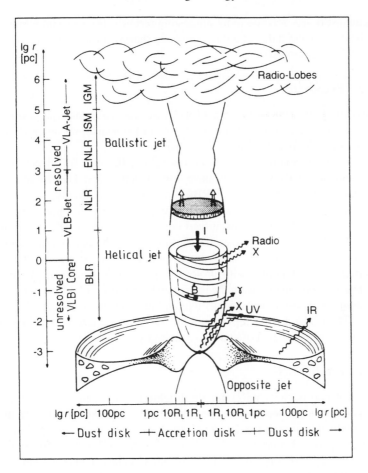

Figure 12.4. Model of the quasar phenomenon and related effects. Depending on the angle of observation, different aspects of this structure can be seen and, therefore, the richness of the phenomenon becomes understandable (from [Qui93]).

processes they accumulate in the core. The annihilation can proceed via

$$\chi + \chi \rightarrow b + \bar{b} \qquad (\text{for } m_\chi < m_W) \qquad (12.4)$$

or

$$\chi + \chi \rightarrow W^+ + W^- \qquad (\text{for } m_\chi > m_W). \qquad (12.5)$$

In both cases neutrinos are finally produced. The ν_μ component might be observed by the detectors described, by looking for the CC reaction. Because the created muon follows the incoming neutrino direction at these energies, it should point towards the Sun or the core of the Earth. Within that context, the galactic centre has become more interesting recently, because simulations show that the dark matter halos of galaxies may be sharply 'cusped' toward a galaxy's centre [Gon00, Gon00a].

12.1.3   Event rates

For experimental detection of any of these sources, three main parameters have to be known: the predicted neutrino flux from the source, the interaction cross section of neutrinos and the detection efficiency.

For various sources like AGNs and GRBs, the flux still depends on the model and, hence, the predictions have some uncertainties. The rate of neutrinos produced by $p\gamma$ interactions in GRBs and AGNs is essentially dictated by the observed energetics of the source. In astrophysical beam dumps, like AGNs and GRBs, typically one neutrino and one photon is produced per accelerated proton [Gai95, Gan96]. The accelerated protons and photons are, however, more likely to suffer attenuation in the source before they can escape. So, a hierarchy of particle fluxes emerges with protons < photons < neutrinos. Using these associations, one can constrain the energy and luminosity of the accelerator from the gamma- and cosmic-ray observations and subsequently anticipate the neutrino fluxes. These calculations represent the basis for the construction of kilometre-scale detectors as the goal of neutrino astronomy.

12.1.4 v from AGNs

Active galactic nuclei (AGN) are the brightest sources in the universe. It is anticipated that the beams accelerated near the central black hole are dumped on the ambient matter in the active galaxy. Typically two jets emerge in opposite directions, perpendicular to the disc of the AGN. An AGN viewed from a position illuminated by the cone of a relativistic jet is called a blazar. Particles are accelerated by Fermi shocks in blobs of matter, travelling along the jet with a bulk Lorentz factor of $\gamma \approx 10$ or higher.

In the estimate (following [Hal98]) of the neutrino flux from a proton blazar, primes will refer to a frame attached to the blob moving with a Lorentz factor γ relative to the observer. In general, the transformation between blob and observer frame is $R' = \gamma R$ and $E' = \frac{1}{\gamma}E$ for distances and energies, respectively. For a burst of 15 min duration, the strongest variability observed in TeV emission, the size of the accelerator, is only

$$R' = \gamma c \Delta t \sim 10^{-4} – 10^{-3} \text{pc} \qquad (12.6)$$

for $\gamma = 10–10^2$. So the jet consists of relatively small structures with short lifetimes. High-energy emission is associated with the periodic formation of these blobs.

Shocked protons in the blob will photoproduce pions on the photons whose properties are known from the observed multi-wavelength emission. From the observed photon luminosity L_γ, the energy density of photons in the shocked region can be deduced:

$$U'_\gamma = \frac{L'_\gamma \Delta t}{\frac{4}{3}\pi R'^3} = \frac{L_\gamma \Delta t}{\gamma} \frac{1}{\frac{4}{3}\pi(\gamma c \Delta t)^3} = \frac{3}{4\pi c^3} \frac{L_\gamma}{\gamma^4 \Delta t^2}.$$ (12.7)

(Geometrical factors of order unity will be ignored throughout.) The dominant photon density is at UV wavelengths, the UV bump. Assume that a luminosity $\mathcal{L}_\gamma$ of 10^{45} erg s^{-1} is emitted in photons with energy $E_\gamma = 10$ eV. Luminosities larger by one order of magnitude have actually been observed. The number density of photons in the shocked region is

$$N'_\gamma = \frac{U'_\gamma}{E'_\gamma} = \gamma \frac{U'_\gamma}{E_\gamma} = \frac{3}{4\pi c^3} \frac{L_\gamma}{E_\gamma} \frac{1}{\gamma^3 \Delta t^2} \sim 6.8 \times 10^{14} - 6.8 \times 10^{11} \text{ cm}^{-3}.$$ (12.8)

From now on the range of numerical values will refer to $\gamma = 10$–10^2, in that order. With such a high density the blob is totally opaque to photons with 10 TeV energy and above. Because photons with such energies have indeed been observed, one must essentially require that the 10 TeV γ are below the $\gamma\gamma \to e^+e^-$ threshold in the blob, i.e.

$$E_{\text{thr}} = \gamma E'_{\gamma \text{ thr}} \geq 10 \text{ TeV}$$ (12.9)

or

$$E_{\text{thr}} > \frac{m_e^2}{E_\gamma}\gamma^2 > 10 \text{ TeV}$$ (12.10)

or

$$\gamma > 10.$$ (12.11)

To be more conservative, the assumption $10 < \gamma < 10^2$ is used.

The accelerated protons in the blob will produce pions, predominantly at the Δ-resonance, in interactions with the UV photons. The proton energy for resonant pion production is

$$E'_p = \frac{m_\Delta^2 - m_p^2}{4} \frac{1}{E'_\gamma}$$ (12.12)

or

$$E_p = \frac{m_\Delta^2 - m_p^2}{4E_\gamma}\gamma^2$$ (12.13)

$$E_p = \frac{1.6 \times 10^{17} \text{ eV}}{E_\gamma}\gamma^2 = 1.6 \times 10^{18} - 1.6 \times 10^{20} \text{ eV}.$$ (12.14)

The secondary ν_μ have energy

$$E_\nu = \frac{1}{4}\langle x_{p\to\pi}\rangle E_p = 7.9 \times 10^{16} - 7.9 \times 10^{18} \text{ eV}$$ (12.15)

for $\langle x_{p\to\pi}\rangle \simeq 0.2$, the fraction of energy transferred, on average, from the proton to the secondary pion produced via the Δ-resonance. The $\frac{1}{4}$ is because each lepton in the decay $\pi \to \mu\nu_\mu \to e\nu_e\nu_\mu\bar{\nu}_\mu$ carries roughly equal energy. The fraction of energy f_π lost by protons to pion production when travelling a distance R' through a photon field of density N'_γ is

$$f_\pi = \frac{R'}{\lambda_{p\gamma}} = R'N'_\gamma\sigma_{p\gamma\to\Delta}\langle x_{p\to\pi}\rangle \tag{12.16}$$

where $\lambda_{p\gamma}$ is the proton interaction length, with $\sigma_{p\gamma\to\Delta\to n\pi^+} \simeq 10^{-28}$ cm^2 resulting in

$$f_\pi = 3.8\text{--}0.038 \qquad \text{for } \gamma = 10\text{--}10^2. \tag{12.17}$$

For a total injection rate in high-energy protons $\dot{E}$, the total energy in ν is $\frac{1}{2}f_\pi t_H\dot{E}$, where $t_H = 10$ Gyr is the Hubble time. The factor $\frac{1}{2}$ accounts for the fact that half of the energy in charged pions is transferred to $\nu_\mu + \bar{\nu}_\mu$, see earlier. The neutrino flux is

$$\Phi_\nu = \frac{c}{4\pi}\frac{(\frac{1}{2}f_\pi t_H\dot{E})}{E_\nu}e^{f_\pi}. \tag{12.18}$$

The last factor corrects for the absorption of the protons in the source, i.e. the observed proton flux is a fraction e^{-f_π} of the source flux which photoproduces pions. We can write this as

$$\Phi_\nu = \frac{1}{E_\nu}\frac{1}{2}f_\pi e^{f_\pi}(E_p\Phi_p). \tag{12.19}$$

For $E_p\Phi_p = 2 \times 10^{-10}$ TeV (cm^2 s sr)$^{-1}$, we obtain

$$\Phi_\nu = 8 \times 10^5 \text{ to } 2 \text{ (km}^2 \text{ yr)}^{-1} \tag{12.20}$$

over 4π sr. (Neutrino telescopes are background free for such high-energy events and should be able to identify neutrinos at all zenith angles.) The detection probability is computed from the requirement that the neutrino has to interact within a distance of the detector which is shorter than the range of the muon it produces. Therefore,

$$P_{\nu\to\mu} \simeq \frac{R_\mu}{\lambda_{\text{int}}} \simeq AE_\nu^n \tag{12.21}$$

where R_μ is the muon range and λ_{int} the neutrino interaction length. For energies below 1 TeV, where both the range and cross section depend linearly on energy, $n = 2$. At TeV and PeV energies $n = 0.8$ and $A = 10^{-6}$, with E in TeV units. For EeV energies $n = 0.47$, $A = 10^{-2}$ with E in EeV [Gai95, Gan96]. The observed neutrino event rate in a detector is

$$N_{\text{events}} = \Phi_\nu P_{\nu\to\mu} \tag{12.22}$$

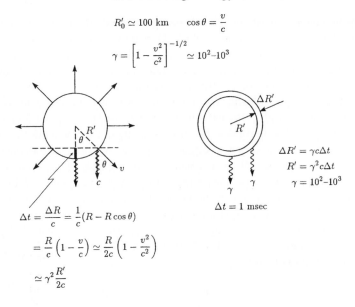

Figure 12.5. Kinematics of GRBs (from [Hal98]).

with

$$P_{\nu \to \mu} \cong 10^{-2} E_{\nu,\text{EeV}}^{0.4} \qquad (12.23)$$

where E_ν is expressed in EeV. Therefore,

$$N_{\text{events}} = (3 \times 10^3 \text{ to } 5 \times 10^{-2}) \text{ km}^{-2} \text{ yr}^{-1} = 10^{1 \pm 2} \text{ km}^{-2} \text{ yr}^{-1} \qquad (12.24)$$

for $\gamma = 10$–10^2. This estimate brackets the range of γ factors considered. Notice, however, that the relevant luminosities for protons (scaled to the high-energy cosmic rays) and the luminosity of the UV target photons are themselves uncertain. The large uncertainty in the calculation of the neutrino flux from AGN is predominantly associated with the boost factor γ.

12.1.5 ν from GRBs

Recently, GRBs may have become the best motivated source for high-energy neutrinos. Their neutrino flux can be calculated in a relatively model-independent way. Although neutrino emission may be less copious and less energetic than that from AGNs, the predicted fluxes can probably be bracketed with more confidence.

In GRBs, a fraction of a solar mass of energy ($\sim 10^{53}$ erg) is released over a time scale of order 1 s into photons with a very hard spectrum. It has been suggested that, although their ultimate origin is a matter of speculation, the same cataclysmic events also produce the highest energy cosmic rays. This association is reinforced by more than the phenomenal energy and luminosity.

The phenomenon consists of three parts. First of all there must be a central engine, whose origin is still under debate. Hypernovae and merging neutron stars are among the candidates. The second part is the relativistic expansion of the fireball. Here, an Earth size mass is accelerated to 99.9% of the velocity of light. The fireball has a radius of 10–100 km and releases an energy of 10^{51-54} erg. Such a state is opaque to light. The observed gamma rays display is the result of a relativistic shock with $\gamma \approx 10^2\text{--}10^3$ which expands the original fireball by a factor of 10^6 over 1 s. For this to be observable, there must be a third condition, namely an efficient conversion of kinetic energy into non-thermal gamma rays.

The production of high-energy neutrinos is a feature of the fireball model because, like in the AGN case, the protons will photoproduce pions and, therefore, neutrinos on the gamma rays in the burst. This is a beam-dump configuration where both the beam and target are constrained by observation: of the cosmic-ray beam and of the photon fluxes at Earth, respectively. Simple relativistic kinematics (see figure 12.5) relates the radius and width R', $\Delta R'$ to the observed duration of the photon burst $c\Delta t$:

$$R' = \gamma^2(c\Delta t) \tag{12.25}$$

$$\Delta R' = \gamma c\Delta t. \tag{12.26}$$

The calculation of the neutrino flux follows the same path as that for AGNs. From the observed GRB luminosity L_γ, we compute the photon density in the shell:

$$U'_\gamma = \frac{(L_\gamma \Delta t/\gamma)}{4\pi R'^2 \Delta R'} = \frac{L_\gamma}{4\pi R'^2 c\gamma^2}. \tag{12.27}$$

The pion production by shocked protons in this photon field is, as before, calculated from the interaction length

$$\frac{1}{\lambda_{p\gamma}} = N_\gamma \sigma_\Delta \langle x_{p\to\pi}\rangle = \frac{U'_\gamma}{E'_\gamma}\sigma_\Delta \langle x_{p\to\pi}\rangle \qquad \left(E'_\gamma = \frac{1}{\gamma}E_\gamma\right) \tag{12.28}$$

σ_Δ is the cross section for $p\gamma \to \Delta \to n\pi^+$ and $\langle x_{p\to\pi}\rangle \simeq 0.2$. The fraction of energy going into π production is

$$f_\pi \cong \frac{\Delta R'}{\lambda_{p\gamma}} \tag{12.29}$$

$$f_\pi \simeq \frac{L_\gamma}{E_\gamma}\frac{1}{\gamma^4 \Delta t}\frac{\sigma_\Delta \langle x_{p\to\pi}\rangle}{4\pi c^2} \tag{12.30}$$

$$f_\pi \simeq 0.14 \left\{\frac{L_\gamma}{10^{51}\ \text{ergs}^{-1}}\right\}\left\{\frac{1\ \text{MeV}}{E_\gamma}\right\}\left\{\frac{300}{\gamma}\right\}^4\left\{\frac{1\ \text{ms}}{\Delta t}\right\}$$

$$\times \left\{\frac{\sigma_\Delta}{10^{-28}\ \text{cm}^2}\right\}\left\{\frac{\langle x_{p\to\pi}\rangle}{0.2}\right\}. \tag{12.31}$$

The relevant photon energy within the problem is 1 MeV, the energy where the typical GRB spectrum exhibits a break. The number of higher energy photons is suppressed by the spectrum and lower energy photons are less efficient at producing pions. Given the large uncertainties associated with the astrophysics, it is an adequate approximation to neglect the explicit integration over the GRB photon spectrum. The proton energy for production of pions via the Δ-resonance is

$$E_p' = \frac{m_\Delta^2 - m_p^2}{4E_\gamma'}. \tag{12.32}$$

Therefore,

$$E_p = 1.4 \times 10^{16} \text{ eV} \left(\frac{\gamma}{300}\right)^2 \left(\frac{1 \text{ MeV}}{E_\gamma}\right) \tag{12.33}$$

$$E_\nu = \tfrac{1}{4}\langle x_{p\to\pi}\rangle E_p \simeq 7 \times 10^{14} \text{ eV}. \tag{12.34}$$

We are now ready to calculate the neutrino flux:

$$\phi_\nu = \frac{c}{4\pi}\frac{U_\nu'}{E_\nu'} = \frac{c}{4\pi}\frac{U_\nu}{E_\nu} = \frac{c}{4\pi}\frac{1}{E_\nu}\left\{\frac{1}{2}f_\pi t_H \dot{E}\right\} \tag{12.35}$$

where the factor 1/2 accounts for the fact that only half of the energy in charged pions is transferred to $\nu_\mu + \bar{\nu}_\mu$. As before, $\dot{E}$ is the injection rate in cosmic rays beyond the 'ankle', a flattening of the cosmic-ray energy spectrum around 10^{18} eV, ($\sim 4 \times 10^{44}$ erg Mpc^{-3} yr^{-1}) and t_H is the Hubble time of $\sim 10^{10}$ Gyr. Numerically,

$$\phi_\nu = 2 \times 10^{-14} \text{ cm}^{-2} \text{ s}^{-1} \text{ sr}^{-1} \left\{\frac{7 \times 10^{14} \text{ eV}}{E_\nu}\right\} \left\{\frac{f_\pi}{0.125}\right\} \left\{\frac{t_H}{10 \text{ Gyr}}\right\}$$

$$\times \left\{\frac{\dot{E}}{4 \times 10^{44} \text{ erg Mpc}^{-3} \text{ yr}^{-1}}\right\}. \tag{12.36}$$

The observed muon rate is

$$N_{\text{events}} = \int_{E_{\text{thr}}}^{E_\nu^{\text{max}}} \Phi_\nu P_{\nu\to\mu} \frac{dE_\nu}{E_\nu} \tag{12.37}$$

where $P_{\nu\to\mu} \simeq 1.7 \times 10^{-6} E_\nu^{0.8}$ (TeV) for TeV energy. Therefore,

$$N_{\text{events}} \cong 26 \text{ km}^{-2} \text{ yr}^{-1} \left\{\frac{E_\nu}{7 \times 10^{14} \text{ eV}}\right\}^{-0.2} \left\{\frac{\Delta\theta}{4\pi}\right\}. \tag{12.38}$$

This number might be reduced by a factor of five due to absorption in the Earth. The result is insensitive to beaming. Beaming yields more energy per burst but less bursts are actually observed. The predicted rate is also insensitive to the neutrino energy E_ν because higher average energy yields less ν but more are detected. Both effects are approximately linear. A compilation of expected fluxes from various diffuse and point sources is shown in figure 12.6. As can be seen, the fluxes at high energies are very low, requiring large detectors. Below about 10 TeV the background due to atmospheric neutrinos will dominate. The rate is expected to be approximately the same in the case of GRBs coming from supernova explosions. The burst only lasts for about 10 s, where the burst is formed by the shock created in the transition of a supernova into a black hole.

12.1.6   Cross sections

Neutrinos can be divided into VHE neutrinos (very high energy) from pp reactions with $E_\nu > 50$ GeV and UHE (ultra-high energy) neutrinos from pγ reactions with $E_\nu > 10^6$ GeV [Ber91]. The reason for this ultra-high energy comes from the high threshold for pion production in photoproduction of nuclei

$$N + \gamma \rightarrow N' + \pi. \tag{12.39}$$

In a collinear collision the threshold is given by $s = (m_N + m_\pi)^2$. Using

$$s = (p_N + p_\gamma)^2 = m_N^2 + 2p_N p_\gamma = m_N^2 + 4E_N E_\gamma \tag{12.40}$$

it follows ($N \equiv p$)

$$E_P^S = \frac{(2m_p + m_\pi)}{4E_\gamma} = 7 \times 10^{16} \frac{\text{eV}}{E_\gamma} \text{ eV}. \tag{12.41}$$

Taking the cosmic microwave background as the photon source (see chapter 13) with $\langle E_\gamma \rangle \approx 7 \times 10^{-4}$ eV, a threshold of $E_P^S = 10^{20}$ eV $= 10^{11}$ GeV follows. This is the well-known Greisen–Zatsepin–Kuzmin (GZK) cutoff [Gre66, Zat66].

The cross section for CC νN interactions has already been given in chapter 4. However, we are dealing now with neutrino energies far beyond the ones accessible in accelerators, implying a few modifications. First of all, at energies of 10^4 GeV, a deviation of the linear rise with E_ν has to be expected because of the W-propagator, leading to a damping, an effect already observed by H1 (see chapter 4). Considering parton distribution functions in the UHE regime the heavier quarks, e.g. charm, bottom and top, have to be included in the sea (see chapter 4). In good approximation the top sea contribution can be neglected and the charm and bottom quarks can be considered as massless. In addition, perturbative QCD corrections are insignificant at these energies. The dominant contribution to the cross section comes from the region $Q^2 \approx m_W^2$ implying that the involved partons have x-values of around $m_W^2/2ME_\nu$. This requires extrapolations towards small x-values, not constrainted by experiments. Data obtained at HERA give important constraints up to energies of 10^8 GeV. Beyond that, one has to rely on the various extrapolations available, causing the main uncertainty in the cross section for higher energetic neutrino interactions

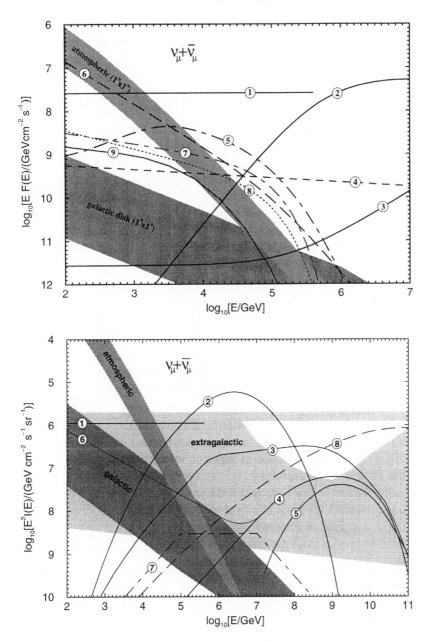

Figure 12.6. Top: Summary plot for the expected $\nu_\mu + \bar{\nu}_\mu$ fluxes from candidate cosmic accelerators. Together with various model predictions the atmospheric neutrino and galactic background are shown. Bottom: Summary of expected $\nu_\mu + \bar{\nu}_\mu$ intensities for diffuse emission from various sources. For details see [Lea00] (from [Lea00]).

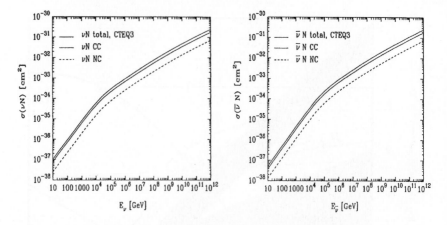

Figure 12.7. Total as well as NC and CC cross sections for high-energy neutrinos (left) and antineutrinos (right) (from [Gan96]).

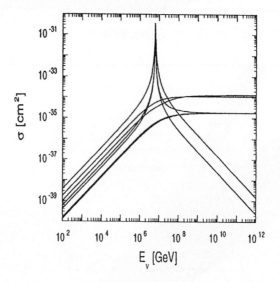

Figure 12.8. Glashow resonance in the $\bar{\nu}_e e$ cross section. The curves correspond, in the low-energy region from highest to lowest, to (i) $\bar{\nu}_e e \to$ hadrons, (ii) $\nu_\mu e \to \mu \nu_e$, (iii) $\nu_e e \to \nu_e e$, (iv) $\bar{\nu}_e e \to \bar{\nu}_\mu \mu$, (v) $\bar{\nu}_e e \to \bar{\nu}_e e$, (vi) $\nu_\mu e \to \nu_\mu e$ and (vii) $\bar{\nu}_\mu e \to \bar{\nu}_\mu e$ (from [Gan96]).

[Gan96, Gan98, Gan01]. Moreover, at $E_\nu > 10^6$ GeV νN and $\bar{\nu}$N cross sections become equal because the $(1 - y)^2$ term from valence quark scattering (4.74) is now of minor importance and the sea-quarks dominate the cross section. A reasonable parametrization of the cross sections in the region $10^{16} < E_\nu <$

10^{21} eV is given within 10% by [Gan98]

$$\sigma_{CC}(\nu N) = 5.53 \times 10^{-36} \text{ cm}^2 \left(\frac{E_\nu}{1 \text{ GeV}}\right)^{0.363} \tag{12.42}$$

$$\sigma_{NC}(\nu N) = 2.31 \times 10^{-36} \text{ cm}^2 \left(\frac{E_\nu}{1 \text{ GeV}}\right)^{0.363} \tag{12.43}$$

$$\sigma_{CC}(\bar{\nu} N) = 5.52 \times 10^{-36} \text{ cm}^2 \left(\frac{E_\nu}{1 \text{ GeV}}\right)^{0.363} \tag{12.44}$$

$$\sigma_{NC}(\bar{\nu} N) = 2.29 \times 10^{-36} \text{ cm}^2 \left(\frac{E_\nu}{1 \text{ GeV}}\right)^{0.363}. \tag{12.45}$$

Below 10^{16} eV basically all the different PDF parametrizations agree and at energies around 10^{21} eV a factor of two uncertainty sounds reasonable. The total cross sections on nucleons are shown in figure 12.7. It should be noted that new physics beyond the standard model might affect these cross sections. The cross section on electrons is, in general, much smaller than on nucleons except in a certain energy range between 2×10^{15}–2×10^{17} eV for $\bar{\nu}_e$. Here the cross section for

$$\bar{\nu}_e + e^- \rightarrow W^- \rightarrow \text{hadrons} \tag{12.46}$$

can dominate. At an energy of 6.3 PeV ($= 6.3 \times 10^{15}$ eV) the cross section shows a resonance behaviour (the Glashow resonance), because here $s = 2m_e E_\nu = m_W^2$ [Gla60] (figure 12.8). At resonance $\sigma(\bar{\nu}_e e) = (3\pi/\sqrt{2})G_F = 3.0 \times 10^{-32} \text{ cm}^2$ while $\sigma(\nu N) \approx 10^{-33} \text{ cm}^2$ at $E_\nu \approx 10^7$ GeV. Another severe effect associated with the rising cross section, is the interaction rate of neutrinos within the Earth. The interaction length L (in water equivalent) defined as

$$L = \frac{1}{\sigma_{\nu N}(E_\nu) N_A} \tag{12.47}$$

in rock is approximately equal to the diameter of the Earth for energies of 40 TeV. At higher energies the Earth becomes opaque for neutrinos. The phenomenon of Earth shielding can be described by a shadow factor S, which is defined to be an effective solid angle divided by 2π for upward-going muons and is a function of the energy-dependent cross section for neutrinos in the Earth:

$$S(E_\nu) = \frac{1}{2\pi} \int_{-1}^{0} d\cos\theta \int d\phi \, \exp[-z(\theta)/L(E_\nu)] \tag{12.48}$$

with z the column-depth, as a function of nadir angle θ and $N_A = 6.022 \times 10^{23} \text{ mol}^{-1} = 6.022 \times 10^{23} \text{ cm}^{-3}$ (water equivalent). The shadowing increases from almost no attenuation to a reduction of the flux by about 93% for the highest energies observed in cosmic rays. For energies above 10^6 GeV the interaction length is about the same for neutrinos and antineutrinos as well as for ν_e- and

ν_μ-type neutrinos. The damping for ν_τ is more or less absent, because with the CC production of a τ-lepton, its decay produces another ν_τ [Hal98a]. Below 100 TeV their interaction is not observable. However, a special situation holds for $\bar{\nu}_e$ because of the previously mentioned resonance. A similar interaction length for interactions with electrons can be defined:

$$L = \frac{1}{\sigma_{\nu e}(E_\nu)(10/18)N_A} \qquad (12.49)$$

where the factor $(10/18)N_A$ is the number of electrons in a mole of water. This interaction length is very small in the resonance region; hence, damping out $\bar{\nu}_e$ in this energy range is very efficient. The high energy of the neutrinos results in a strong correlation of the muon direction from ν_μ CC interactions with the original neutrino direction resulting in a typical angle of $\theta_{\mu\nu} \approx 1.5°/\sqrt{E(\text{TeV})}$. This allows point sources to be sought and identified. VHE neutrinos can only be detected as point sources, because of the overwhelming atmospheric neutrino background.

Combining all numbers the expected event rates for AGNs and GRBs are likely of the order 1–100 events/(km² yr).

12.2   Detection

The most promising way of detection is by looking for upward-going muons, produced by ν_μ CC interactions. Such upward-going muons cannot be misidentified from muons produced in the atmosphere. The obvious detection strategy relies on optical identification with the help of Cerenkov light, producing signals in an array of photomultiplier tubes. A muon can be found by track reconstruction using the timing, amplitude and topology of the hit photomultipliers (figure 12.9). Shower events produced by ν_e NC interactions or CC interactions have a typical extension of less than 10 m, smaller than the typical spacing of the phototubes. They can be considered as point sources of light within the detector.

Water is a reasonable transparent and non-scattering medium available in large quantities. The two crucial quantities are the absorption and scattering lengths, both of which are wavelength dependent. The absorption length should be large because this determines the required spacing of the photomultipliers. Moreover, the scattering length should be long to preserve the geometry of the Cerenkov pattern. The idea is now to equip large amounts of natural water resources like oceans and ice with photomultipliers to measure the Cerenkov light of the produced muons. To get a reasonable event rate, the size of such detectors has to be on the scale of a 1 km³ and cannot be installed in underground laboratories. Even if the experiment is installed deep in the ocean, atmospheric muons dominate neutrino events by orders of magnitude. An irreducible source of background remains in the form of atmospheric neutrinos and their interactions,

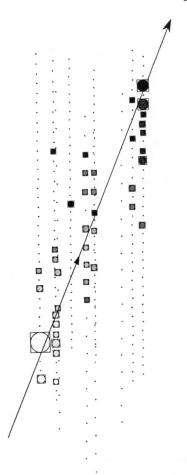

Figure 12.9. Event display of an upward-going muon as observed with AMANDA-B10. The grey scale indicates the flow of time, with early hits at the bottom and the latest hits at the top of the array. The arrival times match the speed of light. The sizes of the circles corresponds to the measured amplitudes.

which is overwhelming at lower energies. However, because of the steeper energy dependence ($\propto E^{-3}$) they fall below predicted astrophysical fluxes starting from around 10–100 TeV because their energy dependence is typically assumed to be more like $\propto E^{-2}$ due to Fermi acceleration [Lon92, 94]. The effective size of a detector is actually enhanced because the range of high-energy muons also allows interactions between the surrounding and muons flying into the detector. Hence, it is not the volume of the detector which is important but the area showing towards the neutrino flux ('effective area A_μ'). The larger this effective area is, the larger

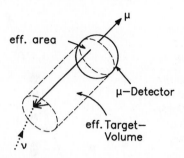

Figure 12.10. Schematic drawing to explain the concept of effective target volume and effective area in a neutrino telescope.

the effective volume $V(E_\mu) \approx A_\mu \times R(E_\mu)$ outside the detector (figure 12.10) will be. If the interesting neutrinos arrive from various directions as in the diffuse case, the detector area in all directions should be large, finally resulting in a sufficiently large detector volume. The energy loss rate of muons with energy E_μ due to ionization and catastrophic losses like bremsstrahlung, pair-production and hadroproduction is given by [Gai90]

$$\left\langle \frac{\mathrm{d}E_\mu}{\mathrm{d}X} \right\rangle = -\alpha - \frac{E_\mu}{\xi}. \tag{12.50}$$

The constant α describing energy loss (Bethe–Bloch formula) is about 2 MeV/g cm^{-2} in rock. The constant ξ describing the catastrophic losses is $\xi \approx 2.5 \times 10^5$ g cm^{-2} in rock. Above a critical energy, $\epsilon = \alpha \xi$, they dominate with respect to ionization. For muons in rock, $\epsilon \approx 500$ GeV. This leads to a change in energy dependence from linear to logarithmic. If α and ξ are energy independent, the range of the average loss for a muon of initial energy E_μ and final energy $E_\mu^{\min}$ is given by

$$R(E_\mu, E_\mu^{\min}) = \int_{E_\mu^{\min}}^{E_\mu} \frac{\mathrm{d}E_\mu}{\langle \mathrm{d}E_\mu/\mathrm{d}X \rangle} \simeq \frac{1}{\xi} \ln \left(\frac{\alpha + \xi E_\mu}{\alpha + \xi E_\mu^{\min}} \right). \tag{12.51}$$

For $E_\mu \ll \epsilon$, the range of muons is correctly reproduced by $R \propto E_\mu$, for higher energies detailed Monte Carlo studies are neccessary to propagate them. For a muon with initial energy larger than 500 GeV, the range exceeds 1 km.

The rate at which upward-going muons can be observed in a detector with effective area A is

$$A \int_{E_\mu^{\min}}^{E_\mu^{\max}} \mathrm{d}E_\nu \, P_\mu(E_\nu, E_\mu^{\min}) S(E_\nu) \frac{\mathrm{d}N}{\mathrm{d}E_\nu} \tag{12.52}$$

with S defined in (12.48) and the probability for a muon arriving in the detector

Table 12.2. Various techniques proposed for large neutrino detectors (after [Lea00])

Radiation	Medium	μ-Det.	E_ν Thres.	Atten. Length	Spectral region
Cerenkov	Filtered H$_2$O	Y	GeV	100 m	300–500 nm
	Natural Lake	Y	GeV	$\simeq$20 m	400–500 nm
	Deep Ocean	Y	GeV	$\simeq$40 m	350–500 nm
	Polar Ice	Y	GeV	$\simeq$15–45 m	300–500 nm
Ceren. Radio	Polar Ice	N	> 5 GeV	$\simeq$1 km	0.1–1 GHz
	Moon	N	>100 EeV		1–2 GHz
Acoustic	Water	N	>1 PeV	$\simeq$5 km	10–20 kHz
	Ice	N	>PeV	?	10–30 kHz
	Salt	N	>PeV	?	10–50 kHz
EAS particles	Air	N	10 PeV	1 km	100 MeV
N$_2$ fluorescence	Air	N	EeV	10 km	337 nm
EAS Radar	Air	N	>EeV	($\simeq$100 km)	30–100 MHz

Figure 12.11. Installation of one of the rods of the NT-96 experiment in Lake Baikal. As the lake freezes over in winter, this season is ideal for installation (with kind permission of Ch Spiering).

with an energy threshold of $E_\mu^{\min}$ is given by

$$P(\mu(E_\nu, E_\mu^{\min}) = N_A \sigma_{CC}(E_\nu) R(E_\mu, E_\mu^{\min}) \quad (12.53)$$

with R given in (12.51). The actual threshold is a compromise between large detector volume (large spacing of the optical modules) and low-energy threshold

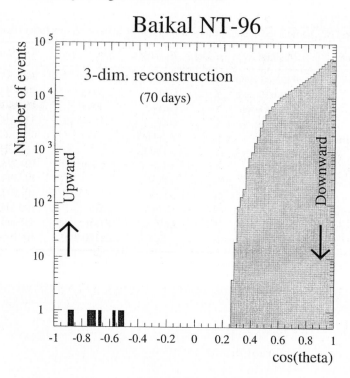

Figure 12.12. Angular distribution of muon tracks in the Baikal NT-96.

for physics reasons (requiring small spacing). In this way various experiments might be complementary as well as by the fact that some are sensitive to the Northern and some to the Southern Sky.

In addition, to such long tracks, cascades might also be detected, e.g. from v_e CC interactions where the electrons produce an electromagnetic shower. Therefore, the effective volume is close to the real geometrical volume of the Cerenkov telescope. For the v_τ interaction a signature has been proposed in the form of a double bang [Lea95]. The pathlength of a τ-lepton in CC interactions is $c\tau E/m_\tau = 86.93\ \mu\text{m}E/1.777$ GeV. At energies of 2 PeV this corresponds to a distance of about 100 m before its decay. This results in two light-emitting processes, its production and decay, where the initial burst shows about half of the energy of the τ-decay burst.

We will now discuss the various Cerenkov detectors—other possible detection methods like acoustic and radio detection will be briefly described later. An overview of techniques is shown in table 12.2.

Figure 12.13. Schematic view of NESTOR (with kind permission of the NESTOR collaboration).

12.2.1 Water Cerenkov detectors

The pioneering effort to build a large-scale neutrino telescope in the ocean was started by DUMAND in the 1970s. However, the project was stopped in the 1990s. The first one to run is Baikal NT-200 in Lake Baikal in Russia.

12.2.1.1 Baikal NT-200

This experiment [Spi96, Dom02, Bal03] is being installed in Lake Baikal (Russia) (see figure 12.11). It is situated at a depth of about 1.1 km. One of the advantages of this experiment is that Lake Baikal is a sweet water lake containing practically no ^{40}K, which could produce a large background. The photomultipliers used for light detection have a diameter of 37 cm and are fastened on rods over a length of about 70 m. The rods are arranged in the form of a heptagon and are attached to an additional rod at the centre. The whole arrangement is supported by an umbrella-like construction, which keeps the rods at a distance of 21.5 m from the centre. The photomultipliers are arranged in pairs, with one facing upwards and the other downwards. The distance between two phototubes with the same orientation is about 7.5 m and between two with opposite orientation about 5 m. The full array with 192 optical modules (OMs) has been operational since April 1998. Figure 12.12 shows the reconstructed upward-going muons in a dataset of 234 days. From a smaller prototype NT-96 and only 70 days of data-taking, an upper limit on a diffuse neutrino flux assuming a $\propto E^{-2}$ shape for the neutrino

Figure 12.14. Schematic view of ANTARES (picture created by F Montanet, with kind permission of the ANTARES collaboration).

spectrum of [Bal03]

$$\frac{\mathrm{d}\Phi_\nu}{\mathrm{d}E}E^2 < 4 \times 10^{-7} \ \mathrm{cm}^{-2} \ \mathrm{s}^{-1} \ \mathrm{sr}^{-1} \ \mathrm{GeV} \qquad (12.54)$$

in the energy range 10–10^4 TeV has been obtained. With a rather small spacing for the OMs the detector is well suited for the detection of WIMPs. An upgraded version of the detector (NT200+) is being considered.

12.2.1.2 NESTOR

The NESTOR experiment [Res94, Nut96, Tra99] is planned for the Mediterranean, off the coast of Pylos (Greece), at a depth of 3.8 km. It will be constructed out of seven rods in the form of a central rod surrounded by a hexagon with a radius of 100–150 m (figure 12.13). In contrast to DUMAND the optical detectors are spaced in groups on the rod rather than linearly along it. Each group consists of a hexagon of radius 16 m, at each apex of which is installed a pair of phototubes (one pointing upwards, one downwards). Twelve such hexagons are attached to each rod with a separation of 20–30 m. The measured attenuation length is about 55 m at 470 nm. The total effective area envisaged is about 10^5 m². A first floor of such hexagons has been deployed recently.

12.2.1.3 ANTARES

Like NESTOR the ANTARES project is located in the Mediterranean about 40 km from the coast near Toulon (France) at a depth of 2400 m [Ant97, Tho01, Mon02]. A first string for a proof of the principle was deployed in November 1999. The optical parameters measured are an absorption length of about 40 m at 467 nm and 20 m at 375 nm and an effective scattering length defined as

$$\Lambda_{\text{eff}} = \frac{\Lambda}{1 - \langle \cos\theta \rangle} \tag{12.55}$$

of about 300 m. The final design will be 10 strings with a separation of about 60 m, each carrying 90 photomultipliers (figure 12.14). The tubes will look sideways and downwards to avoid background from biofouling. The array will work with an energy threshold of 10 GeV.

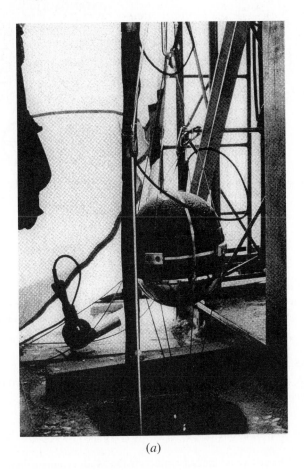

(*a*)

Figure 12.15. (Continues.)

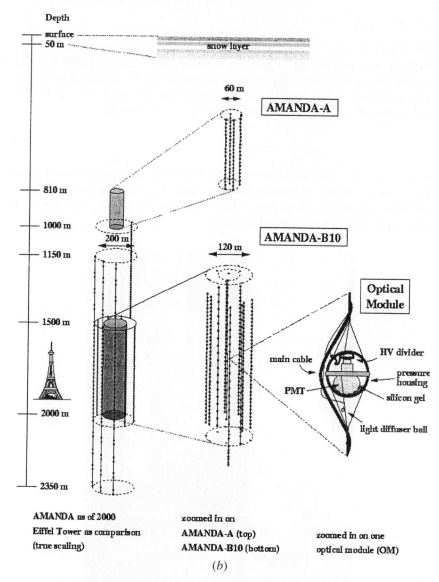

Figure 12.15. (*a*) Deployment of a string of photomultipliers.   (*b*) Positions of AMANDA-A and AMANDA-B10 in the South Pole Ice (with kind permission of the AMANDA collaboration).

On top of these two advanced projects in the Mediterranean there is an Italian initiative called NEMO to study appropriate sites for a 1 km^3 detector near Italy.

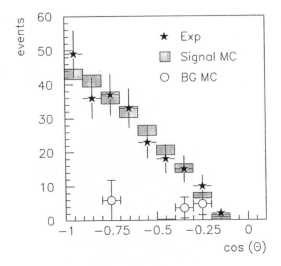

Figure 12.16. Zenith-angle distribution of the atmospheric data of AMANDA-B10 compared to simulated atmospheric neutrinos and a simulated background of downgoing muons produced by cosmic rays. The Monte Carlo prediction is normalized to the experimental data. The error bars report only statistical errors (from [Ahr02b]).

12.2.2   Ice Cerenkov detectors—AMANDA, ICECUBE

Another way of using water is in its frozen form, i.e. building an experiment in the ice of the Antarctica. This is exactly the idea of AMANDA [Wis99, And00]. Photomultipliers of 8-inch diameter are used as OMs and plugged into holes in the ice, obtained by hot-water drilling, along long strings. After an exploratory phase in which AMANDA-A was installed at a depth of 800–1000 m, AMANDA B-10 was deployed between 1995 and 1997 (figure 12.15). It consists of 302 modules at a depth of 1500–2000 m below the surface. The instrumented volume forms a cylinder with an outer diameter of 120 m. In January 2000, AMANDA-II, consisting out of 19 strings with 677 OMs, where the ten strings from AMANDA-B10 form the central core, was completed. The measured absorption length is about 110 m at 440 nm, while the effective scattering length is about 20 m.

Several physics results based on 130.1 days of lifetime have already been obtained with AMANDA-B10, which has an effective area of more than 10 000 m^2 for declinations between 25 and 90 degrees at $E_\mu = 10$ TeV. A proof of the principle understanding and the reliability of Monte Carlo simulations is the observation of atmospheric neutrino events. As described in chapter 9, here also the search is performed with upward-going muons. About 200 neutrino events could be observed [Ahr02b], in good agreement with expectation (figure 12.16). A special case of upward-going muons are nearly vertical muons, which might experience an additional signal due to neutralino dark matter annihilation in the

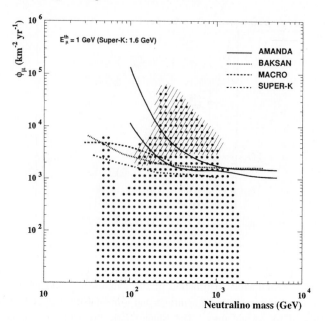

Figure 12.17. Limits on the muon flux from neutralino annihilation for various experiments. The dots represent model predictions from the MSSM, calculated with the DarkSUSY package. The dashed area shows the models disfavoured by direct searches from the DAMA collaboration (from [Ahr02c]).

centre of the Earth. The performed search did not reveal any signal [Ahr02c] and a limit on the muon flux coming from neutralino annihilation together with bounds from other experiments is shown in figure 12.17. Moreover, a search for point sources in the Northern Hemisphere was undertaken [Ahr03]. No obvious excess at any specific point in the sky is seen (figure 12.20). For any source with a differential energy spectrum proportional to E_ν^{-2} and declination larger than +40 degrees a limit of $E^2(dN_\nu/dE) \leq 10^{-6}$ GeV cm^{-2} s^{-1} with a threshold of 10 GeV is obtained. This flux limit for upward-going muons is shown in figure 12.18. Also a search for cascades in the detector, produced by a diffuse flux of neutrinos in the energy range 50–300 TeV, has been performed. The limit on cascades of diffuse neutrinos of all flavours in a 1:1:1 ratio is given by [Nie03]

$$E^2 \frac{d\Phi_\nu}{dE} < 9 \times 10^{-7} \text{ GeV cm}^{-2} \text{ s}^{-1} \text{ sr}^{-1}. \qquad (12.56)$$

This combined with those observed by other experiments is shown in figure 12.19. The final goal will be ICECUBE [Hal01, Spi01, Kar02], a real 1 km^3 detector, consisting of 80 strings spaced by 125 m, each with 60 OMs with a spacing of 17 m, resulting in a total of 4800 photomultipliers.

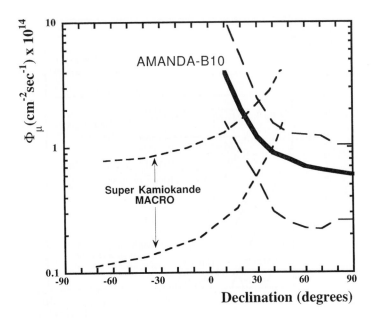

Figure 12.18. Upper limit on the muon flux (90% CL) as a function of declination. The band indicates the range of limits for MACRO and Super-Kamiokande and the statistical fluctuation for AMANDA-B10 (from [Ahr02b]).

12.2.3   Alternative techniques—acoustic and radio detection

Associated with the two previously mentioned projects are alternative experiments using different detection techniques, namely acoustic and radio detection. An electromagnetic shower in matter develops a net charge excess due to the photon and electron scattering pulling additional electrons from the surrounding material in the shower and by positron annihilation [Ash62]. This can result in a 20–30% charge excess. The effect leads to a strong coherent radio Cerenkov emission which has been verified experimentally [Sal01]. This offers a potential detection method of UHE ν_e interactions in matter [Zas92]. The signal will be a radiopulse of several ns with most power emitted along the Cerenkov angle. One of the experiments is RICE [Seu01, Leh02] at the South Pole in close vicinity to AMANDA. A 16-channel array of dipole radio receivers were deployed at a depth of 100–300 m with a bandpass of 200–500 MHz. The first interesting limits could be obtained. Independently, a balloon mission flying an array of 36 antennas over Antarctica (ANITA) has been approved.

In the acoustic case, the shower particles produced in the ν_e interaction lose energy through ionization leading to local heating and a density change localized along the shower. A neutrino interaction with $E_\nu = 10^{20}$ eV creates a hadronic shower, with 90% of the energy in a cylinder of 20 m length and a diameter

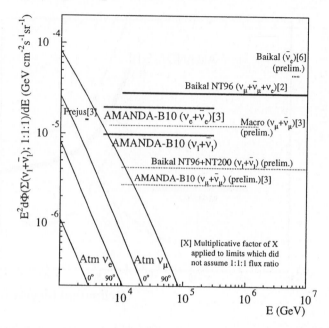

Figure 12.19. Limits on the cascade-producing neutrino flux, summed over the three active flavours, with multiplicative factors applied as indicated to permit comparison of limits derived. Also shown are the predicted horizontal and vertical ν_e and ν_μ atmospheric fluxes (from [Nie03]).

of roughly 20 cm. The density change propagates as sound waves through the medium and can be detected with an array of detectors, e.g. hydrophones. A reconstruction of the event can be performed by measuring the arrival times and amplitudes. The speed of sound in water is about 1.5 km s^{-1}, so the frequency range of interest is between 10–100 kHz. The interesting option of using existing hydrophone arrays exists [Leh02].

A compilation of the various mentioned experiments is given in table 12.3, see also [Lea03].

12.2.4   Horizontal air showers—the AUGER experiment

An alternative method to water Cerenkov detection might be the use of extended air showers (EAS) in the atmosphere. This is a well-established technique to measure the cosmic-ray spectrum by the cosmic rays' interactions with oxygen and nitrogen and, hence, to determine their chemical composition and energy by measuring various shower parameters. In this way, in recent years, events beyond the GZK cutoff have been found. The possible origin of such UHE cosmic rays is discussed in [Bha00, Nag00]. As mentioned, 'beyond GZK' events

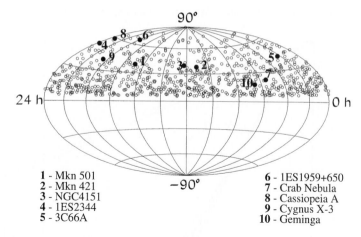

1 - Mkn 501
2 - Mkn 421
3 - NGC4151
4 - 1ES2344
5 - 3C66A

6 - 1ES1959+650
7 - Crab Nebula
8 - Cassiopeia A
9 - Cygnus X-3
10 - Geminga

Figure 12.20. Skymap of 815 events from the point-source analysis of AMANDA-B10. The horizontal coordinates are right ascension and the vertical coordinates are declination. Also shown are the sky coordinates for ten potential high-energy neutrino sources (from [Ahr03]).

Table 12.3. Summary of second-generation initiatives in high-energy neutrino astronomy as of 2000 (after [Lea00]). The key for the table is: N, presently operating; T, testing and deployment; C, construction; P, under discussion or proposed; WC, water Cerenkov; μwv, microwave detection; acoustic, acoustic wave detection.

Detector	Location	Status	μ Area (10^3 m^2)	Depth (mwe)	Technique	Threshold (GeV)
DUMANDII	Hawaii	92–95	20	4760	WC	20
Baikal NT-200	Siberia	96–N	3	1000	WC	10
AMANDA IIB	South Pole	96–N	100	2000	WC in ice	20
NESTOR	Greece	T/C	100	3500	WC	1
SADCO	Greece	T	1000	3500	acoustic	$>10^6$
RICE	South Pole	T/C	1000	1000	μwv	$\simeq 10^6$
ANTARES	France	T/C	?	2000	WC	10
NEMO	Italy	P	?	?	WC	?
RAMAND	US	T	?	Moon	μwv	10^{11}
ICECUBE	South Pole	P	1000	2000	WC in ice	>100
KM3	Ocean	P	1000	>4000	WC	>100?

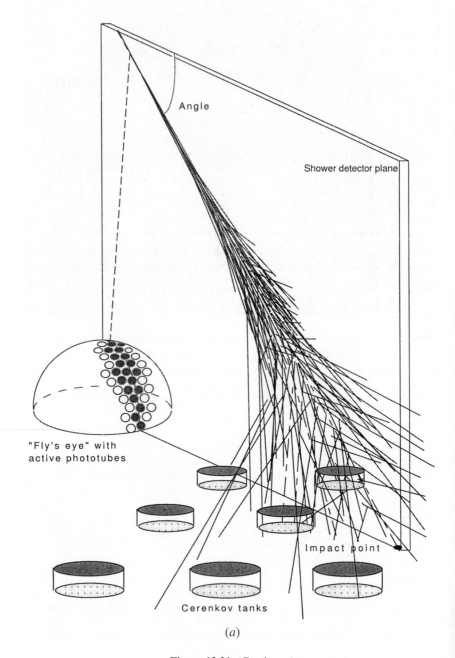

Angle

Shower detector plane

"Fly's eye" with
active phototubes

Impact point

Cerenkov tanks

(*a*)

Figure 12.21. (Continues.)

(b)

Figure 12.21. (a) Principle of the Auger experiment, combining two techniques: the detection of Cerenkov light with a huge array of water tanks and the detection of nitrogen fluorescence with telescopes. (b) One of the Auger Cerenkov tanks in Argentina (with kind permission of the Auger collaboration).

have to come from our cosmological neighbourhood basically a sphere of 50–100 Mpc radius. However, there are no good AGN and GRB sources available. However, UHE neutrinos of the order 10^{21} eV could come from cosmological distances. One interesting explanation combining highest and lowest neutrino energies in the universe is given by [Wei82] (the 'Z-burst' model). Hadrons could be produced from Z-decays created by interactions of UHE neutrinos with low-energy antineutrinos from the relic neutrino background (see chapter 13). The cross section $\sigma(\nu\bar{\nu} \to Z^0)$ shows a resonance at energy

$$E_\nu = \frac{m_Z^2}{2m_\nu} = 4\left(\frac{\text{eV}}{m_\nu}\right) \times 10^{21} \text{ eV}. \tag{12.57}$$

Here the cross section is $\sigma(\nu\bar{\nu}) = (4\pi/\sqrt{2})G_F = 4.2 \times 10^{-32}$ cm^2.

The experimental statistics in the region beyond 10^{18} eV is still limited and larger air shower arrays are needed for improvements, like the Auger experiment [Aug96] currently being built in Argentina. The detection system combines two major techniques (figure 12.21): a fluorescence detector system to measure the longitudinal profile of the EAS; and a surface array of detectors to sample its lateral distribution on the ground. With a detection acceptance larger than $16\,000$ km^2 sr, the Auger experiment should observe each year more than 5000 events above 10^{19} eV, 500 above 5×10^{19} eV and more than 100 above 10^{20} eV. The Auger site will be composed of 1600 Cerenkov stations (our surface detector units) and four fluorescence eyes located at the periphery of the array. Each eye is composed of six $30° \times 30°$ mirror and camera units looking inwards over the surface station network. A counterpart in the Northern Hemisphere is under consideration. Also projects to observe such showers from space (EUSO, OWL) are under consideration.

A striking feature for neutrino detection is their deeper interaction in the atmosphere which allows them to be discriminated from hadrons, interacting high in the atmosphere [Cap98]. Horizontal EAS produced by neutrino interactions are 'young', meaning a shower at its beginning, showing properties like a curved shower front, a large electromagnetic component and a spread in arrival times of the particles larger than 100 ns. None of this is valid for well-advanced showers. These properties could be measured adequately if the interactions happen in the air above the array. In addition, the possibility of measuring a possible ν_τ flux might exist [Ber02, Let03]. τ-leptons produced in the mountains or the ground around the array produce a clear signal if the decay occurs above the detector. Most of them stem from upward-going ν_τ where the CC interactions occur in the ground.

After an overview of the rapidly developing field of high-energy neutrino astrophysics we now want to discuss the role of neutrinos in cosmology.

Chapter 13

Neutrinos in cosmology

It is a reasonable assumption that, on the scales that are relevant to a description of the development of the present universe, of all the interactions only gravity plays a role. All other interactions are neutralized by the existence of opposite charges in the neighbourhood and have an influence only on the detailed course of the initial phase of the development of the universe. The currently accepted theory of gravitation is Einstein's *general theory of relativity*. This is *not* a gauge theory: gravitation is interpreted purely geometrically as the curvature of four-dimensional spacetime. For a detailed introduction to general relativity see [Wei72, Mis73, Sex87]. While general relativity was being developed (1917), the accepted model was that of a stationary universe. In 1922 Friedman examined *non-stationary* solutions of Einstein's field equations. Almost all models based on expansion contain an initial singularity of infinitely high density. From this the universe developed via an explosion (the Big Bang). An expanding universe was experimentally confirmed when Hubble discovered galactic redshifts in 1929 [Hub29] and interpreted their velocity of recession as a consequence of this explosion. With the discovery of the cosmic microwave background in 1964 [Pen65], which is interpreted as the echo of the Big Bang, the Big Bang model was finally established in preference to competing models, such as the steady-state model. The proportions of the light elements could also be predicted correctly over 10 orders of magnitude within this model (see section 13.8). All this has resulted in the Big Bang model being today known as the *standard model of cosmology*. For further literature see [Boe88, Gut89, Kol90, Kol93, Nar93, Pee93, Pee95, Kla97, Bot98a, Pea98, Ber99, Ric01, Dol02].

13.1 Cosmological models

Our present conception of the universe is that of a homogeneous, isotropic and expanding universe. Even though the observable spatial distribution of galaxies seems decidedly lumpy, it is generally assumed that, at distances large enough, these inhomogeneities will average out and an even distribution will exist. At least, this seems to be a reasonable approximation today. The high isotropy of the microwave background radiation (chapter 6) also testifies to the very high isotropy of the universe. These observations are embodied in the so-called *cosmological principle*, which states that there is no preferred observer, which means that the universe looks the same from any point in the cosmos. The spacetime structure is described with the help of the underlying metric. In three-dimensional space the distance is given by the line element ds^2 with

$$ds^2 = dx_1^2 + dx_2^2 + dx_3^2 \tag{13.1}$$

whereas in the four-dimensional spacetime of the *special theory of relativity*, a line element is given by

$$ds^2 = dt^2 - (dx_1^2 + dx_2^2 + dx_3^2) \tag{13.2}$$

which in the general case of non-inertial systems can also be written as

$$ds^2 = \sum_{\mu\nu=1}^{4} g_{\mu\nu} \, dx^\mu \, dx^\nu. \tag{13.3}$$

Here $g_{\mu\nu}$ is the metric tensor which, in the case of the special theory of relativity, takes on the simple diagonal form of

$$g_{\mu\nu} = \text{diag}(1, -1, -1, -1). \tag{13.4}$$

The simplest metric with which to describe a homogeneous isotropic universe in the form of spaces of constant curvature is the *Robertson–Walker metric* [Wei72], in which a line element can be described by

$$ds^2 = dt^2 - a^2(t) \left[\frac{dr^2}{1 - kr^2} + r^2 \, d\theta^2 + r^2 \sin^2\theta \, d\phi^2 \right]. \tag{13.5}$$

Here r, θ and ϕ are the three co-moving spatial coordinates, $a(t)$ is the scale-factor and k characterizes the curvature. A closed universe has $k = +1$, a flat Euclidean universe has $k = 0$ and an open hyperbolic one has $k = -1$. In the case of a closed universe, a can be interpreted as the 'radius' of the universe. The complete dynamics is embodied in this time-dependent scale-factor $a(t)$,[1] which is described by Einstein's field equations

$$R_{\mu\nu} - \tfrac{1}{2} a g_{\mu\nu} = 8\pi G T_{\mu\nu} + \Lambda g_{\mu\nu}. \tag{13.6}$$

In this equation $R_{\mu\nu}$ is the Ricci tensor, $T_{\mu\nu}$ corresponds to the energy–momentum tensor and Λ is the cosmological constant [Wei72, Mis73, Sex87]. If

---

[1] This name implies that the spatial separation of two adjacent 'fixed' space points (with constant r, ϕ, θ coordinates) is scaled in time by $a(t)$.

we look at space only locally, we can assume as a first approximation a flat space, which means the metric is given by the Minkowski metric of the special theory of relativity (13.4). As $g_{\mu\nu}$ is diagonal here the energy–momentum tensor also has to be diagonal. Its spatial components are equal due to isotropy. The dynamics can be described in analogy to the model of a perfect liquid with density $\rho(t)$ and then has the form

$$T_{\mu\nu} = \text{diag}(\rho, -p, -p, -p). \tag{13.7}$$

The cosmological constant acts as a contribution to the energy momentum tensor in the form

$$T^{\Lambda}_{\mu\nu} = \text{diag}(\rho_{\Lambda}, -\rho_{\Lambda}, -\rho_{\Lambda}, -\rho_{\Lambda}) \tag{13.8}$$

with $\rho_{\Lambda} = 3\Lambda/(8\pi G)$. Thus, vacuum energy has a very unusual property, that in the case of a positive ρ_{Λ} it has a negative pressure. In an expanding universe this even accelerates the expansion. From the zeroth component of Einstein's equations, it follows that

$$\frac{\dot{a}^2}{a^2} + \frac{k}{a^2} = \frac{8\pi G}{3}(\rho + \rho_V) \tag{13.9}$$

while the spatial components give

$$2\frac{\ddot{a}}{a} + \frac{\dot{a}^2}{a^2} + \frac{k}{a^2} = -8\pi G p. \tag{13.10}$$

These equations (13.9) and (13.10) are called the *Einstein–Friedmann–Lemaitre equations*. From these equations, it is easy to show that

$$\frac{\ddot{a}}{a} = -\frac{4\pi G}{3}(\rho + 3p - 2\rho_V). \tag{13.11}$$

Since currently $\dot{a} \geq 0$ (i.e. the universe is expanding), and on the assumption that the expression in brackets has always been positive, i.e. $\ddot{a} \leq 0$, it inevitably follows that a was once 0. This singularity at $a = 0$ can be seen as the 'beginning' of the development of the universe. Evidence for such an expanding universe came from the redshift of far away galaxies [Hub29]. The further galaxies are away from us, the more redshifted are their spectral lines, which can be interpreted as a consequence of the velocity of recession v. This can be demonstrated by expanding $a(t)$ as a Taylor series around the value it has today, giving

$$\frac{a(t)}{a(t_0)} = 1 + H_0(t - t_0) - \frac{1}{2}q_0 H_0^2(t - t_0)^2 + \cdots. \tag{13.12}$$

The index 0 represents the current value both here and in what follows. The Hubble constant H_0 is, therefore,

$$H_0 = \frac{\dot{a}(t_0)}{a(t_0)} \tag{13.13}$$

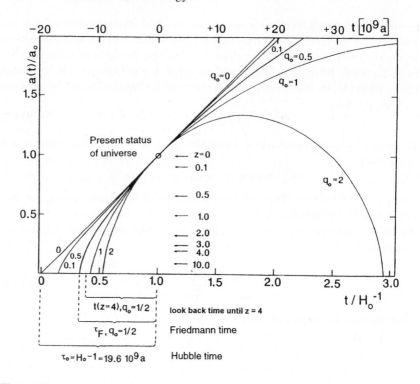

Figure 13.1. Behaviour of the scale factor $a(t)$ for different models of the universe. For all models $\Lambda = 0$ was assumed. Also shown are the various redshifts, as well as the influence of various deceleration parameters q_0. We are currently in an era which allows no discrimination to be made about the model of the universe. A Hubble constant of 50 km s^{-1} Mpc^{-1} has been used (from [Uns92]).

corresponding to the current expansion rate of the universe and the deceleration parameter q_0 is given by

$$q_0 = \frac{-\ddot{a}(t_0)}{\dot{a}^2(t_0)} a(t_0). \tag{13.14}$$

These measurements result for low redshifts in the Hubble relation in

$$v = cz = H_0 r. \tag{13.15}$$

using the redshift z. In general, the Hubble parameter describes the expansion rate at a given time

$$H(t) = \frac{\dot{a}}{a}. \tag{13.16}$$

The behaviour of the scale factor for various cosmological models is shown in figure 13.1.

13.1.1 The cosmological constant Λ

A $\Lambda \neq 0$ would also be necessary if the Hubble time H^{-1} (for $\Lambda = 0$) and astrophysically determined data led to different ages for the universe. Λ has experienced a revival through modern quantum field theories. In these the vacuum is not necessarily a state of zero energy but the latter can have a finite expectation value. The vacuum is only defined as the state of lowest energy. Due to the Lorentz invariance of the ground state it follows that the energy–momentum tensor in every local inertial system has to be proportional to the Minkowski metric $g_{\mu\nu}$. This is the only 4×4 matrix which in special relativity theory is invariant under Lorentz 'boosts' (transformations along a spatial direction). According to this, the cosmological constant can be associated with the energy density ϵ_V of the vacuum to give

$$\epsilon_V = \frac{c^4}{8\pi G}\Lambda = \rho_V c^2. \tag{13.17}$$

All terms contributing in some form to the vacuum energy density also provide a contribution to the cosmological constant. There exist, in principle, three different contributions:

- The static cosmological constant Λ_{geo} impinged by the underlying spacetime geometry. It is identical to the free parameter introduced by Einstein [Ein17].
- Quantum fluctuations Λ_{fluc}. According to Heisenberg's uncertainty principle, virtual particle–antiparticle pairs can be produced at any time even in a vacuum. That these quantum fluctuations really exist was proved clearly via the Casimir effect [Cas48, Lam97].
- Additional contributions of the same type as the previous one due to invisible, currently unknown, particles and interactions Λ_{inv}.

The sum of all these terms is what can be experimentally observed

$$\Lambda_{\text{tot}} = \Lambda_{\text{geo}} + \Lambda_{\text{fluc}} + \Lambda_{\text{inv}}. \tag{13.18}$$

Consider first the *static solutions* ($\dot{a} = \ddot{a} = 0$). The equations are then written (for $p = 0$) as

$$\frac{8\pi G}{3}(\rho + \rho_V) = \frac{k}{a^2} \tag{13.19}$$

$$\rho = 2\rho_V. \tag{13.20}$$

From equation (13.20) it follows that $\rho_V > 0$ and, therefore, equation (13.19) has a solution only for $k = 1$:

$$a^2 = \frac{1}{4\pi G\rho}. \tag{13.21}$$

Equation (13.19) represents the equilibrium condition for the universe. The attractive force due to ρ has to exactly compensate for the repulsive effect of a

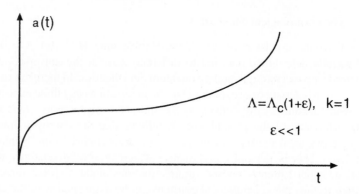

Figure 13.2. The behaviour of the scale factor with a non-vanishing cosmological constant in the case of a Lemaitre universe: a value of Λ only a little larger than Λ_c produces a phase in which the expansion of the universe is almost at a standstill, before a further expansion.

positive cosmological constant in order to produce a static universe. This closed static universe is, however, unstable, since if we increase a by a small amount, ρ decreases, while Λ remains constant. The repulsion then dominates and leads to a further increase in a, so that the solution moves away from the static case. We now consider *non-static* solutions. As can easily be seen, a positive Λ always leads to an acceleration of the expansion, while a negative Λ acts as a brake. Λ always dominates for large a, since ρ_V is constant. A negative Λ, therefore, always implies a contracting universe and the curvature parameter k does not play an important role. For positive Λ and $k = -1, 0$ the solutions are always positive which, therefore, results in a continually expanding universe. For $k = 1$, there exists a critical value

$$\Lambda_c = 4 \left(\frac{8\pi G}{c^2} M \right)^{-2} \tag{13.22}$$

exactly the value of Einstein's static universe, dividing two regimes. For $\Lambda > \Lambda_c$ static, expanding and contracting solutions all exist. A very interesting case is that with $\Lambda = \Lambda_c(1+\epsilon)$ with $\epsilon \ll 1$ (Lemaitre universe). It contains a phase in which the universe is almost stationary, before continuing to expand again (figure 13.2).

Striking evidence for a non-vanishing cosmological constant has arisen in recent years by investigating high redshift supernovae of type Ia [Per97, Rie98, Sch98, Per99, Ton03]. They are believed to behave as standard candles, because the explosion mechanism is assumed to be the same. Therefore, the luminosity as a function of distance scales with a simple quadratic behaviour. By investigating the luminosity distance *versus* redshift relation, equivalent to a Hubble diagram, at high redshift the expected behaviour is sensitive to cosmological parameters. As it turned out [Sch98, Per99, Lei01], the best fit describing the data is a universe with a density Ω (see (13.27)) $\Omega_M = 0.3$ and $\Omega_\Lambda = 0.7$ (figure 13.3). Future satellite missions like Supernova Acceleration Probe (SNAP) will extend the search to

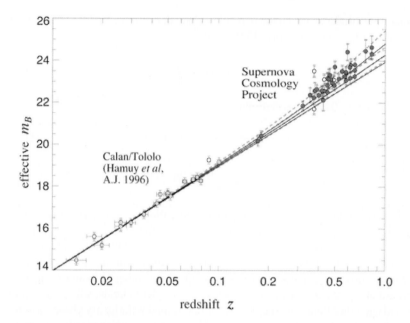

effective m_B

redshift z

Figure 13.3. Hubble diagram in the form of a magnitude–redshift diagram of supernovae type Ia. Shown are the low redshift supernovae of Hamuy *et al* (1995) together with the data from the Supernova Cosmology Project. The full curves are theoretical m_B for $(\Omega_M, \Omega_V) = (0, 0)$ on the top, $(1, 0)$ in the middle, and $(2, 0)$ on the bottom. The dotted curves in (*a*) and (*b*), which are practically indistinguishable from the full curves, represent the flat universe case, with $(\Omega_M, \Omega_V) = (0.5, 0.5)$ on the top, $(1, 0)$ in the middle and $(1.5, -0.5)$ on the bottom (from [Per99]). Similar results are obtained by the high-z redshift team.

even higher redshifts.

It is a striking puzzle that any estimated contribution to ρ_V is 50–100 orders of magnitude larger than the cosmological value [Wei89, Kla97, Dol97]. In addition, there are many phenomenological models with a variable cosmological 'constant' [Sah00]. A special class of them with a generalized equation of state of (13.3)

$$p = w\rho \qquad \text{with } 1- < w < 0 \qquad (13.23)$$

has been named 'quintessence' [Cal98].

13.1.2 The inflationary phase

As mentioned, a positive vacuum energy corresponds to a negative pressure $p_V = -\rho_V$. Should this vacuum energy at some time be the dominant contribution with respect to all matter and curvature terms, new exponential solutions for the time behaviour of the scale factor a result. Consider a universe free of matter and radiation ($T_{\mu\nu} = 0$). Solving (13.9) results in

$$H^2(t) = \frac{\Lambda}{3} \tag{13.24}$$

which for $k = 0$ and $\Lambda > 0$ implies

$$a(t) \propto \exp(Ht) \tag{13.25}$$

where

$$H^2 = \frac{8\pi G \rho_V}{3}. \tag{13.26}$$

Such exponentially expanding universes are called *de Sitter* universes. In the specific case in which the negative pressure of the vacuum is responsible for this, we talk about *inflationary* universes. Such a inflationary phase, where the exponential increase is valid for only a limited time in the early universe, helps to solve several problems within standard cosmology. Inflation is generally generated by scalar fields ϕ, sometimes called inflaton fields, which only couple weakly to other fields. As the period for the limited inflationary phase, in general the GUT phase transition is considered. Here a new vacuum ground state emerged due to spontaneous symmetry breaking (see chapter 3). For more detailed reviews on inflation see [Gut81, Alb82, Lin82, Lin84, Kol90, Lin02, Tur02]. The extension of the Big Bang hypothesis through an inflationary phase 10^{-35} s after the Big Bang has proven to be very promising and successful. This is the reason why today the combination of the Big Bang model with inflation is often called the standard model of cosmology.

13.1.3   The density in the universe

From equation (13.9) it is clear that a flat universe ($k = 0$) is only reached for a certain density, the so-called *critical density*. This is given as [Kol90]

$$\rho_{c0} = \frac{3H_0^2}{8\pi G} \approx 18.8h^2 \times 10^{-27} \text{ kg m}^{-3} \approx 11h^2 \text{ H-atoms m}^{-3} \tag{13.27}$$

where $h = H_0/100 \text{ km s}^{-1}\text{Mpc}^{-1}$. It is convenient to normalize to this density and, therefore, a *density parameter* Ω is introduced, given by

$$\Omega = \frac{\rho}{\rho_c}. \tag{13.28}$$

$\Omega = 1$, therefore, means a Euclidean universe. This is predicted by inflationary models. An $\Omega > 1$ implies a closed universe, which means that at some time the gravitational attraction will stop the expansion and the universe will collapse again (the 'Big Crunch'). An $\Omega < 1$, however, means a universe which expands

Table 13.1. Current experimental values of the most important cosmological parameters (after [Fre02]) and new results from the WMAP satellite (after [Spe03]).

Quantity	Experimental value	WMAP
Ω_0	1.03 ± 0.06	1.02 ± 0.02
Ω_M	0.3 ± 0.1	$0.135^{+0.005}_{-0.009}$
Ω_Λ	0.7 ± 0.3	0.73
H_0	$72 \pm 8 \text{ km s}^{-1} \text{ Mpc}^{-1}$	$72 \pm 5 \text{ km s}^{-1} \text{ Mpc}^{-1}$
t_0	$13 \pm 2 \times 10^9 \text{ yr}$	$13.7 \pm 0.2 \times 10^9 \text{ yr}$

forever. If the Friedmann equation (13.9) is solved for the $\mu = 0$ component, the first law of thermodynamics results:

$$d(\rho a^3) = -p d(a^3). \tag{13.29}$$

This means simply that the change in energy in a co-moving volume element is given by the negative product of the pressure and the change in volume. Assuming a simple equation of state $p = k\rho$, where k is a time-independent constant, it immediately follows that

$$\rho \sim a^{-3(1+k)} \tag{13.30}$$

$$a \sim t^{\frac{2}{3}(1+k)}. \tag{13.31}$$

The dependence of the density on a can, hence, be derived for different energy densities using the known thermodynamic equations of state. For the two limiting cases—relativistic gas (the early radiation-dominated phase of the cosmos, particle masses negligible) and cold, pressure-free matter (the later, matter-dominated phase)—we have:

$$\text{Radiation} \rightarrow p = 1/3\rho \rightarrow \rho \sim a^{-4} \tag{13.32}$$

$$\text{Matter} \rightarrow p = 0 \rightarrow \rho \sim a^{-3}. \tag{13.33}$$

Hence, in the considered Euclidean case, a simple time dependence for the scale parameter (see figure 13.1) follows:

$$a \sim t^{\frac{1}{2}} \qquad \text{radiation dominated} \tag{13.34}$$

$$a \sim t^{\frac{2}{3}} \qquad \text{matter dominated.} \tag{13.35}$$

For the vacuum energy which is associated with the cosmological constant Λ, one has

$$\text{vacuum energy} \rightarrow p = -\rho \rightarrow \rho \sim \text{constant.} \tag{13.36}$$

The current experimental numbers of cosmological parameters is shown in table 13.1.

13.2   The evolution of the universe

13.2.1   The standard model of cosmology

In this section we consider how the universe evolved from the Big Bang to what
we see today. We start from the assumption of thermodynamic equilibrium for
the early universe, which is a good approximation, because the particle number
densities n were so large, that the rates of reactions $\Gamma \propto n\sigma$ (σ being the
cross section of the relevant reactions) were much higher than the expansion
rate $H = \dot{a}/a$. A particle gas with g internal degrees of freedom, number
density n, energy density ρ and pressure p obeys the following thermodynamic
relations [Kol90]:

$$n = \frac{g}{(2\pi)^3} \int f(\boldsymbol{p})\,\mathrm{d}^3 p \tag{13.37}$$

$$\rho = \frac{g}{(2\pi)^3} \int E(\boldsymbol{p}) f(\boldsymbol{p})\,\mathrm{d}^3 p \tag{13.38}$$

$$p = \frac{g}{(2\pi)^3} \int \frac{|\boldsymbol{p}|^2}{3E} f(\boldsymbol{p})\,\mathrm{d}^3 p \tag{13.39}$$

where $E^2 = |\boldsymbol{p}|^2 + m^2$. The phase space partition function $f(\boldsymbol{p})$ is given,
depending on the particle type, by the Fermi–Dirac (+ sign in equation (13.40))
or Bose–Einstein (− sign in equation (13.40)) distribution

$$f(\boldsymbol{p}) = [\exp((E - \mu)/kT) \pm 1]^{-1} \tag{13.40}$$

where μ is the chemical potential of the corresponding type of particle. In the case
of equilibrium, the sum of the chemical potentials of the initial particles equals
that of the end products and particles and antiparticles have equal magnitude in μ
but opposite sign. Consider a gas at temperature T. Since non-relativistic particles
($m \gg T$) give an exponentially smaller contribution to the energy density than
relativistic ($m \ll T$) particles, the former can be neglected and, thus, for the
radiation-dominated phase, we obtain:

$$\rho_R = \frac{\pi^2}{30} g_{\mathrm{eff}} T^4 \tag{13.41}$$

$$p_R = \frac{\rho_R}{3} = \frac{\pi^2}{90} g_{\mathrm{eff}} T^4 \tag{13.42}$$

where g_{eff} represents the sum of all effectively contributing massless degrees of
freedom and is given by [Kol90]

$$g_{\mathrm{eff}} = \sum_{i=\mathrm{bosons}} g_i \left(\frac{T_i}{T}\right)^4 + \frac{7}{8} \sum_{i=\mathrm{fermions}} g_i \left(\frac{T_i}{T}\right)^4, \tag{13.43}$$

In this relation the equilibrium temperature T_i of the particles i is allowed to differ
from the photon temperature T. The statistical weights are $g_\gamma = 2$ for photons,

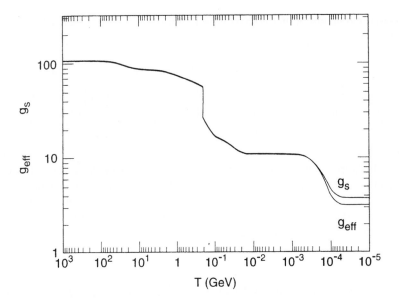

Figure 13.4. The cosmological standard model: behaviour of the summed effective degrees of freedom g_{eff} and g_S as a function of decreasing temperature. Only the particles of the standard model have been taken into consideration. One can see that both g_{eff} and g_S are identical over a wide range (from [Kol90]).

$g_e = 4$ for e^+, e^- and $g_\nu = 6$ for ν_α with $\alpha \equiv e, \mu, \tau$. This is valid for Dirac neutrinos contributing with one helicity or Majorana neutrinos contributing with two helicity states. If, indeed, four helicity states for Dirac neutrinos were to exist, the weight would be $g_\nu = 12$. Figure 13.4 illustrates the behaviour of g_{eff}. Starting at 106.75 at high energies where all particles of the standard model contribute, it decreases down to 3.36 if only neutrinos are participating.

In addition to the temperature, the entropy also plays an important role. The entropy is given by

$$S = \frac{R^3(\rho + p)}{T} \tag{13.44}$$

or, in the specific case of relativistic particles, by [Kol90]

$$S = \frac{2\pi^2}{45} g_s T^3 a^3 \tag{13.45}$$

where

$$g_s = \sum_{i=\text{bosons}} g_i \left(\frac{T_i}{T}\right)^3 + \frac{7}{8} \sum_{i=\text{fermions}} g_i \left(\frac{T_i}{T}\right)^3. \tag{13.46}$$

For the major part of the evolution of the universe, the two quantities g_{eff} and g_s were identical [Kol90]. The entropy per co-moving volume element is a conserved quantity in thermodynamic equilibrium, which together with constant g_s leads to the condition

$$T^3 a^3 = \text{constant} \Rightarrow a \sim T^{-1}. \tag{13.47}$$

The adiabatic expansion of the universe is, therefore, clearly connected with cooling. In the radiation-dominated phase, it leads to a dependence of (see equations (13.34) and (13.47))

$$t \sim T^{-2}. \tag{13.48}$$

With the help of equation (13.48) the evolution can now be discussed in terms of either times or energies. During the course of the evolution at certain temperatures particles which were until then in thermodynamic equilibrium ceased to be so. In order to understand this we consider the relation between the reaction rate per particle Γ and the expansion rate H. The former is

$$\Gamma = n\langle\sigma v\rangle \tag{13.49}$$

with a suitable averaging of relative speed v and cross section σ [Kol90]. The equilibrium can be maintained as long as $\Gamma > H$ for the most important reactions. For $\Gamma < H$ the corresponding particle is decoupled from equilibrium. This is known as *freezing out*. Let us assume a temperature dependence of the reaction rate of the form $\Gamma \sim T^n$. Consider two interactions mediated either by massless bosons such as the photon or by massive bosons with a mass m_M as the Z^0. In the first case for the scattering of two particles a cross section of

$$\sigma \sim \frac{\alpha^2}{T^2} \quad \text{with} \quad g = \sqrt{4\pi\alpha} = \text{gauge coupling strength} \tag{13.50}$$

results. In the second case the same behaviour can be expected for $T \gg m_M$. For $T \leq m_M$,

$$\sigma \sim G_M^2 T^2 \quad \text{with } G_M = \frac{\alpha}{m_M^2} \tag{13.51}$$

holds. With a thermal number density, i.e. $n \sim T^3$, for the case of massless exchange particles it follows that

$$\Gamma \sim \alpha^2 T. \tag{13.52}$$

For reactions involving the exchange of massive particles the corresponding relation is

$$\Gamma \sim G_M^2 T^5. \tag{13.53}$$

During the early radiation-dominated phase, the Hubble parameter can be written as [Kol90]

$$H = 1.66 g_{\text{eff}}^{1/2} \frac{T^2}{m_{Pl}} \tag{13.54}$$

The Planck mass m_{Pl} is given by

$$m_{Pl} = \left(\frac{\hbar c}{G}\right)^{\frac{1}{2}} = 1.221 \times 10^{19} \text{GeV}/c^2. \tag{13.55}$$

For massless particles, it then follows that

$$\frac{\Gamma}{H} \propto \frac{\alpha^2 m_{Pl}}{T} \tag{13.56}$$

As long as $T > \alpha^2 m_{Pl} \approx 10^{16}$ GeV, the reactions occur rapidly: in the opposite case they 'freeze out'. For massive particles, the analogous relation is

$$\frac{\Gamma}{H} \propto G_M^2 m_{Pl} T^3. \tag{13.57}$$

This means that as long as

$$m_M \geq T \geq G_M^{-\frac{2}{3}} m_{Pl}^{-\frac{1}{3}} \approx \left(\frac{m_M}{100 \text{ GeV}}\right)^{\frac{4}{3}} \text{MeV} \tag{13.58}$$

holds, such processes remain in equilibrium. If a particle freezes out its evolution is decoupled from the general thermal evolution of the universe.

We will now discuss the evolution of the universe step by step (see figure 13.5). The earliest moment to which our present description can be applied is the Planck time. Planck time t_{Pl} and Planck length l_{Pl} are given by

$$l_{Pl} = \left(\frac{\hbar G}{c^3}\right)^{\frac{1}{2}} = 1.6 \times 10^{-33} \text{ cm} \tag{13.59}$$

$$t_{Pl} = \left(\frac{\hbar G}{c^5}\right)^{\frac{1}{2}} = 5.4 \times 10^{-44} \text{ s}. \tag{13.60}$$

Here, the Schwarzschild radius and Compton wavelength are of the same order. Before this point, a quantum mechanical description of gravity is necessary which does not exist currently. All particles are highly relativistic and the universe is radiation dominated. At the moment at which energies drop to around 10^{16} GeV GUT symmetry breaking takes place, where the heavy gauge bosons X and Y (see chapter 5) freeze out. At about 300 GeV a second symmetry breaking occurs, which leads to the interactions that can be observed in today's particle accelerators. At about 10^{-6} s, the quarks and antiquarks annihilate and the surplus of quarks represents the whole of today's observable baryonic matter. The slight surplus of quarks is reflected in a baryon–photon ratio of about 10^{-10}. After about 10^{-5} s equivalent to 100–300 MeV, characterized by Λ_{QCD}, a further phase transition takes place. This is connected with the breaking of the chiral symmetry of the strong interaction and the transition from free quarks in the form of a

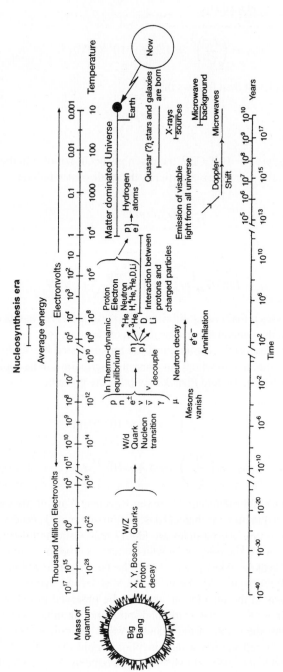

Figure 13.5. The chronological evolution of the universe since the Big Bang (from [Wil93]).

Table 13.2. GUT cosmology (from [Gro90]).

	Time t (s)	Energy $E = kT$ (GeV)	Temperature T (K)	'Diameter' of the universe a (cm)
Planck time t_{Pl}	10^{-44}	10^{19}	10^{32}	10^{-3}
GUT SU(5) breaking M_X	10^{-36}	10^{15}	10^{28}	10
$SU(2)_L \otimes U(1)$ breaking M_W	10^{-10}	10^2	10^{15}	10^{14}
Quark confinement $p\bar{p}$ annihilation	10^{-6}	1	10^{13}	10^{16}
ν decoupling, e^+e^- annihilation	1	10^{-3}	10^{10}	10^{19}
light nuclei form	10^2	10^{-4}	10^9	10^{20}
γ decoupling, transition from radiation-dominated to matter-dominated universe, atomic nuclei form, stars and galaxies form	10^{12} ($\approx 10^5$ yr)	10^{-9}	10^4	10^{25}
Today, t_0	$\approx 5 \times 10^{17}$ ($\approx 2 \times 10^{10}$ yr)	3×10^{-13}	3	10^{28}

quark–gluon plasma to quarks confined in baryons and mesons. At temperatures of about 1 MeV several things happen simultaneously. During the period 1–10^2 s, the process of primordial nucleosynthesis takes place. Therefore, the observation of the lighter elements provides the furthest look back into the history of the universe. Around the same time or, more precisely, a little before, the neutrinos decouple and develop further independently. As a result, a cosmic neutrino background is produced, which has, however, not yet been observed. Also the almost total destruction of electrons and positrons happens at this time. The resulting annihilation photons make up part of the cosmic microwave background. The next crucial stage only takes place about 300 000 years later. By then the temperature has sunk so far that nuclei can recombine with the electrons. As Thomson scattering (scattering of photons from free electrons) is strongly reduced, the universe suddenly becomes transparent and the radiation decouples from matter. This can still be detected today as 3 K background radiation. Starting at this time density fluctuations can increase and, therefore, the creation of large-scale structures which will finally result in galaxies can begin. At that time the universe also passes from a radiation-dominated to a matter-dominated state. This scenario, together with the discussed characteristics, is called the *standard model of cosmology* (see table 13.2).

13.3   The cosmic microwave background (CMB)

The cosmic microwave background is one of the most important supports for the Big Bang theory. Gamov, Alpher and Herman already predicted in the 1940s that if the Big Bang model was correct, a remnant noise at a temperature of about 5 K should still be present [Gam46, Alp48]. For extensive literature concerning this cosmic microwave background we refer to [Par95, Ber02, Sil02, Hu02, Hu03].

13.3.1   Spectrum and temperature

During the radiation-dominated era, radiation and matter were in a state of thermodynamic equilibrium. Thompson scattering on free electrons resulted in an opaque universe. As the temperature continued to fall, it became possible for more and more of the nucleons and electrons to recombine to form hydrogen. As most of the electrons were now bound, the mean free path of photons became much larger (of the order c/H) and they decoupled from matter. As the photons were in a state of thermodynamic equilibrium at the time of decoupling, their intensity distribution $I(\nu)\,d\nu$ corresponds to a black-body spectrum:

$$I(\nu)\,d\nu = \frac{2h\nu^3}{c^2} \frac{1}{\exp\left(\frac{h\nu}{kT}\right) - 1}\,d\nu. \tag{13.61}$$

The black-body form in a homogeneous Friedmann universe remains unchanged despite expansion. The maximum of this distribution lies, according to Wien's law, at a wavelength of

$$\lambda_{\max} T = 2.897 \times 10^{-3}\ \text{mK} \tag{13.62}$$

which for 5 K radiation corresponds to about 1.5 mm. Indeed, in 1964 Penzias and Wilson of the Bell Laboratories discovered an isotropic radiation at 7.35 cm, with a temperature of (3.5 ± 1) K [Pen65]. The energy density of the radiation is found by integrating over the spectrum (Stefan–Boltzmann law):

$$\rho_\gamma = \frac{\pi^2 k^4}{15 h^3 c^3} T_\gamma^4 = a T_\gamma^4. \tag{13.63}$$

From equation (13.37) we obtain the following relationship:

$$n_\gamma = \frac{30 \zeta(3) a}{\pi^4 k} T_\gamma^3 \approx 20.3 T_\gamma^3\ \text{cm}^{-3} \tag{13.64}$$

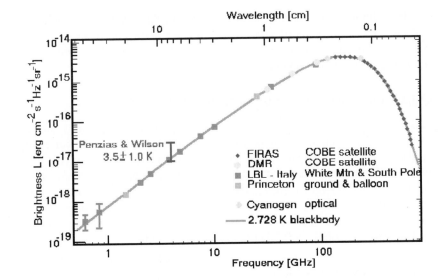

Figure 13.6. Spectrum of the cosmic background radiation, measured with the FIRAS and DMR detectors on the COBE satellite. It shows a perfect black-body behaviour. The smooth curve is the best-fit black-body spectrum with a temperature of 2.728 K. Also shown are the original data point of Penzias and Wilson as well as further terrestrial measurements (from [Fix96]).

for the number density of photons. Here $\zeta(3)$ is the Riemann ζ function of 3, which is $1.202\,06\ldots$.

The probability distribution of events at the time of last scattering, the so called last scattering surface (LSS), is approximately Gaussian with a mean at a redshift of $z = 1070$ and a standard deviation of 80. This means that roughly half of the last scattering events took place at redshifts between 990 and 1150. This redshift interval today corresponds to a length scale of $\lambda \simeq 7(\Omega h^2)^{1/2}$ Mpc, and an angle of $\theta \simeq 4\Omega^{1/2}$ [arcmin]. Structures on smaller angular scales are smeared out.

13.3.2 Measurement of the spectral form and temperature of the CMB

The satellite COBE (cosmic background explorer) brought a breakthrough in the field [Smo90]. It surveyed the entire sky in different wavelengths. In previous measurements only a few wavelengths had been measured and these were different in every experiment. The measured spectrum shows a perfect black-body form at a temperature of (2.728 ± 0.004) K [Wri94a, Fix96] (figure 13.6). No deviations whatsoever are seen in the spectral form. From that the number density of photons can be determined as $n_\gamma = (412 \pm 2)$ cm^{-3}. The number density is particularly interesting for the photon–baryon ratio η.

In addition to the spectral form and its distortions, the homogeneity and isotropy are also of extraordinary interest, as they allow conclusions as to the expansion of the universe and are an extremely important boundary condition for all models of structure formation.

13.3.3   Anisotropies in the 3 K radiation

Anisotropies in the cosmic background radiation are of extraordinary interest, on one hand for our ideas about the formation of large-scale structures and galaxies in the universe and, on the other hand, for our picture of the early universe. The former reveals itself through anisotropies on small angular scales (arc minutes up to a few degrees), while the latter is noticeable on larger scales (up to 180 degrees). We only consider the small angular scales in more detail [Whi94], because this part is important for neutrino physics. For an overview see [Rea92, Hu95, Ber02, Sil02, Hu02, Hu03, Wri03, Zal03].

13.3.3.1   Measurement of the anisotropy

The temperature field of the CMB can be expanded into its spherical harmonics Y^{lm}

$$\frac{\Delta T}{T}(n) = \sum_{l=2}^{\infty} \sum_{m=-l}^{l} a_{lm} Y^{lm}(n). \tag{13.65}$$

By definition the mean value of a_{lm} is zero. The correlation function $C(\theta)$ of the temperature field is the average across all pairs of points in the sky separated by an angle θ:

$$C(\theta) = \left\langle \frac{\Delta T}{T}(n_1) \frac{\Delta T}{T}(n_2) \right\rangle = \frac{1}{4\pi} \sum_l (2l+1) C_l P_l(\cos\theta) \tag{13.66}$$

with the Legendre polynomials $P_l(\cos\theta)$ and C_l as a cosmological ensemble average

$$C_l = \langle |a_{lm}^2| \rangle. \tag{13.67}$$

In the case of random phases the power spectrum is related to a temperature difference ΔT via

$$\Delta T_l = \sqrt{C_l \frac{l(l+1)}{2\pi}}. \tag{13.68}$$

The harmonic index ℓ is associated with an angular scale θ via $\ell \approx 180°/\theta$. The main anisotropy observed is the dipole component due to the Earth movement with respect to the CMB. It was measured by COBE to be $\Delta T/T = 3.353 \pm 0.024$ mK [Ben96] and, more recently, by WMAP as $\Delta T/T = 3.346 \pm 0.017$ mK [Ben03]. After subtracting the dipole component, anisotropies are observed by various experiments on the level of 10^{-5}. These result mainly from thermo-acoustic oscillations of baryons and photons [Hu95, Smo95, Teg95, Ber02, Sil02,

Wri03]. It produces a series of peaks in the power spectrum whose positions, heights and numbers depend critically on various cosmological parameters, which is of major importance. The position of the first acoustic peak depends on the total density in the universe as $l \approx 200\sqrt{\Omega_0}$. In addition, Ω_b will increase the odd peaks with respect to even ones. Both h and Ω_Λ will change the height and location of the various peaks. The existence of such acoustic peaks has been shown by various experiments [Tor99, Mau00, Mel00, Net02, Hal02a, Lee01, Pea03, Sco02, Ben03, Ruh02, Gra02, Kuo02, Ben03] and is shown for the WMAP data in figure 13.7. The radiation power spectrum shows two characteristic angular scales. A prominent peak occurs at $l \approx 220$ or about 30 arcmin. This is the angular scale that corresponds to the horizont at the moment of last scattering of radiation. The corresponding comoving scale is about 100 Mpc. The second scale is the damping scale of about 6 arcmin, equivalent to the thickness of the last scattering surface of about 10 Mpc.

It will be one of the main goals of the next generation satellite missions like WMAP (whose first results were recently published [Ben03, Spe03]) and PLANCK to determine these cosmological quantities even more precisely. How both high-z supernova observations and CMB peak position measurements, restrict cosmological parameters in a complementary way is shown in figure 13.8.

13.3.3.2 Anisotropies on small scales

Anisotropies have to be divided into two types, depending on the horizon size at the time of decoupling. Fluctuations outside the event horizon are independent of the microphysics present during decoupling and so reflect the primordial perturbation spectrum, while the sub-horizontal fluctuations depend on the details of the physical conditions at the time of decoupling. The event horizon at the time of decoupling today corresponds to an angular size of [Kol90]

$$\Theta_{\text{dec}} = 0.87\Omega_0^{1/2} \left(\frac{z_{\text{dec}}}{1100} \right)^{-1/2} \quad \text{[deg]}. \tag{13.69}$$

Below about $1°$, therefore, the fluctuations mirror those which show up in structure formation (see section 13.6). There is a correlation between the mass scale and the corresponding characteristic angular size of the anisotropies (see, e.g., [Nar93]):

$$(\delta\theta) \simeq 23 \left(\frac{M}{10^{11}M_\odot} \right) (h_0 q_0^2)^{1/3} \quad \text{[arcsec]}. \tag{13.70}$$

Typical density fluctuations that led to the formation of galaxies, therefore, correspond today to anisotropies on scales of 20 arcsec. Assuming that density fluctuations $\delta\rho/\rho$ develop adiabatically, the temperature contrast in the background radiation should be given by

$$\left(\frac{\delta T}{T} \right)_R = \frac{1}{3} \left(\frac{\delta\rho}{\rho} \right)_R \tag{13.71}$$

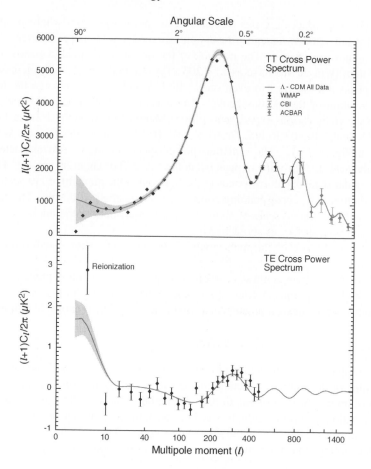

Figure 13.7. Compilation of the anisotropy power spectrum as a function of the multipole order l as observed in various experiments. The first acoustic peak is clearly visible at $l \approx 200$, also higher order peaks are visible (from [Ben03]).

where the subscript R stands for 'at the recombination time'. In order to produce the density currently observed in galaxies, observable temperature anisotropies of $\delta T / T \approx 10^{-3}$–$10^{-4}$ would be expected. However, this has not been observed. The density fluctuations in the baryon sector are damped and additional terms are needed in the density fluctuations. The damping results from the period of recombination. Due to the suddenly increasing mean free path for photons, these can also effectively flow away from areas of high density. Their spreading, due to frequent collisions, does, however, correspond to diffusion rather than to a free flow. This kind of damping is called collisional damping or *Silk-damping* [Sil67, Sil68, Efs83]. Here smaller scales are effectively smeared out as, due to

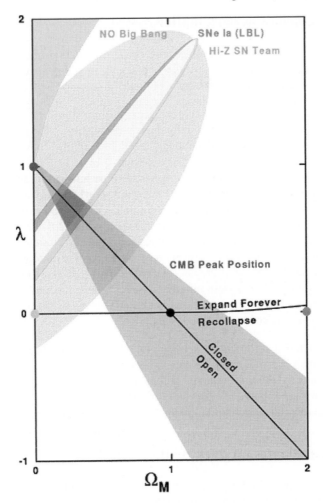

Figure 13.8. Ω_Λ *versus* Ω_m plot. The determination of cosmological parameters using high-z supernovae type Ia and the acoustic peak positions in the CMB power spectrum result in quite complementary information. The combined parameter values show with high significance a non-vanishing cosmological constant and an $\Omega_\Lambda \approx 0.7$ and $\Omega_m \approx 0.3$.

the frequent interaction of the photons, inhomogeneities in the photon–baryon plasma are damped. A significant amount of dark matter in the form of WIMPs could, for example, produce a similar effect. They already form a gravitational potential before recombination which the baryons experience later.

After having briefly discussed the basic picture of cosmology we now want to discuss some special topics which are influenced by neutrinos in more detail.

13.4   Neutrinos as dark matter

Among the most interesting problems of modern particle astrophysics is dark matter. Roughly speaking, the problem is the realization that there seems to be a great deal more gravitationally interacting matter in the universe than is luminous. It shows up on various scales:

- Rotational curves $v(r)$ of spiral galaxies. They show a flat behaviour even in regions far out where the optical detection, according to Newton's law, a Keplerian $v(r) \sim r^{-1/2}$ should be expected.
- Dark matter in galaxy clusters. It has been established by the kinematics of clusters (virial theorem), x-ray emission (gravitational binding of a hot electron gas) and gravitational lensing that a large fraction of the mass of clusters is dark. Estimates of cluster masses result in $\Omega \approx 0.3$.
- Cosmology: Big Bang nucleosynthesis and galaxy clusters estimates on matter densities in the universe are only consistent if assuming non-baryonic dark matter. Also the discrepancy from the observed value to the theoretical prediction of $\Omega = 1$ from inflation requires a large amount of dark matter. However, the latter argument was weakened recently by the observation of a non-vanishing cosmological constant.

For more detailed information, see [Jun96, Kla97, Ber99]. In summary, it is clear that the observed visible matter is insufficient to close the universe. An explanation of the rotation curves of galaxies and the behaviour of galaxy clusters also does not seem possible. Baryonic forms of dark matter seem to be able to explain the rotation curves but fail in the large-scale problems.

13.5   Candidates for dark matter

Having shortly presented the evidence for the existence of dark matter we discuss some particle physics candidates among them are neutrinos.

13.5.1   Non-baryonic dark matter

The possible candidates for this are limited not so much by physical boundary conditions as by the human imagination and the resulting theories of physics. Consider first the abundance of relics (such as massive neutrinos) from the early period of the universe, which was then in thermodynamic equilibrium. For temperatures T very much higher than the particle mass m, their abundance is similar to that of photons, while at low temperatures ($m > T$) the abundance is exponentially suppressed (Boltzmann factor). How long a particle remains in equilibrium depends on the ratio between the relevant reaction rates and the Hubble expansion. Pair production and annihilation determine the abundance of long-lived or stable particles. The particle density n is then determined by the

Boltzmann equation [Kol90]

$$\frac{dn}{dt} + 3Hn = -\langle \sigma v \rangle_{\text{ann}}(n^2 - n_{\text{eq}}^2) \tag{13.72}$$

where H is the Hubble constant, $\langle \sigma v \rangle_{\text{ann}}$ the thermally averaged product of the annihilation cross section and velocity and n_{eq} is the equilibrium abundance. The annihilation cross section of a particle results from a consideration of all of its decay channels. It is useful to parametrize the temperature dependence of the reaction cross section as follows: $\langle \sigma v \rangle_{\text{ann}} \sim v^p$, where in this partial wave analysis $p = 0$ corresponds to an s-wave annihilation, $p = 2$ to a p-wave annihilation, etc. As, furthermore, $\langle v \rangle \sim T^{1/2}$, it follows that $\langle \sigma v \rangle_{\text{ann}} \sim T^n$, with $n = 0$ s-wave, $n = 1$ p-wave, etc. This parametrization is useful in the calculation of abundances for Dirac and Majorana particles. While the annihilation of Dirac particles only occurs via s-waves, i.e. independent of velocity, Majorana particles also have a contribution from p-wave annihilation, leading to different abundances.

There is a lower mass limit on any dark matter particle candidate, which relies on the fact of conservation of phase space (Liouville theorem). The evolution of dark matter distributions is collisionless; therefore, they accumulate in the centre of astronomical objects. Consider hot dark matter and an initial particle density distribution given by Fermi–Dirac statistics:

$$dN = g\frac{V}{(2\pi\hbar)^3}\exp(E/kT) \pm 1]^{-1}\,d^3p \tag{13.73}$$

Having an average occupation number of $\bar{n} = 1/2$, this results in a phase space density ρ of

$$\rho_i < (2\pi\hbar)^{-3}g/2. \tag{13.74}$$

Assuming that the velocity dispersion has relaxed to a Maxwellian

$$dp = (2\pi\sigma^2)^{3/2}\exp(-v^2/2\sigma^2)\,d^3v \tag{13.75}$$

resulting in

$$\rho_f < \left(\frac{\rho}{m}\right)(2\pi\sigma^2)^{-3/2}\,\text{m}^{-3}. \tag{13.76}$$

Then the condition that the maximum phase space density has not increased results in

$$m^4 > \frac{\rho}{Ng}\left(\frac{\sqrt{2\sigma}\hbar}{\sigma}\right)^3 \tag{13.77}$$

assuming a number N of neutrinos with mass m. This bound is known as the Tremaine–Gunn limit [Tre79]. Take, as an example, a simple isothermal sphere with radius r, having a density according $\rho = \sigma^2/2\pi Gr^2$ and the simple case $Ng = 1$. For galaxy clusters with $\sigma = 1000$ km s^{-1}, $r = 1$ Mpc it follows that

$m_\nu > 1.5$ eV and there is no problem in fitting in neutrinos there. However, for a galactic halo with $\sigma = 150$ km s^{-1}, $r = 5$ kpc, it follows that $m_\nu > 33$ eV already close to a value necessary for the critical density. Therefore, neutrinos cannot be the dark matter in small objects like regular galaxies—this is further supported by the observation that dwarf galaxies also contain a significant amount of dark matter.

13.5.1.1 Hot dark matter, light neutrinos

Light neutrinos remain relativistic and freeze out at about 1 MeV (see (13.11)), so that their density is given by

$$\rho_\nu = \sum_i m_{\nu i} n_{\nu i} = \Omega_\nu \rho_c. \tag{13.78}$$

From this it follows that for $\Omega \approx 1$ a mass limit for light neutrinos (masses smaller than about 1 MeV) is given by [Cow72]

$$\sum_i m_{\nu i} \left(\frac{g_\nu}{2}\right) = 94 \text{ eV}\Omega_\nu h^2. \tag{13.79}$$

Given the experimentally determined mass limits mentioned in chapter 6, and the knowledge that there are only three light neutrinos, ν_e is already eliminated as a dominant contribution. Recent CMB data from WMAP, CBI and ACBAR in combination with data from large-scale structures and Ly α systems have been used to give a neutrino density contributions of [Spe03]

$$\Omega_\nu h^2 < 0.0076 \qquad (95\% \text{ CL}). \tag{13.80}$$

In the case of degenerated neutrinos this would imply $m_\nu < 0.23$ eV. However there should be some caution because a bound on the neutrino mass is strongly correlated with other cosmological quantities like the Hubble parameter (see section 13.6). Therefore, conclusions that the observed evidence in double β-decay (chapter 7) and the LSND oscillation evidence (chapter 8) are ruled out is premature. Instead, one should use the future KATRIN β-decay result (chapter 6) as input for the determination of other cosmological parameters.

13.5.1.2 Cold dark matter, heavy particles, WIMPs

The freezing-out of non-relativistic particles with masses of GeV and higher has the interesting characteristic that their abundance is inversely proportional to the annihilation cross section. This follows directly from the Boltzmann equation and implies that the weaker particles interact, the more abundant they are today. Such 'weakly interacting massive particles' are generally known as WIMPs. If

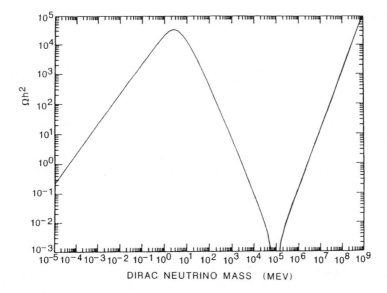

Ωh^2

DIRAC NEUTRINO MASS (MEV)

Figure 13.9. The contribution of stable neutrinos of mass m to the matter density of the universe. Only neutrino masses smaller than 100 eV and heavier than several GeV to TeV are cosmologically acceptable. Otherwise neutrinos have to be unstable (from [Tur92a]).

we assume a WIMP with mass m_{WIMP} smaller than the Z^0 mass, the cross section is roughly equal to $\langle \sigma v \rangle_{\text{ann}} \approx G_F^2 m_{\text{WIMP}}^2$ [Kol90], i.e.

$$\Omega_{\text{WIMP}} h^2 \approx 3 \left(\frac{m_{\text{WIMP}}}{\text{GeV}} \right)^{-2}. \qquad (13.81)$$

Above the Z^0 mass the annihilation cross section decreases as m_{WIMP}^{-2}, due to the momentum dependence of the Z^0 propagator and, hence, a correspondingly higher abundance results. Figure 13.9 shows, as an example, the contribution of massive neutrinos to the mass density in the universe. Neutrinos between 100 eV and about 2 GeV as well as beyond the TeV region should, if they exist, be unstable according to these cosmological arguments [Lee77a]. In order to be cosmologically interesting, that is to produce a value of $\Omega \approx 1$, stable neutrinos must either be lighter than 100 eV as mentioned before or heavier than about 2 GeV (Dirac neutrinos) or 5 GeV (Majorana neutrinos). However, this does not exclude other particles like sneutrinos in this mass range.

With heavy neutrinos a little bit out of fashion, currently the most preferred class of possible candidates for dark matter are supersymmetric particles, especially the neutralino as a possible lightest supersymmetric particle. A calculation of their relic abundance depends on various assumptions, due to the many free parameters in SUSY models (see [Jun96]).

13.5.2 Direct and indirect experiments

The experimental search for dark matter is currently one of the most active fields in particle astrophysics. Two basic strategies are followed: either the direct detection of dark matter interactions mostly via elastic scattering of WIMPs on nuclei or indirect detection by looking mostly for their annihilation products. Recent reviews of the direct detection efforts can be found in [Smi90, Jun96, Kla97, Ram03]. Indirect experiments do not detect the interaction of dark matter in the laboratory but the products of reactions of particles of dark matter taking place extraterrestrially or inside the Earth. For dark matter this is mainly particle–antiparticle annihilation. Two main types of annihilation are considered:

- annihilation inside the Sun or Earth and
- annihilation within the galactic halo.

13.5.2.1 *Annihilation inside the Sun or Earth*

It is possible that dark matter may accumulate gravitationally in stars and annihilate there with anti-dark matter. One signal of such an indirect detection would be high-energy solar neutrinos in the GeV–TeV range. These would be produced through the capture and annihilation of dark matter particles within the Sun [Pre85, Gou92]. An estimate of the expected signal due to photino (neutralino) annihilation results in about two events per kiloton detector material and year. These high-energy neutrinos would show up in the large water detectors via both charged and neutral weak currents. The charged weak interactions are about three times as frequent as the neutral ones. So far no signal has been found in detectors like Super-Kamiokande and AMANDA (see chapter 12). The capture of particles of dark matter in the Earth and their annihilation, have also been discussed. Neutrinos from neutralino–antineutralino annihilation in the Earth are being sought by looking for vertical upward-going muons (see chapter 12). Again no signal has been observed yet.

13.6 Neutrinos and large-scale structure

One assumption in describing our universe is homogeneity. However, even this seems to be justified on very large scales, observations have revealed a lot of structure on scales going beyond 100 Mpc. Galaxies group themselves into clusters and the clusters into superclusters, separated by enormous regions with low galaxy density, the so-called voids. The existence of large-scale structure (LSS) depends on the initial conditions of the Big Bang and about how physics processes have operated subsequently. The general picture of structure formation as the ones described is gravitational instability, which amplifies the growth of density fluctuations, produced in the early universe. The most likely source for producing density perturbations are quantum zero-point fluctuations during the inflationary era. Initial regions of higher density, which after the recombination era can concentrate through gravity, thereby form the starting points for the formation of structure. Hence, defining an initial spectrum of perturbation, the growth of the various scales has to be explored. The most viable framework of describing structure formation is the self-gravitating fluid which experiences a critical instability, as already worked out in classical mechanics by Jeans. We refer to further literature with respect to a more detailed treatment of structure formation [Pee80, Pad93, Pee93, Bah97, Pea98].

For the theoretical description of the development of the fluctuations, it is convenient to introduce the dimensionless density perturbation field or density contrast

$$\delta(x) = \frac{\rho(x) - \langle \rho \rangle}{\langle \rho \rangle}. \tag{13.82}$$

The correlation function of the density field is given by

$$\xi(r) = \langle \delta(x) \delta(x + r) \rangle \tag{13.83}$$

with brackets indicating an averaging over the normalization volume V. The density contrast can be decomposed into its Fourier coefficients:

$$\xi(r) = \frac{V}{(2\pi)^3} \int \delta_k e^{-ik \cdot r} \, d^3 k. \tag{13.84}$$

It can be shown that

$$\xi(r) = \frac{V}{(2\pi)^3} \int |\delta_k|^2 e^{-ik \cdot r} \, d^3 k. \tag{13.85}$$

The quantity $|\delta_k|^2$ is known as the *power spectrum*. The correlation function is, therefore, the Fourier transform of the power spectrum. If we assume an isotropic correlation function, integration over the angle coordinates gives

$$\xi(r) = \frac{V}{(2\pi)^3} \int |\delta_k|^2 \frac{\sin kr}{kr} 4\pi k^2 \, dk. \tag{13.86}$$

The aim is to predict this power spectrum theoretically, in order to describe the experimentally determined correlation function. Theory suggests a spectrum with no preferred scale, called the Harrison–Zeldovich spectrum, equivalent to a power law

$$|\delta_k|^2 \sim k^n. \tag{13.87}$$

For Gaussian-like, and, therefore, uncorrelated fluctuations, there is a connection between the mean square mass fluctuation and the power spectrum $|\delta_k|^2$, which contains all the information about the fluctuation [Kol90]:

$$\langle \delta^2 \rangle_\lambda \simeq V^{-1} (k^3 |\delta_k|^2 / 2\pi^2)_{k \approx 2\pi/\lambda}. \tag{13.88}$$

with

$$\lambda = \frac{2\pi}{k} a(t). \tag{13.89}$$

Weakly interacting particles, such as light neutrinos, can escape without interaction from areas of high density to areas of low density, which can erase small scale perturbations entirely. This process of free streaming, or collisionless damping, is important before the Jeans instability becomes effective. Light, relativistic neutrinos travel approximately with the speed of light, so any perturbation that has entered the horizon will be damped. The relevant scale is the redshift of matter radiation equality [Bon80, Bon84]

$$\lambda_{FS} \simeq 1230 \left(\frac{m_\nu}{\text{eV}}\right)^{-1} \text{Mpc} \tag{13.90}$$

corresponding to a mass scale of

$$M_{FS} \simeq 1.5 \times 10^{17} \left(\frac{m_\nu}{\text{eV}}\right)^{-2} M_\odot. \tag{13.91}$$

Such masses are the size of galaxy superclusters and such objects are the first to form in a neutrino dominated universe. Below this scale perturbations are completely erased or at least strongly damped by neutrinos, resulting in a suppression of small scales in the matter power spectrum by roughly [Hu98]

$$\frac{\Delta P_M}{P_m} \sim -\frac{8\Omega_\nu}{\Omega_M}. \tag{13.92}$$

The larger m_ν and Ω_ν, the stronger is the suppression of density fluctuations at small scales. The effect of neutrino masses in the LSS power spectra can be seen in figure 13.11. While massive neutrinos have little impact on the CMB power spectrum, it is still necessary to include the CMB data to determine the other cosmological parameters and to normalize the matter power spectrum. Ω_M has been recently restricted by WMAP, so the measurements of the power spectrum obtained by large scale galaxy surveys can now be normalized to the CMB data. The most recent large-scale surveys are the 2dF galaxy redshift survey (2dFGRS) [Elg02] and the Sloan Digital Sky Survey (SDSS) [Dor03]. The galaxy distribution as observed in the 2dFGRS is shown in figure 13.10. To obtain a bound on neutrino masses, the correlation with other cosmological parameters has to be taken into account accordingly, especially the Hubble parameter H, the matter density Ω_M and the bias parameter b [Han03]. The bias parameter relates the matter spectrum to the observable galaxy–galaxy correlation function ξ_{gg} in large-scale structure surveys via

$$b^2(k) = \frac{P_g(k)}{P_M(k)} \tag{13.93}$$

In this way it defines a threshold for how light traces matter. The current data seem to suggest (according to the assumptions made) an upper limit in the region of [Han03a]

$$\sum m_\nu < 0.7\text{–}2.2 \text{ eV}. \tag{13.94}$$

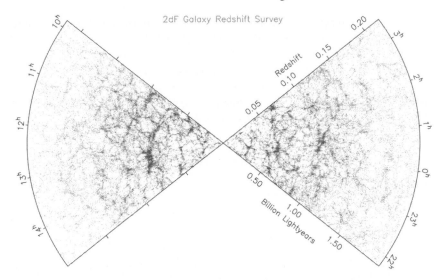

Figure 13.10. The distribution of galaxies (wedge diagram) in part of the 2dFGRS with a slice thickness of 4 degrees. In total 213 703 galaxies are drawn. The filament-like structures, i.e. areas of very high density (superclusters), as well as voids, can clearly be seen (from [Pea02a]).

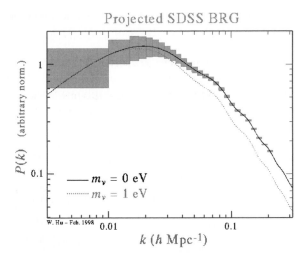

Figure 13.11. The projected power spectrum of the Sloan Digital Sky Survey and its sensitivity to neutrino masses. The effect of massive neutrinos on small scales is clearly visible and, hence, a mass sensitivy on the level of less than 1 eV seems feasible (from [Hu98]).

The optimistic sensitivity estimated of the SDSS survey together with future PLANCK data results in a possible bound of [Han03]

$$\sum m_\nu < 0.12 \text{ eV}. \tag{13.95}$$

This result is obtained by the assumption that all other relevant cosmological parameters would be measured with a precision of 1%.

13.7   The cosmic neutrino background

Analogous to the photon background, a cosmic neutrino background should also exist. At temperatures above 1 MeV, neutrinos, electrons and photons are in thermal equilibrium with each other via reactions such as $e^+e^- \leftrightarrow \gamma\gamma$ or $e^+e^- \leftrightarrow \nu\bar{\nu}$. As the temperature drops further to less than the rest mass of the electron, all energy is transferred to the photons via pair annihilation, thereby increasing their temperature. Next consider the entropy of relativistic particles which is given by (see equations (13.45) and (13.46)):

$$S = \frac{4}{3}k_B \frac{R^3}{T}\rho \tag{13.96}$$

where ρ represents the energy density. As $\rho \sim T^4$, according to the Stefan–Boltzmann law, it follows that, for constant entropy,

$$S = (Ta)^3 = \text{constant}. \tag{13.97}$$

From the relations mentioned in section 13.2.1, it follows that

$$\rho_{\nu_i} = \tfrac{7}{16}\rho_\gamma \tag{13.98}$$

and, for $kT \gg m_e c^2$,

$$\rho_{e^\pm} = \tfrac{7}{8}\rho_\gamma. \tag{13.99}$$

Using the appropriate degrees of freedom and entropy conservation, we obtain:

$$(T_\gamma a)_B^3 (1 + 2\tfrac{7}{8}) + (T_\nu a)_B^3 \sum_{i=1}^{6}\rho_{\nu_i} = (T_\gamma a)_A^3 + (T_\nu a)_A^3 \sum_{i=1}^{6}\rho_{\nu_i} \tag{13.100}$$

where B ('before') represents times $kT > m_e c^2$ and A ('after') times $kT < m_e c^2$. Since the neutrinos had already decoupled, their temperature developed proportional to a^{-1} and, therefore, the last terms on both sides cancel. Hence,

$$(T_\gamma a)_B^3 \frac{11}{4} = (T_\gamma a)_A^3 \tag{13.101}$$

However, since prior to the annihilation phase of the e^+e^--pairs $T_\gamma = T_\nu$, this means that

$$\left(\frac{T_\gamma}{T_\nu}\right)_B = \left(\frac{11}{4}\right)^{1/3} \simeq 1.4. \tag{13.102}$$

On the assumption that no subsequent significant changes to these quantities have taken place, the following relation between the two temperatures exists today:

$$T_{\nu,0} = (\tfrac{4}{11})^{1/3} T_{\gamma,0}. \tag{13.103}$$

A temperature of $T_{\gamma,0} = 2.728$ K corresponds then to a neutrino temperature of 1.95 K. If the photon background consists of a number density of $n_{\gamma,0} = 412$ cm^{-3}, the particle density of the neutrino background is

$$n_{\nu_0} = \frac{3}{4}\frac{g_\nu}{g_\gamma}\frac{4}{11}n_{\gamma_0} = \frac{3g_\nu}{22}n_{\gamma_0} = 336\,\text{cm}^{-3} \tag{13.104}$$

with $g_\nu = 6$ ($g_\gamma = 2$). The energy density and average energy are:

$$\rho_{\nu_0} = \frac{7}{8}\frac{g_\nu}{g_\gamma}\left(\frac{4}{11}\right)^{\frac{4}{3}}\rho_{\gamma_0} = 0.178\,\text{eV cm}^{-3} \qquad \langle E_\nu\rangle_0 = 5.28 \times 10^{-4}\,\text{eV}. \tag{13.105}$$

These relations remain valid even if neutrinos have a small mass. It is also valid for Majorana neutrinos and light left-handed Dirac neutrinos because, in both cases, $g_\nu = 2$. For heavy Dirac neutrinos ($m > 300$ keV) with a certain probability, the 'wrong' helicity states can also occur and $g_\nu = 4$ [Kol90].

The very small cross section of relic neutrinos has so far thwarted any experimental attempt to obtain evidence for their existence. However, it plays an important role in the Z-burst model mentioned in chapter 12 to explain UHE cosmic-ray events.

13.8 Primordial nucleosynthesis

In this chapter we turn our attention to another very important support of the Big Bang model, namely the synthesis of the light elements in the early universe. These are basically H, D, ^{3}He, ^{4}He and ^{7}Li. Together with the synthesis of elements in stars and the production of heavy elements in supernova explosions (see chapter 11), this is the third important process in the formation of the elements. The fact that their relative abundances are predicted correctly over more than ten orders of magnitude can be seen as one of the outstanding successes of the standard Big Bang model. Studying the abundance of ^{4}He allows statements to be made about the number of possible neutrino flavours in addition to the precise measurements made at LEP. For a more detailed discussion we refer the reader to [Yan84, Boe88, Den90, Mal93, Pag97, Tyt00, Oli00, Oli02, Ste03].

13.8.1   The process of nucleosynthesis

The synthesis of the light elements took place in the first three minutes after the Big Bang, which means at temperatures of about 0.1–10 MeV. The first step begins at about 10 MeV, equivalent to $t = 10^{-2}$ s. Protons and neutrons are in thermal equilibrium through the weak interaction via the reactions

$$p + e^- \longleftrightarrow n + \nu_e \qquad (13.106)$$

$$p + \bar{\nu}_e \longleftrightarrow n + e^+ \qquad (13.107)$$

$$n \longleftrightarrow p + e^- + \bar{\nu}_e \qquad (13.108)$$

and the relative abundance is given in terms of their mass difference $\Delta m = m_n - m_p$ (neglecting chemical potentials) as

$$\frac{n}{p} = \exp\left(-\frac{\Delta m c^2}{kT}\right). \qquad (13.109)$$

The weak interaction rates are (see (13.50) and (13.52))

$$\Gamma \propto G_F^2 T^5 \qquad T \gg Q, m_e. \qquad (13.110)$$

If this is compared with the expansion rate H, it follows that

$$\frac{\Gamma}{H} = \frac{T^5 G_F^2}{T^2/m_{Pl}} \approx \left(\frac{T}{0.8 \text{ MeV}}\right)^3 \qquad (13.111)$$

implying that from about 0.8 MeV the weak reaction rate becomes less than the expansion rate and freezes out. The neutron–proton ratio begins to deviate from the equilibrium value. One would expect a significant production of light nuclei here, as the typical binding energies per nucleon lie in the region of 1–8 MeV. However, the large entropy, which manifests itself in the very small baryon–photon ratio η, prevents such production as far down as 0.1 MeV.

The second step begins at a temperature of about 1 MeV or, equivalently, at 0.02 s. The neutrinos have just decoupled from matter and, at about 0.5 MeV, the electrons and positrons annihilate. This is also the temperature region in which these interaction rates become less than the expansion rate, which implies that the weak interaction freezes out, which leads to a ratio of

$$\frac{n}{p} = \exp\left(-\frac{\Delta m c^2}{kT_f}\right) \simeq \frac{1}{6}. \qquad (13.112)$$

The third step begins at 0.3–0.1 MeV, corresponding to about 1–3 min after the Big Bang. Here practically all neutrons are converted into ^{4}He via the reactions shown in figure 13.14 beginning with

$$n + p \leftrightarrow D + \gamma. \qquad (13.113)$$

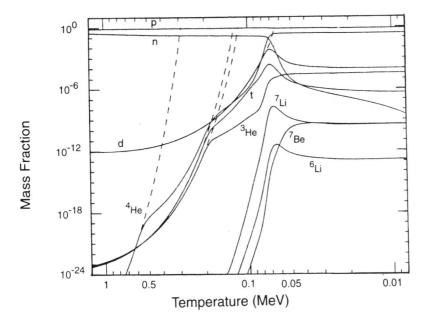

Figure 13.12. Development of the abundances of the light elements during primordial nucleosynthesis (from [Ree94]).

This is a certain bottleneck, because high energetic photons dissociate the deuteron. Once it builds up, the reaction chain continues via

$$D + D \leftrightarrow He + \gamma \tag{13.114}$$
$$D + p \leftrightarrow {}^3He + \gamma \tag{13.115}$$
$$D + n \leftrightarrow {}^3H + \gamma. \tag{13.116}$$

The amount of primordial helium γ is then

$$Y = \frac{2n_n}{n_n + n_p}. \tag{13.117}$$

Meanwhile the initial n/p fraction has fallen to about $1/7$, due to the decay of the free neutrons. The equilibrium ratio, which follows from an evolution according to equation (13.109), would be $n/p = 1/74$ at 0.3 MeV. The non-existence of stable nuclei of mass 5 and 8, as well as the now essential Coulomb barriers, very strongly inhibit the creation of ${}^7Li$, and practically completely forbid that of even heavier isotopes (see figure 13.12). Because of the small nucleon density, it is also not possible to get over this bottleneck via 3α reactions, as stars do. Therefore, BBN comes to an end if the temperature drops below about 30 keV, when the Universe was about 20 min old.

The process of nucleosynthesis

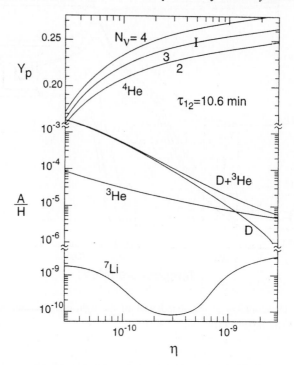

Figure 13.13. The primordial abundances of the light elements, as predicted by the standard model of cosmology, as a function of today's baryon density n_B or of $\eta = n_B/n_\gamma$. The ^{4}He fraction is shown under the assumptions $N_\nu = 2, 3$ and 4 where N_ν is the number of light neutrino flavours. A consistent prediction is possible over 10 orders of magnitude (from [Tur92a]).

Current experimental numbers of the elemental abundances are [Oli02, Ste03]:

$$Y = 0.238 \pm 0.002 \pm 0.005 \qquad (13.118)$$

$$\mathrm{Li/H} = 1.23 \pm 0.01 \times 10^{-10} \qquad (13.119)$$

$$\mathrm{D/H} = 2.6 \pm 0.4 \times 10^{-5}. \qquad (13.120)$$

Using the D/H value this corresponds to

$$\eta = 6.1^{+0.7}_{-0.3} \times 10^{-10}. \qquad (13.121)$$

it can be converted into a baryonic density ($\eta \times 10^{10} = 274\Omega_b h^2$)

$$\Omega_b h^2 = 0.022^{+0.003}_{-0.002}. \qquad (13.122)$$

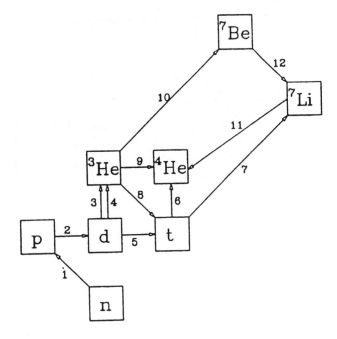

Figure 13.14. The 12 fundamental reactions in the chain of synthesis of the light elements, illustrating which elements can be built up in this way (from [Smi93b]). Labels indicate the following reactions: (1) $n \leftrightarrow p$, (2) $p(n, \gamma)d$, (3) $d(p, \gamma)\,^3He$, (4) $d(d, n)\,^3He$, (5) $d(d, p)t$, (6) $t(d, n)\,^4He$, (7) $t(\alpha, \gamma)\,^7Li$, (8) $^3He(n, p)t$, (9) $^3He(d, p)\,^4He$, (10) $^3He(\alpha, \gamma)\,^7Be$, (11) $^7Li(p, \alpha)\,^4He$, (12) $^7Be(n, p)\,^7Li$.

Recent WMAP results imply a $\Omega_b h^2 = 0.024 \pm 0.001$ which results in the rather high value of $\eta = 6.5^{+0.4}_{-0.3} \times 10^{-10}$ [Spe03]. While both numbers are in good agreement, there is a certain tension in the ^4He measurement. The given Ω_b implies $Y = 0.248 \pm 0.001$, higher than the value of (13.118). Further studies will show whether there is reasonable agreement. Therefore, according to primordial nucleosynthesis, it is not possible to produce a closed universe from baryons alone. However, if Ω_b has a value close to the upper limit, a significant fraction can be present in dark form, as the luminous part is significantly less ($\Omega_b^L \lesssim 0.02$) than that given by equation (13.122). Some of it could be 'Massive Compact Halo Objects' (MACHOs) searched for by gravitational microlensing in the Milky Way halo. It is at least possible to use baryonic matter to explain the rotation curves of galaxies.

The predicted abundances (especially ^4He) depend on a number of parameters (figure 13.13). These are, in principle, three: the lifetime of the neutron τ_n, the fraction of baryons to photons $\eta = n_B/n_\gamma$ and the number of relativistic degrees of freedom g_{eff}, where neutrinos contribute. In the following we only investigate the latter, for more details see [Kla97].

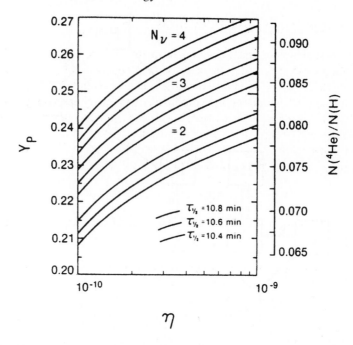

Figure 13.15. A more detailed illustration of the ^{4}He abundance as a function of the baryon/photon ratio $\eta = n_B/n_\gamma$. The influence of the number of neutrino flavours and the neutron lifetime on the predicted ^{4}He abundance can clearly be seen (from [Yan84]).

13.8.2 The relativistic degrees of freedom g_{eff} and the number of neutrino flavours

The expansion rate H is proportional to the number of relativistic degrees of freedom of the available particles (13.54). According to the standard model, at about 1 MeV these are photons, electrons and three neutrino flavours. The dependence of the freeze-out temperature on the number of degrees of freedom then results using equation (13.111) in

$$H \sim g_{\text{eff}}^{1/2} T^2 \Rightarrow T_F \sim g_{\text{eff}}^{1/6}. \tag{13.123}$$

Each additional relativistic degree of freedom (further neutrino flavours, axions, majorons, right-handed neutrinos, etc) therefore means an increase in the expansion rate and, therefore, a freezing-out of these reactions at higher temperatures. This again is reflected in a higher ^{4}He abundance. The maximum number of neutrino flavours which can be determined by observation is less than 4.2 (95% CL) for rather low η (see figure 13.15) [Oli02]. Using the information from the CMB, a range of values of $N_\nu = 1.8$–3.3 can be obtained [Oli02].

13.9 Baryogenesis via leptogenesis

Under the assumption of equal amounts of matter and antimatter at the time of the Big Bang we observe today an enormous preponderance of matter compared with antimatter. If we assume that antimatter is not concentrated in regions which are beyond the reach of current observation, this asymmetry has to originate from the earliest phases of the universe. Here matter and antimatter destroy themselves almost totally except for a small excess of matter, leading to the current baryon asymmetry in the universe (BAU) of

$$Y_B = \frac{n_B - n_{\bar{B}}}{n_\gamma} \approx 10^{-10}. \tag{13.124}$$

In order to accomplish this imbalance, three conditions have to be fulfilled [Sac67b]:

(i)   both a C and a CP violation of one of the fundamental interactions,
(ii) non-conservation of baryon number and
(iii) thermodynamic non-equilibrium.

The production of the baryon asymmetry is usually associated with the GUT transition. The violation of baryon number is not unusual in GUT theories, since in these theories leptons and quarks are situated in the same multiplet, as discussed in chapter 5. That a CP violation is necessary can be seen in the following illustrative set of reactions:

$$X \xrightarrow{r} u + u \qquad X \xrightarrow{1-r} \bar{d} + e^+ \tag{13.125}$$

$$\bar{X} \xrightarrow{\bar{r}} \bar{u} + \bar{u} \qquad \bar{X} \xrightarrow{1-\bar{r}} d + e^-. \tag{13.126}$$

In the case of CP violation $r \neq \bar{r}$. A surplus of u, d, e over $\bar{u}$, $\bar{d}$ and e^+ would follow therefore for $r \geq \bar{r}$. This is, however, only possible in a situation of thermodynamic non-equilibrium, as a higher production rate of baryons will otherwise also lead to a higher production rate of antibaryons. In equilibrium, the particle number is independent from the reaction dynamics. Theoretical estimates show that the CP-violating phase δ in the CKM matrix (see chapter 3) is not sufficient to generate the observed baryon asymmetry and other mechanisms have to be at work.

13.9.0.1 Leptogenesis

The leptonic sector offers a chance for baryogenesis. In the case of massive neutrinos we have the MNS matrix (see chapter 5) in analogy to the CKM matrix. Moreover, for Majorana neutrinos with three flavours two additional CP-violating phases exist. Associated with Majorana neutrinos is lepton number violation. How this can be transformed into a baryon number violation is

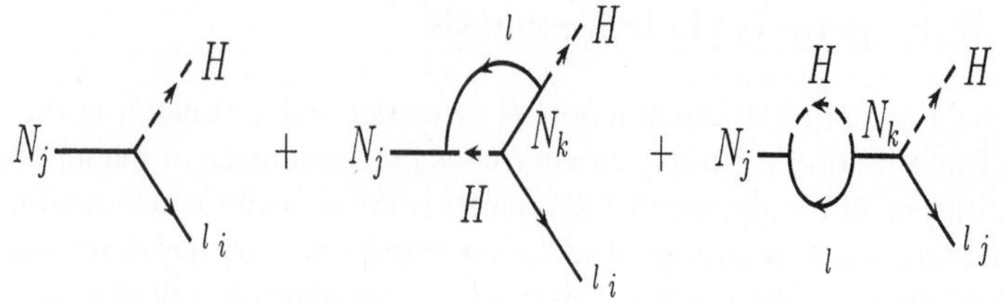

Figure 13.16. Feynman diagrams (tree level and radiative corrections) for heavy Majorana decays.

described later. Moreover, the seesaw mechanism (see chapter 5) requires the existence of a heavy Majorana neutrino, which can be a source for leptogenesis. The idea to use lepton number violation to produce baryon number violation was first discussed in [Fuk86]. The argument is that radiative corrections (figure 13.16) to

$$\mathcal{L} = \mathcal{L}_{\mathcal{EW}} + M_{Rij}\bar{N}_i^c N_j + \frac{(m_D)_{i\alpha}}{v}\bar{N}_i l_\alpha \phi^\dagger + h.c. \tag{13.127}$$

lead to a decay asymmetry of the heavy Majorana neutrino N_i of

$$
\begin{aligned}
\epsilon_i &= \frac{\Gamma(N_i \to \phi l^c) - \Gamma(N_i \to \phi^\dagger l)}{\Gamma(N_i \to \phi l^c) + \Gamma(N_i \to \phi^\dagger l)} \\
&= \frac{1}{8\pi v^2} \frac{1}{(m_D^\dagger m_D)_{ii}} \times \sum_{j \neq i} \mathrm{Im}((m_D^\dagger m_D)_{ij}^2) f(M_j^2/M_i^2).
\end{aligned} \tag{13.128}
$$

To fulfil the observations $\epsilon \approx Y_L \propto Y_B \approx 10^{-10}$. Unfortunately there is no chance to explore the heavy Majorana neutrino sector directly, its only connection to experiment is via the seesaw mechanism to light neutrinos. Numerous models for neutrino masses and heavy Majorana neutrinos have been presented to reproduce the low energy neutrino observations together with Y_B [Fla96, Rou96, Buc96, Pil97, Buc98, Pil99, Buc00]. As it turned out, there is no direct connection between low- and high-energy (meaning the heavy Majorana neutrino scale, normally related to the GUT scale) CP violation, unless there is a symmetry relating the light and heavy sectors [Bra01]. As a general tendency of most models, a strong dependence on the Majorana phases is observed [Rod02], while a possible Dirac CP-phase in the MNS matrix seems to play a minor role. This fact makes the investigation of neutrinoless double β-decay (see chapter 7) very important, because this is the only known process, where these phases can be explored.

The moment in the evolution of the universe at which lepton number violation is converted into baryon number violation is the electroweak phase transition. Its scale is characterized by the vacuum expectation value of the Higgs

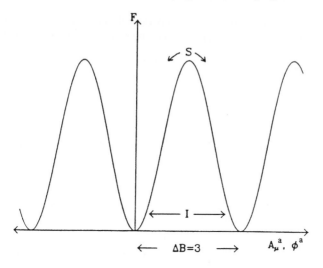

Figure 13.17. Schematic illustration of the potential with different vacua, which appear as different possible vacuum configurations of the fields A_μ, Φ in non-Abelian gauge theories. The possibility of the instanton tunnelling (I) *through* the barrier of height T_C, as well as the sphaleron (S) jumping *over* the barrier are indicated. In the transition $B + L$ changes by $2N_F$, and B by N_F, where N_F represents the number of families (in this case 3) (from [Kol90]).

boson in the electroweak standard model and, therefore, lies around 200 GeV. It has been shown that non-Abelian gauge theories have non-trivial vacuum structures and with a different number of left- and right-handed fermions can produce baryon and lepton number violation [t'Ho76, Kli84]. Figure 13.17 shows such vacuum configurations, which are characterized by different topological winding numbers and are separated by energy barriers of height T_C. In the case of $T = 0$ a transition through such a barrier can only take place by means of quantum mechanical tunnelling (*instantons*) and is, therefore, suppressed by a factor $\exp(2\pi/\alpha_w) \approx 10^{-86}$ (α_w = weak coupling constant). However, this changes at higher temperatures [Kuz85]. Now thermal transitions are possible and for $T \gg T_C$ the transition is characterized by a Boltzmann factor $\exp(-E_{\text{sph}}(T)/T)$. Here E_{sph} represents the sphaleron energy. The *sphaleron* is a saddle point in configuration space which is classically unstable. This means that the transition takes place mainly via this configuration. The sphaleron energy E_{sph} is equivalent to the height of the barrier T_C and, therefore, is also temperature dependent. If one proceeds from one vacuum to the next via the sphaleron configuration the combination of $B + L$ changes by $2 \times N_F$, where N_F is the number of families, of which three are currently known. As, in addition, $B - L$ is free of gauge anomalies, which means that $B - L$ is conserved, the vacuum transition leads to $\Delta B = 3$. Calculations show that in the transition roughly half of the lepton number violation Y_L is converted into baryon number violation Y_B. In this way an elegant solution for baryogenesis could be found resulting from CP-violating phases in the neutrino sector.

Chapter 14

Summary and outlook

Neutrino physics has experienced quite a significant boost in recent years with the establishment of a non-vanishing rest mass of the neutrino. As discussed in more detail in the corresponding chapters, all evidence stems from neutrino oscillation searches. There are three pieces of evidence:

- A deficit in upward-going muons produced by atmospheric neutrinos. This can be explained by ν_μ oscillations with $\Delta m^2 \approx 3 \times 10^{-3}$ eV^2 and $\sin^2 2\theta \approx 1$ very likely into ν_τ.
- The LSND evidence for $\bar{\nu}_\mu$–$\bar{\nu}_e$ oscillations where parts of the allowed parameter sets are in disagreement with KARMEN and NOMAD. The preferred allowed region is about $\Delta m^2 \approx 1$ eV^2 and $\sin^2 2\theta \approx 10^{-3}$.
- The observation of active neutrinos from the Sun besides ν_e. This observation performed by SNO together with the other solar neutrino data result in a best global fit value of $\Delta m^2 \approx 7 \times 10^{-5}$ eV^2 and $\tan^2 \theta = 0.34$, showing that the LMA solution is the most preferred one. This has been confirmed in a completely different way by the KamLAND experiment using nuclear power plants resulting in best-fit values of $\Delta m^2 \approx 7 \times \times 10^{-5}$ eV^2 and $\sin^2 2\theta = 1$.

A graphical representation of all current results is shown in figure 14.1. An incredible number of papers fitting the observed oscillation experiments exists, e.g. see [Bil03, Fer02, Pak03]. Within the context of mass models or underlying theories, a determination of the matrix elements is performed. If all three pieces of evidence turn out to be correct, there has to be at least one additional neutrino because of the unitarity of the MNS matrix (chapter 5). Such a neutrino would not take part in the known weak interaction and is called sterile. How many of them exist or whether we need them at all might be clarified by new experiments like MiniBOONE, currently probing the LSND evidence (chapter 8). But even without sterile neutrinos a precise determination of the mixing matrix elements is neccessary. As part of this programme, a large accelerator-based long-baseline programme has been launched to investigate the atmospheric neutrino anomaly

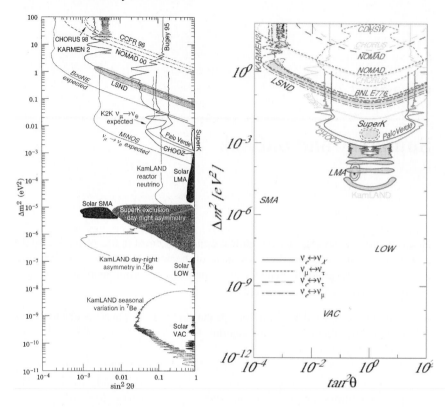

Figure 14.1. Compilation of most neutrino oscillation results in two different presentations: left, Δm^2 *versus* $\sin^2 2\theta$ plot of oscillation searches before KamLAND and SNO; right, a more modern version plotting Δm^2 *versus* $\tan^2 \theta$ more reliable to include matter effects. The latest KamLAND and SNO data are also implemented.

over the next few years (chapter 9). The first indications of ν_μ disappearance observed by K2K, using a beam from KEK to Super-Kamiokande, and future experiments like MINOS (from Fermilab to the Soudan mine), ICARUS and OPERA (using a beam from CERN to Gran Sasso) will sharpen our understanding of the atmospheric neutrinos anomaly. The MNS-matrix can be written in a suggestive way:

$$
\begin{pmatrix}
\cos\theta_{12} & \sin\theta_{12} & 0 \\
-\sin\theta_{12} & \cos\theta_{12} & 0 \\
0 & 0 & 1
\end{pmatrix}
\begin{pmatrix}
\cos\theta_{13} & 0 & \sin\theta_{13}e^{i\delta} \\
0 & 1 & 0 \\
-\sin\theta_{13}e^{i\delta} & 0 & \cos\theta_{13}
\end{pmatrix}
$$
$$
\times
\begin{pmatrix}
1 & 0 & 0 \\
0 & \cos\theta_{23} & \sin\theta_{23} \\
0 & -\sin\theta_{23} & \cos\theta_{23}
\end{pmatrix}.
\tag{14.1}
$$

As can be seen, a possible observation of a CP-violating phase in the leptonic sector relies on the fact that $\sin\theta_{13}$ is definitely non-zero and not too small to account for a reasonable effect in experiments. A whole chain of accelerator-based activities are currently under investigation and culminate in the concept of a neutrino factory (chapter 4). However, the first step to prove is a non-vanishing $\sin\theta_{13}$. A promising way to investigate this is to use off-axis neutrino beams. One reasonable effort is JHF including SK and an upgrade to about 1 Mt called Hyper-K using the concept of superbeams (chapter 4). These are rather low-energy neutrino beams (below about 1 GeV to reduce kaon contaminations) but with a much higher intensity than beams used in the past. The ultimate goal would be a neutrino factory, using muons from pion decay instead of the neutrinos directly. Muon decay has the advantage of being precisely described by theory and the flavour contents of the beam are well known. The finite lifetime of muons and the necessary event rate to explore CP violation necessitates a sophisticated accelerator system. For CP violation and matter effects to show up, long baselines of a few thousand kilometres are required. In a three-flavour scenario the involved Δm^2 has a sign, so it can be determined as well with this type of experiment.

The Majorana phases are unobservable in oscillation experiments. In the case of three Majorana neutrinos this would add two more CP-violating phases. They may play a significant role in leptogenesis (chapter 13), the explanation of the baryon asymmetry in the universe with the help of lepton number violation, which will partly be transformed to a baryon number violation via the electroweak phase transition. Neutrinoless double β-decay is the most preferred process having sensitivity to their existence (chapter 7). This process is only valid if neutrinos are massive Majorana particles. So, in addition to a sensitivity for the CP phases, its main purpose is to probe the fundamental character of the neutrino and its mass. If we take the current oscillation results seriously, a measurement in the region down to 50 meV would have discovery potential and could discriminate among the various neutrino mass models (normal or inverted mass hierarchies). This requires a scaling up of existing experiments by more than an order of magnitude into the region of hundreds of kilograms of the isotope of interest. Various ideas are currently considered to reach this goal. In addition, the KATRIN experiment plans to improve the direct measurement of a neutrino mass in tritium β-decay by an order of magnitude (chapter 6). In the case of the non-observation of a signal both types of experiments would imply a strong bound on the absolute mass scale of neutrinos and severely restrict the parameter space of various mass models (figure 14.2).

A rapidly expanding field is that of neutrino astrophysics. The solution of the solar neutrino problem by SNO (chapter 10) being due to neutrino oscillations and independently confirmed by KamLAND (chapter 8) using nuclear reactors is one of the major milestones in recent history. Various global fits are applied to all solar neutrino observations to disentangle the exact oscillation parameters [Fog03, Bah02, deH02]. However, from typically about 80 fit parameters all but three rely

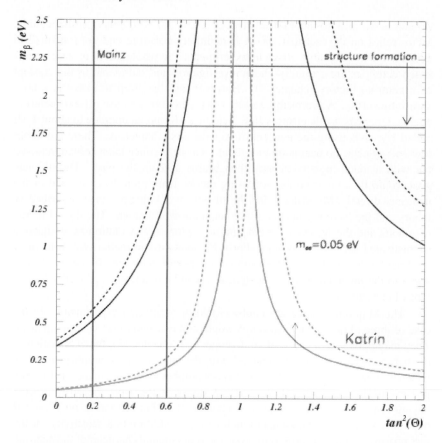

Figure 14.2. Neutrino mass *versus* $\tan^2\theta$. The current Mainz and Troitzk sensitivity as well as the expected KATRIN bound are shown. Also shown are current double beta bounds (black curves) and a possible limit of 50 meV in future experiments (grey curves).

on only 2% of the total solar neutrino flux. Only the gallium (GALLEX/GNO and SAGE) and chlorine rate include the low-energy neutrino sources like pp and ^{7}Be. From this unsatisfactory point and to allow astrophysicists the best view into the interior of the Sun, a real-time low-energy neutrino experiment is highly desirable. With several experiments running in this energy range, we are pretty well equipped to observe nearby supernova explosions (chapter 11). This will open a new window to neutrino astrophysics. An extension to higher neutrino energies arising from various astrophysical sources like AGNs and GRBs is on its way (chapter 12). The expected flux is much smaller which has to be compensated by a larger detector volume. Here, the way to go is to use natural water resources like lakes, oceans or Antarctic ice. Several of these neutrino telescopes (Baikal, NESTOR, ANTARES, AMANDA, ICECUBE) for observing these high-energy neutrinos are in preparation or are already taking data. The very high end of the astrophysical neutrino spectrum might be explored by giant cosmic air shower arrays like the Auger experiment currently under construction in Argentina. The investigations of the horizontal air showers produced by neutrinos and the creation of ultra-high-energy cosmic rays, may be produced by ultra-high-energy neutrinos with the help of the cosmic relic neutrino background are exciting options. This might be one of the very few options to detect the 1.9 K relic neutrino background (chapter 13), whose detection might be the ultimate challenge for experimental neutrino physics. But as the field has proven in the past, the vital excitement of neutrino physics stems from the fact that you always have to expect the unexpected.

References

[Aal02]   Aalseth C E *et al* 2002 *Mod. Phys. A* **17** 1475
[Aal99]   Aalseth C E *et al* 1999 *Phys. Rev. C* **59** 2108
            Aalseth C E *et al* 2002 *Phys. Rev. D* **65** 092007
[Aba01]   Abashian A *et al* 2001 *Phys. Rev. Lett.* **86** 2509
[Abd96]   Abdurashitov J N *et al* 1996 *Phys. Rev. Lett.* **77** 4708
[Abd99]   Abdurashitov J N *et al* (SAGE Collaboration) 1999 *Phys. Rev. C* **59** 2246
[Abe02]   Abele H 2002 *Talk Presented at XXXVII Recontres de Moriond, Electroweak Interactions and Unified Theories (Les Arcs)*
[Abr82]   Abramowicz H *et al* 1982 *Phys. Lett. B* **109** 115
[Ack98]   Ackerstaff K *et al* 1998 *Eur. Phys. J. C* **C5** 229
[Ack98a]   Ackerstaff K *et al* 1998 *Phys. Rev. Lett.* **81** 5519
[Ada00]   Adams T *et al* 2000 *Phys. Rev. D* **61** 092001
[Ade86]   Aderholz M *et al* 1986 *Phys. Lett. B* **173** 211
[Adl00]   Adler S *et al* 2000 *Phys. Rev. Lett.* **84** 3768
[Adl66]   Adler S L 1966 *Phys. Rev.* **143** 1144
[Adl98]   Adloff C *et al* 1998 *Eur. Phys. J. C* **5** 575
[Adl98a]   Adelberger E *et al* 1998 *Rev. Mod. Phys.* **70** 1265
[Adl99]   Adelberger E G *et al* 1999 *Phys. Rev. Lett.* **83** 1299
            Adelberger E G *et al* 1999 *Phys. Rev. Lett.* **83** 3101 (erratum)
[Aga01]   Agasa ICRC01
[Agl87]   Aglietta M *et al* (Frejus Collaboration) 1987 *Europhys. Lett.* **3** 1315
[Agl89]   Aglietta M *et al* 1989 *Europhys. Lett.* **8** 611
[Agr96]   Agraval V *et al* 1996 *Phys. Rev. D* **53** 1314
[Agu01]   Aguilar A *et al* 2001 *Phys. Rev. D* **64** 112007
[Ahm01]   Ahmad Q R *et al* 2001 *Phys. Rev. Lett.* **87** 071301
[Ahm02]   Ahmad Q R *et al* 2002 *Phys. Rev. Lett.* **89** 011301
[Ahm02a]   Ahmad Q R *et al* 2002 *Phys. Rev. Lett.* **89** 011302
[Ahm03]   Ahmed S N *et al* Preprint nucl-ex/0309004
[Ahn03]   Ahn M H *et al* 2003 *Phys. Rev. Lett.* **90** 041801
[Ahr02a]   Ahrens J *et al* 2002 *Preprint* astro-ph/0206487
[Ahr02b]   Ahrens J *et al* 2002 *Phys. Rev. D* **66** 012005
[Ahr02c]   Ahrens J *et al* 2002 *Phys. Rev. D* **66** 032006
[Ahr03]   Ahrens J *et al* 2003 *Ap. J.* **583** 1040
[Ahr87]   Ahrens L A *et al* 1987 *Phys. Rev. D* **35** 785
[Ahr90]   Ahrens L A *et al* 1990 *Phys. Rev. D* **41** 3297

[Ait89]	Aitchison I J R and Hey A J G 1989 *Gauge Theories in Particle Physics* (Bristol: Adam Hilger)
[Ait89]	Aitchison I J R and Hey A J G 1989 *Gauge Theories in Particle Physics* (Bristol: Adam Hilger)
[Akh03]	Akhmedov E Kh and Pulido J 2003 *Phys. Lett.* B **553** 7
[Akh88]	Akhmedov E Kh and Khlopov M Yu 1988 *Mod. Phys. Lett.* A **3** 451
[Akh97]	Akhmedov E Kh 1997 *Preprint* hep-ph/9705451, Proc. 4th Int. Solar Neutrino Conference, Heidelberg
[Alb00]	Albright C *et al* Fermilab-FN-692
[Alb82]	Albrecht A and Steinhardt P J 1982 *Phys. Rev. Lett.* **48** 1220
[Alb92]	Albrecht H *et al* 1992 *Phys. Lett.* B **292** 221
[Alc00]	Alcaraz J *et al* 2000 *Phys. Lett.* B **472** 215
[Ale87]	Alekseev E N, Alexeyeva L N, Krivosheina I V and Volchenko V I 1987 *JETP Lett.* **45** 589
[Ale87a]	Alekseev E N, Alexeyeva L N, Krivosheina I V and Volchenko V I 1987 *Proc. ESO Workshop SN1987A, (Garching near Munich, 6–8 July 1987)* p 237
[Ale88]	Alekseev E N Alexeyeva L N, Krivosheina I V and Volchenko V I 1988 *Phys. Lett.* B **205** 209
[Ale88a]	Alekseev E N, Alexeyeva I N, Krivosheina I V and Volchenko V I 1988 *Neutrino Physics* ed H V Klapdor and B Povh (Heidelberg: Springer) p 288
[Ale94]	Alexander D R and Ferguson J W 1994 *Astrophys. J.* **437** 879
[Ale96]	Alexander G *et al* 1996 *Z. Phys.* C **72** 231
[Ale97]	Fiorini E 1997 *Proc. Neutrino'96* (Singapore: World Scientific) p 352
[Ale98]	Alessandrello A *et al* 1998 *Phys. Lett.* B **420** 109
[Ale99]	Alessandrello A *et al* 1999 *Phys. Lett.* B **457** 253
[All84]	Allasia D *et al* 1984 *Nucl. Phys.* B **239** 301
[All85a]	Allasia *et al* 1985 *Z. f. Phys.* C **28** 321
[All87]	Allaby J V *et al* 1987 *Z. Phys.* C **36** 611
[All88]	Allaby J V *et al* 1988 *Z. Phys.* C **38** 403
[All93]	Allen R C *et al* 1993 *Phys. Rev.* D **47** 11
[All97]	Allen C *et al* 1997 *Preprint* astro-ph/9709223
[Alp48]	Alpher R A and Herman R C 1948 *Nature* **162** 774
[Als02]	Alsharoa M M *et al* 2002 *Preprint* hep-ex/0207031
[Alt02]	Altarelli G and Feruglio F 2002 *Preprint* hep-ph/0206077
[Alt77]	Altarelli G and Parisi G 1977 *Nucl. Phys.* B **126** 298
[Alt98]	Altegoer J *et al* 1998 *Nucl. Instrum. Methods* A **404** 96
[Ama87]	Amaldi U *et al* 1987 *Phys. Rev.* D **36** 1385
[Ama91]	Amaldi U, deBoer W and Färstenau H 1991 *Phys. Lett.* B **260** 447
[Amb01]	Ambrosio M *et al* 2001 *Astrophys. J.* **546** 1038
[Amb02]	Ambrosio M *et al* 2002 *Nucl. Instrum. Methods* A **486** 663
[Amb99]	Ambrosini G *et al* 1999 *Eur. Phys. J.* C **10** 605
[Ams97]	Amsler C *et al* 1997 *Nucl. Instrum. Methods* A **396** 115
[And00]	Andres E *et al* 2000 *Ann. Phys., Paris* **13** 1
[And01]	Andres E *et al* 2001 (AMANDA-coll.) *Nucl. Phys. B (Proc. Suppl.)* **91** 423
[And03]	Ando S and Sato K 2003 *Phys. Rev.* D **67** 023004
[Ang98]	Angelopoulos A *et al* 1998 *Phys. Lett.* B **444** 43

[Ans92]   Anselmann P *et al* (GALLEX Collaboration) 1992 *Phys. Lett.* B **285** 376
[Ans92a]   Anselmann P *et al* (GALLEX Collaboration) 1992 *Phys. Lett.* B **285** 390
[Ans95a]   Anselmann P *et al* (GALLEX Collaboration) 1995 *Phys. Lett.* B **342** 440
[Ans95b]   Anselmann P *et al* (GALLEX Collaboration) 1995 *Nucl. Phys. B (Proc. Suppl.)* **38** 68
            Anselmann P *et al* (GALLEX Collaboration) 1994 *Proc. 16th Int. Conf. on Neutrino Physics and Astrophysics, NEUTRINO '94* ed A Dar, G Eilam and M Gronau (Amsterdam: North-Holland)
[Ant97]   Antares Proposal 1997 *Preprint* astro-ph/9707136
[Ant99]   Antonioli P *et al* 1999 *Nucl. Instrum. Methods* A **433** 104
[Aok03]   Aoki M *et al* 2003 *Phys. Rev.* D **67** 093004
[Apo02]   Apollonio M *et al* 2002 *Preprint* hep-ph/0210192
[Apo98]   Apollonio M *et al* 1998 *Phys. Lett.* B **420** 397
[Apo99]   Apollonio M *et al* 1999 *Phys. Lett.* B **466** 415
[App00]   Appel R *et al* 2000 *Phys. Rev. Lett.* **85** 2877
[Ara97]   Arafune J and Sato J 1997 *Phys. Rev.* D **55** 1653
[Arm02]   Armbruster B *et al* 2002 *Phys. Rev.* D **65** 112001
[Arm98]   Armbruster B *et al* 1998 *Phys. Rev. Lett.* **81** 520
[Arn03]   Arnaboldi C *et al* 2003 *Preprint* hep-ex/0302006
[Arn77]   Arnett W D 1977 *Astrophys. J.* **218** 815
            Arnett W D 1977 *Astrophys. J. Suppl.* **35** 145
[Arn78]   Arnett W D 1978 *The Physics and Astrophysics of Neutron Stars and Black Holes* ed R Giacconi and R Ruffins (Amsterdam: North-Holland) p 356
[Arn80]   Arnett W D 1980 *Ann. NY Acad. Sci.* **336** 366
[Arn83]   Arnison G *et al* 1983 UA1-coll. *Phys. Lett.* B **122** 103
            Arnison G *et al* 1983 *Phys. Lett.* B **126** 398
[Arn89]   Arnett W D *et al* 1989 *Ann. Rev. Astron. Astrophys.* **27** 629
[Arn91]   Arnett W D 1991 *Astrophys. J.* **383** 295
[Arn94]   Arneodo M *et al* 1994 *Phys. Rev.* D **50** R1
[Arn98]   Arnold R *et al* 1998 *Nucl. Phys.* A **636** 209
[Arn99]   Arnold R *et al* 1999 *Nucl. Phys.* A **658** 299
[Arp96]   Arpesella C *et al* (LUNA Collaboration) 1996 *Nucl. Phys. B (Proc. Suppl.)* **48** 375
            Arpesella C *et al* (LUNA Collaboration) 1995 *Proc. 4th Int. Workshop on Theoretical and Phenomenological Aspects of Underground Physics (Toledo, Spain)* ed A Morales, J Morales and J A Villar (Amsterdam: North-Holland)
[Arp98]   Arpesella C *et al* (LUNA Collaboration) 1998 *Phys. Rev.* C **57** 2700
[Arr94]   Arroyo C G *et al* 1994 *Phys. Rev. Lett.* **72** 3452
[Ash62]   Asharyan G A 1962 *Sov. J. JETP* **14** 441
            Asharyan G A 1965 *Sov. J. JETP* **21** 658
[Ass96]   Assamagan K *et al* 1996 *Phys. Rev.* D **53** 6065
[Ast00]   Astier P *et al* 2000 *Nucl. Phys.* B **588** 3
[Ast00a]   Astier P *et al* 2000 *Phys. Lett.* B **486** 35
[Ast01]   Astier P *et al* 2001 *Nucl. Phys.* B **605** 3
[Ast01]   Astier P *et al* 2001 *Nucl. Phys.* B **611** 3
[Ast02]   Astier P *et al* , 2002 *Phys. Lett.* B **526** 278
[Ast03]   Astier P *et al* 2003 *Preprint* hep-ex/0306037

[Ath80]	Atherton H W *et al* 1980 *Preprint* CERN 80-07
[Ath97]	Athanassopoulos C *et al* 1997 *Nucl. Instrum. Methods* A **388** 149
[Ath98]	Athanassopoulos C *et al* 1998 *Phys. Rev.* C **58** 2489
[Aub01]	Aubert B *et al* 2001 *Phys. Rev. Lett.* **86** 2515
[Aue01]	Auerbach L B *et al* 2001 *Phys. Rev.* D **63** 112001
[Auf94]	Aufderheide M *et al* 1994 *Phys. Rev.* C **49** 678
[Aug96]	The Pierre Auger Design Report, Fermilab-PUB-96-024
[Aut99]	Autin B, Blondel A and Ellis J (ed) 1999 *Preprint* CERN 99-02
[Ayr02]	Ayres D *et al* 2002 *Preprint* hep-ex/0210005
[Ays01]	Aysto J *et al* 2001 *Preprint* hep-ph/0109217
[Bab03]	Baby L T *et al* 2003 *Phys. Rev. Lett.* **90** 022501
	Baby L T *et al* 2003 *Phys. Rev.* C **67** 065805
[Bab87]	Babu K S and Mathur V S 1987 *Phys. Lett.* B **196** 218
[Bab88]	Babu K S 1988 *Phys. Lett.* B **203** 132
[Bab95]	BaBar experiment 1995 *Tech. Design Report* SLAC-R-95-437
[Bad01]	Int. Workshop on neutrino masses in the sub-eV range, Bad Liebenzell 2001, http://www-ik1.fzk.de/tritium/liebenzell
[Bag83]	Bagnaia P *et al* 1983 UA2-coll. *Phys. Lett.* B **129** 130
[Bah01]	Bahcall J N, Pinnseaunault M H and Basu S 2001 *Astrophys. J.* **550** 990
[Bah01a]	Bahcall J N and Waxman E 2001 *Phys. Rev.* D **64** 023002
[Bah02]	Bahcall J N, Gonzale-Garcia M C and Penya-Garay C 2002 *Preprint* hep-ph/0212147
[Bah03]	Bahcall J N 2003 *Nucl. Phys.* B *(Proc. Suppl.)* **118** 77
[Bah03a]	Bahcall J N, Gonzalez-Garcia M C and Pena-Garay C 2003 *J. High Energy Phys.* **0302** 009
[Bah88]	Bahcall J N and Ulrich R K 1988 *Rev. Mod. Phys.* **60** 297
[Bah89]	Bahcall J N 1989 *Neutrino Astrophysics* (Cambridge: Cambridge University Press)
[Bah92]	Bahcall J N and Pinsonneault M H 1992 *Rev. Mod. Phys.* **64** 885
[Bah95]	Bahcall J N and Pinsonneault M H 1995 *Rev. Mod. Phys.* **67** 781
[Bah97]	Bahcall J N and Ostriker J P (ed) 1997 *Unsolved Problems in Astrophysics* (Princeton, NJ: Princeton University Press)
[Bal00]	Balkanov V *et al* 2000 *Ann. Phys., Paris* **14** 61
[Bal01]	Balkanov V A *et al* 2001 (Baikal-coll.) *Nucl. Phys.* B *(Proc. Suppl.)* **91** 438
[Bal03]	Balkanov V *et al* 2003 *Nucl. Phys.* B *(Proc. Suppl.)* **118** 363
[Bal94]	Baldo-Ceolin M *et al* 1994 *Z. Phys.* C **63** 409
[Ban83]	Banner M *et al* 1983 UA2-coll. *Phys. Lett.* B **122** 476
[Bar01]	Barish B C 2001 *Nucl. Phys.* B *(Proc. Suppl.)* **91** 141
[Bar02]	Barger V, Marfatia D and Whisnant K 2002 *Phys. Rev.* D **65** 073023
[Bar76]	Barnett R M 1976 *Phys. Rev. Lett.* **36** 1163
[Bar83]	Baruzzi V *et al* 1983 *Nucl. Instrum. Methods* A **207** 339
[Bar84]	Bardin G *et al* 1984 *Phys. Lett.* B **137** 135
[Bar85]	Baron E, Cooperstein J and Kahana S 1985 *Phys. Rev. Lett.* **55** 126
[Bar87]	Barbiellini G and Cocconi G 1987 *Nature* **329** 21
[Bar88]	Barbieri R and Mohapatra R N 1988 *Phys. Rev. Lett.* **61** 27
[Bar89]	Barr G *et al* 1989 *Phys. Rev.* D **39** 3532
[Bar96]	Barabanov I *et al* 1996 *Astropart. Phys.* **5** 159
[Bar96a]	Barabash A S 1995 *Proc. ECT Workshop on Double Beta Decay and*

Related Topics (Trento) ed H V Klapdor-Kleingrothaus and S Stoica (Singapore: World Scientific) p 502

[Bar97b]   Barabash A S 1997 *Proc. Neutrino'96* (Singapore: World Scientific) p 374

[Bar98]   Barate R *et al* 1998 ALEPH-coll. *Eur. Phys. J.* C **2** 395

[Bar99]   Barger V *et al* 1999 *Phys. Lett.* B **462** 109

[Bat03]   Battistoni G *et al* 2003 *Astropart. Phys.* **19** 269

            Battistoni G *et al* 2003 *Astropart. Phys.* **19** 291 (erratum)

[Bau97]   Baudis L *et al* 1997 *Phys. Lett.* B **407** 219

[Baz01]   Bazarko A O 2001 *Nucl. Phys. B (Proc. Suppl.)* **91** 210

[Baz95]   Bazarko A O *et al* 1995 *Z. Phys.* C **65** 189

[Bea02]   Beacom J F, Farr W M and Vogel P 2002 *Phys. Rev.* D **66** 033001

[Bea98]   Beacom J F and Vogel P 1998 *Phys. Rev.* D **58** 093012

[Bea99]   Beacom J F 1999 *Preprint* hep-ph/9909231

[Bec92]   Becker-Szendy R *et al* 1992 *Phys. Rev. Lett.* **69** 1010

[Bed99]   Bednyakov V, Faessler A and Kovalenko S 1999 *Preprint* hep-ph/9904414

[Bel01]   Bellotti E 2001 *Nucl. Phys. B (Proc. Suppl.)* **91** 44

[Bel03]   Bellotti E 2003 *TAUP 2003 Conference, Seattle*

[Bel95b]   Belle experiment 1995 *Tech. Design Report* KEK-Rep 95-1

[Bem02]   Bemporad C, Gratta G and Vogel P 2002 *Rev. Mod. Phys.* **74** 297

[Ben03]   Bennett C L *et al* 2003 *Preprint* astro-ph/0302207

[Ben03]   Benoit A *et al* 2003 *Astron. Astrophys.* **398** L19

[Ben96]   Bennett C L *et al* 1996 *Astrophys. J.* **464** L1

[Ber02]   Bertou X *et al* 2002 *Ann. Phys., Paris* **17** 183

[Ber02]   Berzanelli M, Maino D and Mennella 2002 *Preprint* astro-ph/0209215

            Berzanelli M, Maino D and Mennella 2003 *Riv. Nuovo Cim.* **25** 1

[Ber02a]   Bernard V, Elouadrhiri L and Meissner U G 2002 *J. Phys.* G **28**

[Ber87]   Berge J P *et al* 1987 *Z. Phys.* C **35** 443

[Ber90]   Berger C *et al* 1990 *Phys. Lett.* B **245** 305

[Ber90a]   Berger C *et al* 1990 *Phys. Lett.* B **245** 305

            Berger C *et al* 1989 *Phys. Lett.* B **227** 489

[Ber90b]   Berezinskii V S *et al* 1990 *Astrophysics of Cosmic Rays* (Amsterdam: North-Holland)

[Ber91]   Berezinksy V S 1991 *Nucl. Phys. B (Proc. Suppl.)* **19** 187

[Ber91a]   van den Bergh S and Tammann G A 1991 *Ann. Rev. Astron. Astrophys.* **29** 363

[Ber92]   Bernatowicz T *et al* 1993 *Phys. Rev.* C **47** 806

[Ber93a]   Berthomieu G *et al* 1993 *Astron. Astrophys.* **268** 775

[Ber98]   Bergstrom L, Edsjo J and Gondolo P 1998 *Phys. Rev.* D **58** 103519

[Ber99]   Bergström L and Goobar A 1999 *Cosmology and Particle Astrophysics* (New York: Wiley)

[Bet00]   Bethke S 2000 *J. Phys. G: Nucl. Part. Phys.* **26** R27

[Bet38]   Bethe H A and Critchfield C L 1938 *Phys. Rev.* **54** 248, 862

[Bet39]   Bethe H A 1939 *Phys. Rev.* **55** 434

[Bet79]   Bethe H A, Brown G E, Applegate J and Lattimer J M 1979 *Nucl. Phys.* A **324** 487

[Bet82]   Bethe H A 1982 *Essays in Nuclear Astrophysics* ed C A Barnes, D D Clayton and D N Schramm (Cambridge: Cambridge University Press)

[Bet85]	Bethe H A and Wilson J F 1985 *Astrophys. J.* **295** 14
[Bet85]	Bethe H A and Wilson J F 1985 *Astrophys. J.* **295** 14
[Bet86]	Bethe H A 1986 *Phys. Rev. Lett.* **56** 1305
[Bet86a]	Bethe H A 1986 *Proc. Int. School of Physics 'Enrico Fermi', Course XCI (1984)* ed A Molinari and R A Ricci (Amsterdam: North-Holland) p 181
[Bet88]	Bethe H A 1988 *Ann. Rev. Nucl. Part. Sci.* **38** 1
[Bha00]	Bhattachargjee P and Sigl G 2000 *Phys. Rep.* **327** 109
[Bie64]	Bienlein J K *et al* 1964 *Phys. Lett.* **13** 80
[Bil03]	Bilenky S M *et al* 2003 *Phys. Rep.* **379** 69
[Bil78]	Bilenky S M and Pontecorvo B 1978 *Phys. Rev.* **41** 225
[Bil80]	Bilenky S M, Hosek J and Petcov S T 1980 *Phys. Lett.* B **94** 495
[Bil87]	Bilenky S and Petcov S T 1987 *Rev. Mod. Phys.* **59** 671
[Bil94]	Bilenky S M 1994 *Basics of Introduction to Feynman Diagrams and Electroweak Interactions Physics* (Gif sur Yvette: Editions Frontières)
[Bil99]	Bilenky S M, Giunti C and Grimus W 1999 *Prog. Nucl. Part. Phys.* **43** 1
[Bio87]	Bionta R M *et al* (IMB Collaboration) 1987 *Phys. Rev. Lett.* **58** 1494
[Bir91]	Bird D J *et al* 1991 *Phys. Rev. Lett.* **71** 3401
[Bir97]	Birnbaum C *et al* 1997 *Phys. Lett.* B **397** 143
[Bjo64]	Bjorken J D and Drell S D 1964 *Relativistic Quantum Mechanics* (New York: McGraw-Hill)
[Bjo67]	Bjorken J D 1967 *Phys. Rev.* **163** 1767
	Bjorken J D 1969 *Phys. Rev.* **179** 1547
[Bla97]	Blanc F *et al* 1997 ANTARES Proposal *Preprint* astro-ph/9707136
[Blo90]	Blondel A *et al* 1990 *Z. Phys.* C **45** 361
[Blu92]	Bludman S *et al* 1992 *Phys. Rev.* D **45** 1810
[Blu92a]	Bludman S A 1992 *Phys. Rev.* D **45** 4720
[Blu95]	Blundell S A, Johnson W R, Sapirstein J 1995 *Precision Test of the Standard Model* ed P Langacker (Singapore: World Scientific)
[Bod94]	Bodmann B E *et al* 1994 *Phys. Lett.* B **339** 215
[Boe00]	Boehm F 2000 *Current Aspects of Neutrinophysics* ed D O Caldwell (Springer)
[Boe01]	Boehm F *et al* 2001 *Phys. Rev.* D **64** 112001
[Boe88]	Boerner G 1988 *The Early Universe* (Berlin: Springer)
[Boe92]	Boehm F and Vogel P 1992 *Physics of Massive Neutrinos* (Cambridge: Cambridge University Press)
[Boe99]	Boezio M *et al* 1999 *Astrophys. J.* **518** 457
[Bog00]	Boger J *et al* 2000 SNO Collaboration *Nucl. Instrum. Methods* A **449** 172
[Boh75]	Bohr A N and Mottelson B R 1975 *Nuclear Structure* 1st edn (Reading, MA: Benjamin)
	Bohr A N and Mottelson B R 1998 *Nuclear Structure* 2nd edn (Singapore: World Scientific)
[Bol90]	Bolton T *et al* Fermilab Proposal P-815
[Bol97]	Boliev M *et al* Proc. *DARK96* HVKK
[Bon80]	Bond J R, Efstathiou G and Silk J 1980 *Phys. Rev. Lett.* **45** 1980
[Bon84]	Bond J R and Efstathiou G 1984 *Ap. J.* **285** L45
[Bor91]	BOREXINO Collaboration 1991 *Proposal for a Solar Neutrino Detector at Gran Sasso*
[Bos95]	Bosted P E 1995 *Phys. Rev.* C **51** 409

[Bot98]	Boettcher M and Dermer C D 1998 *Preprint* astro-ph/9801027
[Bot98a]	Bothun G 1998 *Modern Cosmological Observations and Problems* (London: Taylor and Francis)
[Bra01]	Branco G *et al* 2001 *Nucl. Phys.* B **617** 475
[Bra02]	Brash E J *et al* 2002 *Phys. Rev.* C **65** 051001
[Bra88]	Bratton C B *et al* 1988 *Phys. Rev.* D **37** 3361
[Bra88a]	Bratton C B *et al* (IMB Collaboration) 1988 *Phys. Rev.* D **37** 3361
[Bre69]	Breidenbach M *et al* 1969 *Phys. Rev. Lett.* **23** 935
[Bri92]	Britton D I *et al* 1992 *Phys. Rev.* D **46** R885
	Britton D I *et al* 1992*Phys. Rev. Lett.* **68** 3000
[Bro03]	Broggini C 2003 *Preprint* astro-ph/0308537
[Bru85]	Bruenn S W 1985 *Astrophys. J. Suppl.* **58** 771
[Bru87]	Bruenn S W 1987 *Phys. Rev. Lett.* **59** 938
[Buc00]	Buchmüller W and Plümacher M 2000 *Int. J. Mod. Phys.* **A15** 5047
[Buc96]	Buchmüller W and Plümacher M 1996 *Phys. Lett.* B **389** 73
[Buc98]	Buchmüller W and Plümacher M 1998 *Phys. Lett.* B **431** 354
[Bud03]	Budd M, Bodek A and Arrington J 2003 *Preprint* hep-ex/0308005
[Bur00]	Burrows A *et al* 2000 *Astrophys. J.* **539** 865
[Bur01]	Burguet-Castell J *et al* 2001 *Nucl. Phys.* B **608** 301
[Bur02]	Burrows A and Thompson T A 2002 *Preprint* astro-ph/0210212
[Bur57]	Burbidge E M *et al* 1957 *Rev. Mod. Phys.* **29** 547
[Bur57]	Burbidge E M *et al* 1957 *Rev. Mod. Phys.* **29** 547
[Bur86]	Burrows A and Lattimer J M 1986 *Astrophys. J.* **307** 178
[Bur92]	Burrows A, Klein D and Gandhi R 1992 *Phys. Rev.* D **45** 3361
[Bur94]	Burgess C P and Cline J M 1994 *Phys. Rev.* D **49** 5925
[Bur95]	Burrows A and Hayes J 1995 *Ann. NY Acad. Sci.* **759** 375
[Bur95a]	Burrows A, Hayes J and Fryxell B A 1995 *Astrophys. J.* **450** 830
[Bur97]	Burrows A 1997 *Preprint* astro-ph/9703008
[Cal69]	Callan C G and Gross D J 1969 *Phys. Rev. Lett.* **22** 156
[Cal98]	Caldwell R R, Dave R and Steinhardt P J 1998 *Phys. Rev. Lett.* **80** 1582
[Cap98]	Capelle K S *et al* 1998 *Ann. Phys., Paris* **8** 321
[Car00]	Carey R M *et al* Proposal to BNL PAC 2000
[Car93]	Carone C D 1993 *Phys. Lett.* B **308** 85
[Car99]	Carey R M *et al Proposal* PSI R-99-07
[Cas48]	Casimir H G B 1948 *Proc. Kon. Akad. Wet.* **51** 793
[Cau96]	Caurier E *et al* 1996 *Phys. Rev. Lett.* **77** 1954
[Cer97]	Abbaneo D *et al* 1997 *Preprint* CERN-PPE/97-154
[Cha14]	Chadwick J 1914 *Verh. der Deutschen Physikalischen Ges.* **16** 383
[Cha32]	Chadwick J 1932 *Nature* **129** 312
[Cha39, 67]	Chandrasekhar S 1939 and 1967 *An Introduction to the Study of Stellar Structure* (New York: Dover)
[Che02]	Chen S 2002 *Talk Presented at XXXVII Recontres de Moriond, Electroweak Interactions and Unified Theories (Les Arcs)*
[Che92]	Chevalier R A 1992 *Nature* **355** 691
[Chi80]	Chikashige Y, Mohapatra R H and Peccei R D 1980 *Phys. Rev. Lett.* **45** 1926
[Chr64]	Christenson J H *et al* 1964 *Phys. Rev. Lett.* **13** 138
[Chu02]	Church E D *et al* 2002 *Phys. Rev.* D **66** 013001

[Chu89]	Chupp E L, Vestrand W T and Reppin C 1989 *Phys. Rev. Lett.* **62** 505
[Cin98]	Ammar R *et al* 1998 *Phys. Lett.* B **431** 209
[Cla68]	Clayton D D 1968 *Principles of Stellar Evolution* (New York: McGraw-Hill)
[Cle98]	Cleveland B T *et al* 1998 *Ap. Y* **496** 505
[Cli92]	Cline D 1992 *Proc. 4th Int. Workshop on Neutrino Telescopes (Venezia, 10–13 March 1992)* ed M Baldo-Ceolin (Padova University) p 399
[Cli94]	Cline D B *et al* 1994 *Phys. Rev.* D **50** 720
[Clo79]	Close F E 1979 *An Introduction to Quarks and Partons* (New York: Academic)
[Cno78]	Cnops A M *et al* 1978 *Phys. Rev. Lett.* **41** 357
[Col90]	Colgate S A 1990 *Supernovae, Jerusalem Winter School* vol 6, ed J C Wheeler, T Piran and S Weinberg (Singapore: World Scientific) p 249
[Col90a]	Colgate S A 1990 *Supernovae* ed S E Woosley (Berlin: Springer) p 352
[Com83]	Commins E D and Bucksham P H 1983 *Weak Interaction of Leptons and Quarks* (Cambridge: Cambridge University Press)
[Com83]	Commins E D and Bucksham P H 1983 *Weak Interaction of Leptons and Quarks* (Cambridge: Cambridge University Press)
[Con98]	Conrad J M, Shaevitz M H and Bolton T 1998 *Rev. Mod. Phys.* **70** 1341
[Coo84]	Cooperstein J, Bethe H A and Brown G E 1984 *Nucl. Phys.* A **429** 527
[Cos88]	Costa G *et al* 1988 *Nucl. Phys.* B **297** 244
[Cos95]	Cosulich E *et al* 1995 *Nucl. Phys.* A **592** 59
[Cou02]	Couvidat S, Turck-Chieze S and Kosovichev A G 2002 *Preprint* astro-ph/0203107
[Cow72]	Cowsik R and McClelland J 1972 *Phys. Rev. Lett.* **29** 669
[Cox68]	Cox J P and Guili R T 1968 *Stellar Structure and Evolution* (New York: Gordon and Breach)
[Cre03]	Cremonesi O 2003 *Nucl. Phys. B (Proc. Suppl.)* **118** 287
[Dan00]	Danevich F A *et al* 2000 *Phys. Rev.* C **62** 045501
[Dan01]	Danevich F A *et al* 2001 *Nucl. Phys.* A **694** 375
[Dan62]	Danby G *et al* 1962 *Phys. Rev. Lett.* **9** 36
[Dar03]	Daraktchieva Z *et al* 2003 *Phys. Lett.* B **564** 190
[Dau95]	Daum K *et al* 1995 *Z. Phys.* C **66** 417
[Dav55]	Davis R 1955 *Phys. Rev.* **97** 766
[Dav64]	Davis R 1964 *Phys. Rev. Lett.* **12** 303
[Dav83]	Davis M and Peebles P J E 1983 *Astrophys. J.* **267** 465
[Dav94a]	Davis R 1994 *Prog. Part. Nucl. Phys.* **32** 13
[Dav94b]	Davis R 1994 *Proc. 6th Workshop on Neutrino Telescopes (Venezia, Italy)* ed M Baldo-Ceolin (Padova University)
[Dav96]	Davis R 1996 *Nucl. Phys. B (Proc. Suppl.)* **48** 284
[Ded97]	Dedenko L G *et al* 1997 *Preprint* astro-ph/9705189
[Ded98]	Beda A G *et al* 1998 *Phys. Atom. Nucl.* **61** 66
[deG00]	de Gouvea A, Friedland A and Murayama H 2000 *Phys. Lett.* B **490** 125
[deH02]	deHolanda P C and Smirnov Y A 2002 *Preprint* hep-ph/0212270
[Del03]	Della Valle M *et al* 2003 *Astron. Astroph.* **406** L33
[Den90]	Denegri D, Sadoulet B and Spiro M 1990 *Rev. Mod. Phys.* **62** 1
[Der81]	de Rujula A 1981 *Nucl. Phys.* B **188** 414

414   *References*

[Der93]   Derbin A I *et al* 1993 *JETP* **57** 768
[Des97]   DeSilva A *et al* 1997 *Phys. Rev.* C **56** 2451
[Die00]   Dienes K R 2001 *Nucl. Phys. B (Proc. Suppl.)* **91** 321
[Die91]   Diemoz M 1991 *Neutrino Physics* ed K Winter (Cambridge: Cambridge University Press)
[Dig00]   Dighe A S and Smirnov A Y 2000 *Phys. Rev.* D **62** 033007
[Doi83]   Doi M, Kotani T and Takasugi E 1983 *Prog. Theor. Phys.* **69** 602
[Doi85]   Doi M, Kotani T and Takasugi E 1985 *Prog. Theo. Phys. Suppl.* **83** 1
[Doi88]   Doi M, Kotani T and Takasugi E 1988 *Phys. Rev.* D **37** 2575
[Doi92]   Doi M and Kotani T 1992 *Prog. Theor. Phys.* **87** 1207
[Doi93]   Doi M and Kotani T 1993 *Prog. Theor. Phys.* **89** 139
[Dok77]   Dokshitser Y L 1977 *Sov. J. JETP* **46** 641
[Dol02]   Dolgov A D 2002 *Phys. Rep.* **370** 333
[Dol97]   Dolgov A 1997 *Proc. 4th Paris Cosmology Meeting* (Singapore: World Scientific)
[Dom02]   Domogatsky G 2002 *Nucl. Phys. B (Proc. Suppl.)* **110** 504
[Dom78]   Domogatskii G V and Nadyozhin 1978 *Sov. Ast.* **22** 297
[Dom93]   Domokos G, Elliott B and Kovesi-Domokos S J 1993 *J. Phys. G: Nucl. Phys.* **19** 899
[Don92]   Donoghue J F, Golowich F and Holstein B R 1992 *Dynamics of the Standard Model* (Cambridge: Cambridge University Press)
[Dor03]   Doroshkevich A *et al* 2003 *Preprint* astro-ph/0307233
[Dor89]   Dorenbosch J *et al* 1989 *Z. Phys.* C **41** 567
             Dorenbosch J *et al* 1991 *Z. Phys.* C **51** 142(E)
[Dra87]   Dragon N, Ellwanger U and Schmidt M G 1987 *Prog. Nucl. Part. Phys.* **18** 1
[Dre94]   Drexlin G *et al* 1994 *Prog. Nucl. Part. Phys.* **32** 351
[Dyc93]   van Dyck R S *et al* 1993 *Phys. Rev. Lett.* **70** 2888
[Dyd03]   Dydak F 2003 *Nucl. Phys. B (Proc. Suppl.)* **118** 233
[Efs83]   Efstathiou G and Silk J 1983 *Fund. Cosmic Phys.* **9** 1
[Egu03]   Eguchi K *et al* 2003 *Phys. Rev. Lett.* **90** 021802
[Ein17]   Einstein A 1917 *Sitzungsberichte Preuß. Akad. Wiss.* **142**
[Eis86]   Eisele F 1986 *Rep. Prog. Phys.* **49** 233
[Eit01]   Eitel K 2001 *Nucl. Phys. B (Proc. Suppl.)* **91** 191
[Eji00]   Ejiri H 2000 *Phys. Rep.* **338** 265
[Eji00a]   Ejiri H *et al* 2000 *Phys. Rev. Lett.* **85** 2917
[Eji01]   Ejiri H *et al* 2001 *Phys. Rev.* C **63** 065501
[Eji97]   Ejiri H 1997 *Proc. Neutrino'96* (Singapore: World Scientific) p 342
[Elg02]   Elgaroy O *et al* 2002 *Phys. Rev. Lett.* **89** 061301
[Elg03]   Elgaroy O and Lahav O 2003 *Preprint* astro-ph/0303089
[Ell02]   Elliott S R and Vogel P 2002 *Annu. Rev. Nucl. Part. Phys.* **52** 115
[Ell27]   Ellis C D and Wooster W A 1927 *Proc. R. Soc.* A **117** 109
[Ell87]   Elliott S R, Hahn A A and Moe M 1987 *Phys. Rev. Lett.* **59** 1649
[Ell91b]   Ellis J 1991 *Int. School of Astroparticle Physics (Woodlands, TX)* ed D V Nanopoulos and R Edge (Singapore: World Scientific)
[Ell98]   Ellis J 1998 *Preprint* hep-ph/9812235, Lecture presented at 1998 European School for High Energy Physics

[Els98]	Elsener K (ed) 1998 *NGS Conceptual Technical Design Report* CERN 98-02 INFN/AE-98/05
[Eng88]	Engel J, Vogel P and Zirnbauer M R 1988 *Phys. Rev.* C **37** 731
[Eri94]	Ericson M, Ericson T and Vogel P 1994 *Phys. Lett.* B **328** 259
[Esk00]	Eskut E *et al* 2000 *Preprint* CERN-EP-2000-154
[Esk97]	Eskut E *et al* 1997 *Nucl. Instrum. Methods* A **401** 7
[Fae95]	Faessler A 1995 *Proc. ECT Workshop on Double Beta Decay and Related Topics (Trento)* ed H V Klapdor-Kleingrothaus and S Stoica (Singapore: World Scientific) p 339
[Fae99]	Faessler A and Simkovic F 1998 *J. Phys. G: Nucl. Part. Phys.* **24** 2139–78
[Fai78]	Faissner H *et al* 1978 *Phys. Rev. Lett.* **41** 213
[Far97]	Farine J 1997 *Proc. Neutrino'96* (Singapore: World Scientific) p 347
[Fei88]	von Feilitzsch F 1988 *Neutrinos* ed H V Klapdor (Berlin: Springer)
[Fer02]	Feruglio F, Strumia A and Vissani F 2002 *Nucl. Phys.* B **637** 345
[Fer34]	Fermi E 1934 *Z. Phys.* **88** 161
[Fey58]	Feynman R P and Gell-Mann M 1958 *Phys. Rev.* **109** 193
	Feynman R P and Gell-Mann M 1958 *Phys. Rev.* **111** 362
[Fey69]	Feynman R P 1969 *Phys. Rev. Lett.* **23** 1415
[Fie97]	Fields B D, Kainulainen K and Olive K A 1997 *Astropart. Phys.* **6** 169
[Fil83]	Filippone B W *et al* 1983 *Phys. Rev. Lett.* **50** 412
	Filippone B W *et al* 1983 *Phys. Rev.* C **28** 2222
[Fio01]	Fiorini E 2001 Int. Workshop on neutrino masses in the sub-eV range, Bad Liebenzell 2001, http://www-ik1.fzk.de/tritium/liebenzell
[Fio01a]	Fiorini E 2001 *Nucl. Phys. B (Proc. Suppl.)* **91** 262
[Fio67]	Fiorini E *et al* 1967 *Phys. Lett.* B **45** 602
[Fio95]	Fiorentini G, Kavanagh R W and Rolfs C 1995 *Z. Phys.* A **350** 289
[Fio98]	Fiorini E 1998 *Phys. Rev.* **307** 309
[Fix96]	Fixsen D J *et al* 1996 *Ap. J.* **473** 576
[Fla00]	Flanz M, Rodejohann W and Zuber K 2000 *Eur. Phys. J.* C **16** 453
[Fla00a]	Flanz M, Rodejohann W and Zuber K 2000 *Phys. Lett.* B **473** 324
[Fla96]	Flanz M *et al* 1996 *Phys. Lett.* B **389** 693
[Fle01]	Fleming B T *et al* 2001 *Phys. Rev. Lett.* **86** 5430
[Fog03]	Fogli G L *et al* 2003 *Phys. Rev.* D **67** 073002
[Fog99]	Fogli G Z *et al* 1999 hep-ph/9912237
[Fow75]	Fowler W A, Caughlan G R and Zimmerman B A 1975 *Ann. Rev. Astron. Astrophys.* **13** 69
[Fra95]	Franklin A 1995 *Rev. Mod. Phys.* **67** 457
[Fre02]	Freedman W 2002 *Int. J. Mod. Phys.* A **1751** 58
[Fre74]	Freedman D Z 1974 *Phys. Rev.* D **9** 1389
[Fri75]	Fritzsch H and Minkowski P 1975 *Ann. Phys.* **93** 193
[Fuk00]	Fukuda S *et al* 2000 *Phys. Rev. Lett.* **85** 3999
[Fuk02]	Fukuda S *et al* 2002 *Phys. Lett.* B **539** 179
[Fuk03]	Fukuda S *et al* 2003 *Nucl. Instrum. Meth.* A **501** 418
[Fuk03a]	Fukugita M and Yanagida T 2003 *Physics of Neutrinos* (Berlin: Springer)
[Fuk86]	Fukugita M and Yanagida T 1986 *Phys. Lett.* B **174** 45
[Fuk87]	Fukugida M and Yanagida T 1987 *Phys. Rev. Lett.* **58** 1807
[Fuk88]	Fukugita M *et al* 1988 *Phys. Lett.* B **212** 139
[Fuk94]	Fukuda Y *et al* 1994 *Phys. Lett.* B **235** 337

[Ful01]   Fuller G M 2001 *Current Aspects of Neutrino Physics* ed D Caldwell (Berlin: Springer)
[Ful92]   Fuller G M *et al* 1992 *Astrophys. J.* **389** 517
[Ful98]   Fuller G M, Haxton W C and McLaughlin G C 1998 *Phys. Rev.* D **59** 085055
[Ful99]   Fulgione W 1999 *Nucl. Phys. B (Proc. Suppl.)* **70** 469
[Fur39]   Furry W H 1939 *Phys. Rev.* **56** 1184
[Fut99]   Futagami T *et al* 1999 *Phys. Rev. Lett.* **82** 5192
[Gai02]   Gaisser T K and Honda M 2002 *Ann. Rev. Nucl. Part. Sci.* **52** 153
[Gai90]   Gaisser T K 1990 *Cosmic Rays and Particle Physics* (Cambridge: Cambridge University Press)
[Gai95]   Gaisser T K, Halzen F and Stanev T 1995 *Phys. Rep.* **258** 173
[Gai98]   Gaisser T K and Stanev T 1998 *Phys. Rev.* D **57** 1977
[Gal98]   Galeazzi M 1998 *Nucl. Phys. B (Proc. Suppl.)* **66** 203
[Gam36]   Gamow G and Teller E 1936 *Phys. Rev.* **49** 895
[Gam38]   Gamow G 1938 *Phys. Rev.* **53** 595
[Gam46]   Gamow G 1946 *Phys. Rev.* **70** 572
[Gan01]   Gandhi R 2001 *Nucl. Phys. B (Proc. Suppl.)* **91** 453
[Gan03]   Gando Y *et al* 2003 *Phys. Rev. Lett.* **90** 171302
[Gan96]   Gandhi R *et al* 1996 *Ann. Phys., Paris* **5** 81
[Gan98]   Gandhi R 1998 *Phys. Rev.* D **58** 093009
[Gar57]   Garwin R L, Ledermann L M, Weinrich M 1957 *Phys. Rev.* **105** 1415
[Gat01]   Gatti F 2001 *Nucl. Phys. B (Proc. Suppl.)* **91** 293
[Gat97]   Gatti F *et al* 1997 *Phys. Lett.* B **398** 415
[Gat99]   Gatti F *et al* 1999 Nature **397** 137
[Gav03]   Gavrin V N 2003 *Nucl. Phys. B (Proc. Suppl.)* **118**
[Gav03a]   Gavrin V 2003 *TAUP 2003 Conference, Seattle*
[Gee98]   Geer S 1998 *Phys. Rev.* D **57** 6989
            Geer S 1999 *Phys. Rev.* D **59** 039903 (erratum)
[Gei90]   Geiregat G *et al* 1990 *Phys. Lett.* B **245** 271
[Gei93]   Geiregat D *et al* 1993 *Nucl. Instrum. Methods* A **325** 91
[Gel64]   Gell-Mann M 1964 *Phys. Lett.* **8** 214
[Gel78]   Gell-Mann M, Ramond P and Slansky R 1978 *Supergravity* ed F van Nieuwenhuizen and D Freedman (Amsterdam: North-Holland) p 315
[Gel81]   Gelmini G B and Roncadelli M 1981 *Phys. Lett.* B **99** 411
[Geo74]   Georgi H and Glashow S L 1974 *Phys. Rev. Lett.* **32** 438
[Geo75]   Georgi H 1975 *Particles and Fields* ed C E Carloso (New York: AIP)
[Geo76]   Georgi H and Politzer H D 1976 *Phys. Rev.* D **14** 1829
[Geo95]   Georgadze A S 1995 *Phys. Atom. Nucl.* **58** 1093
[Ger56]   Gershtein S S and Zeldovich Ya B 1956 *Sov. J. JETP* **2** 596
[Gio84]   Giovanetti K L *et al* 1984 *Phys. Rev.* D **29** 343
[Giu98]   Giunti C, Kim C W and Monteno M 1998 *Nucl. Phys.* B **521** 3
[Gla60]   Glashow S L 1960 *Phys. Rev.* **118** 316
[Gla61]   Glashow S L 1961 *Nucl. Phys.* **22** 579
[Gle00]   Glenzinski D A and Heintz U 2000 *Annu. Rev. Nucl. Part. Phys.* **50** 207
[Glu95]   Glück M, Reya E and Vogt A 1995 *Z. Phys.* C **67** 433
[Gni00]   Gninenko S N and Krasnikov N V 2000 *Phys. Lett.* B **490** 9

[Goe35]   Goeppert-Mayer M 1935 *Phys. Rev.* **48** 512

[Gol58]   Goldhaber M, Grodzin L and Sunyar A W 1958 *Phys. Rev.* **109** 1015

[Gol80]   Goldstein H 1980 *Classical Mechanics* (New York: Addison-Wesley)

[Gon00]   Gondolo P *et al* 2000 *Preprint* astro-ph/0012234

[Gon00a]   Gondolo P and Silk J 2000 *Nucl. Phys. B (Proc. Suppl.)* **87** 87

[Gon99]   Gonzalez-Garcia M C *et al* 1999 *Phys. Rev. Lett.* **82** 3202

[Got67]   Gottfried K 1967 *Phys. Rev. Lett.* **18** 1174

[Gou92]   Gould A 1992 *Astrophys. J.* **388** 338

[Gra02]   Grainge K *et al* 2002 *Preprint* astro-ph/0212495

[Gre66]   Greisen K 1966 *Phys. Rev.* **118** 316

[Gre86]   Greiner W and Müller B 1986 *Theoretische Physik Bd 8 Eichtheorie der Schwachen Wechselwirkung* (Frankfurt: Deutsch)

[Gre86a]   Greiner W and Müller B 1986 *Theoretische Physik Bd 8 Eichtheorie der Schwachen Wechselwirkung* (Frankfurt: Harri Deutsch)

[Gre93]   Grevesse N 1993 *Origin and Evolution of Elements* ed N Prantzos, E Vangioni Flam and M Casse (Cambridge: Cambridge University Press) pp 15–25

[Gre93a]   Grevesse N and Noels A 1993 *Phys. Scr.* T **47** 133

[Gre94]   Greife U *et al* 1994 *Nucl. Instrum. Methods* A **350** 326

[Gri01]   Grieder P K F 2001 *Cosmic Rays on Earth* (Amsterdam: Elsevier Science)

[Gri72]   Gribov V N and Lipatov L N 1972 *Sov. J. Nucl. Phys.* **15** 438, 675

[Gro69]   Gross D J and Llewellyn-Smith C H 1969 *Nucl. Phys.* B **14** 337

[Gro73]   Gross D J and Wilczek F 1973 *Phys. Rev. Lett.* **30** 1343

[Gro79]   deGroot J G H *et al* 1979 *Z. Phys.* C **1** 143

[Gro90]   Grotz K and Klapdor H V 1990 *The Weak Interaction in Nuclear, Particle and Astrophysics* (Bristol: Hilger)

[Gro97]   Grossman Y and Lipkin H J 1997 *Phys. Rev.* D **55** 2760

[Gue97]   Guenther M *et al* 1997 *Phys. Rev.* D **55** 54

[Gul00]   Guler M *et al* 2000 *OPERA Proposal* LNGS P25/2000 CERN SPSC 2000-028

[Gun90]   Gunion G F, Haber H E, Kane G and Dawson S 1990 *The Higgs–Hunter Guide, Frontiers in Physics* vol 80 (London: Addison-Wesley)

[Gut81]   Guth A H 1981 *Phys. Rev.* D **23** 347

[Gut89]   Guth A H 1989 *Bubbles, Voids and Bumps in Time: The New Cosmology* ed J Cornell (Cambridge: Cambridge University Press)

[Gyu95]   Gyuk G and Turner M S 1995 *Nucl. Phys. B (Proc. Suppl.)* **38** 13

[Hab01]   Habig A 2001 *Preprint* hep-ex/0106026, ICRC01

[Hab85]   Haber H E and Kane G L 1985 *Phys. Rep.* **117** 75

[Hai88]   Haidt D and Pietschmann H 1988 *Landolt-Boernstein I/10* (Berlin: Springer)

[Hal01]   Halzen F 2001 *AIP Conf. Proc.* **558** 43

[Hal02]   Halzen F and Hooper D 2002 *Rep. Prog. Phys.* **65** 1025

[Hal02a]   Halverson N W *et al* 2002 *Ap. J.* **568** 38

[Hal83]   Halprin A, Petcov S T and Rosen S P 1983 *Phys. Lett.* B **125** 335

[Hal84]   Halzen F and Martin A D 1984 *Quarks and Leptons* (London: Wiley)

[Hal96]   Halprin A, Leung C N and Pantaleone J 1996 *Phys. Rev.* D **53** 5356

[Hal97]   Halzen F and Zas E 1997 *Astrophys. J.* **488** 669

[Hal98]   Halzen F 1998 *Preprint* astro-ph/9810368, Lecture presented at TASI
               School, 1998
[Hal98a]   Halzen F and Saltzberg D 1998 *Phys. Rev. Lett.* **81** 4305
[Ham03]   Hamuy M 2003 *Preprint* astro-ph/0301006
[Ham93]   Hampel W 1993 *J. Phys. G: Nucl. Part. Phys.* **19** 209
[Ham96a]   Hampel W *et al* (GALLEX Collaboration) 1996 *Phys. Lett.* B **388** 384
[Ham98]   Hammache F *et al* 1998 *Phys. Rev. Lett.* **80** 928
[Ham99]   Hampel W *et al* (GALLEX Collaboration) 1999 *Phys. Lett.* B **447** 127
[Han03]   Hannestad S 2002 *Preprint* astro-ph/0208567
               Hannestad S 2003 *Nucl. Phys. B (Proc. Suppl.)* **118** 315
[Han03]   Hannestadt S 2003 *Phys. Rev.* D **67** 085017
[Han03a]   Hannestadt S 2003 *Preprint* astro-ph/0303076
[Han98]   Hannestadt S and Raffelt G G 1998 *Astrophys. J.* **507** 339
[Har02]   Harney H L 2002 *Mod. Phys.* A **16** 2409
[Har99]   Catanesi M G *et al* 1999 *Preprint* CERN-SPSC-99-35
[Has73]   Hasert F J *et al* 1973 *Phys. Lett.* B **46** 121
[Has73a]   Hasert F J *et al* 1973 *Phys. Lett.* B **46** 138
[Has74]   Hasert F J *et al* 1974 *Nucl. Phys.* B **73** 1
[Hax84]   Haxton W C and Stephenson G J 1984 *Prog. Nucl. Part. Phys.* **12** 409
[Hay99]   Hayato Y *et al* 1999 *Phys. Rev. Lett.* **83** 1529
[Heg03]   Heger A *et al* 2003 *Preprint* astro-ph/0307546
[Hei80]   Heisterberg R H *et al* 1980 *Phys. Rev. Lett.* **44** 635
[Her94]   Herant M *et al* 1994 *Astrophys. J.* **435** 339
[Her94]   1994 *Proposal* HERA-B, DESY/PRC 94-02
[Her95]   Herczeg P 1995 *Precision Tests of the Standard Model* ed P Langacker
               (Singapore: World Scientific)
[Heu95]   Heusser G 1995 *Annu. Rev. Nucl. Part. Phys.* **45** 543
[Hig64]   Higgs P W 1964 *Phys. Lett.* **12** 252
[Hil72]   Hillas A M *Cosmic Rays* (Oxford: Pergamon)
[Hil88]   Hillebrandt W 1988 *NEUTRINOS* ed H V Klapdor (Berlin: Springer) p 285
[Hir02]   Hirsch M *et al* 2002 *Preprint* hep-ph/0202149
[Hir87]   Hirata K S *et al* (KAMIOKANDE Collaboration) 1987 *Phys. Rev. Lett.* **58**
               1490
[Hir88]   Hirata K S *et al* (KAMIOKANDE Collaboration) 1988 *Phys. Rev.* D **38**
               448
[Hir94]   Hirsch M *et al* 1994 *Z. Phys.* A **347** 151
[Hir95]   Hirsch M, Klapdor-Kleingrothaus H V and Kovalenko S 1995 *Phys. Rev.
               Lett.* **75** 17
               Hirsch M, Klapdor-Kleingrothaus H V and Kovalenko S 1995 *Phys. Lett.*
               B **352** 1
[Hir96]   Hirsch M *et al* 1996 *Phys. Lett.* B **374** 7
[Hir96a]   Hirsch M, Klapdor-Kleingrothaus H V and Kovalenko S 1995 *Phys. Lett.*
               B **352** 1
               Hirsch M, Klapdor-Kleingrothaus H V and Kovalenko S 1996 *Phys. Rev.*
               D **53** 1329
[Hir96b]   Hirsch M, Klapdor-Kleingrothaus H V and Kovalenko S 1996 *Phys. Lett.*
               B **378** 17

	Hirsch M, Klapdor-Kleingrothaus H V and Kovalenko S 1996 *Phys. Rev.* D **54** 4207
[Hir96c]	Hirsch M *et al* 1996 *Phys. Lett.* B **372** 8
[Hir96d]	Hirsch M, Klapdor-Kleingrothaus H V and Kovalenko S G 1996 *Phys. Rev.* D **53** 1329
[Hjo03]	Hjorth J *et al* 2003 *Nature* **423** 847
[Hol78]	Holder M *et al* 1978 *Nucl. Instrum. Methods* **148** 235
	Holder M *et al* 1978 *Nucl. Instrum. Methods* **151** 69
[Hol92]	Holzschuh E 1992 *Rep. Prog. Phys.* **55** 851
[Hol99]	Holzschuh E *et al* 1999 *Phys. Lett.* B **451** 247
[Hon01]	Honda M *et al* 2001 *27th Int. Cosmic Ray Conf. (ICRC01) (Hamburg)* **3** 1162
[Hon95]	Honda M *et al* 1995 *Phys. Rev.* D **52** 4985
[Hor82]	Horstkotte J *et al* 1982 *Phys. Rev.* D **25** 2743
[Hu02]	Hu W and Dodelson S 2002 *Ann. Rev. Astron. Astroph.* **40** 171
[Hu03]	Hu W 2003 *Annals. Phys.* **303** 203
[Hu95]	Hu W, Sugiyama N and Silk J 1995 *Preprint* astro-ph/9504057
[Hu98]	Hu W, Eisenstein D J and Tegmark M 1998 *Phys. Rev. Lett.* **80** 5255
[Hub02]	Huber P, Lindner N and Winter W 2002 *Nucl. Phys.* B **645** 3
[Hub29]	Hubble E 1929 *Proc. Nat. Acad.* **15** 168
[Igl96]	Iglesias C A and Rogers F J 1996 *Astrophys. J.* **464** 943
[Ing96]	Ingelman G and Thunman T 1996 *Phys. Rev.* D **54** 4385
[Ing96a]	Ingelman G and Thunman T 1996 *Preprint* hep-ph/9604286
[Ito01]	Itow Y *et al* 2001 *Preprint* hep-ex/0106019
[Jan89a]	Janka H T and Hillebrandt W 1989 *Astron. Astrophys. Suppl. Series* **78** 375
[Jan89b]	Janka H T and Hillebrandt W 1989 *Astron. Astrophys.* **224** 49
[Jan95]	Janka H T and Müller E 1995 *Astrophys. J. Lett.* **448** L109
[Jan95a]	Janka H T and Müller E 1995 *Ann. NY Acad. Sci.* **759** 269
[Jan96]	Janka H T and Müller E 1996 *Astron. Astrophys.* **306** 167
[Jec95]	Jeckelmann B *et al* 1995 *Phys. Lett.* B **335** 326
[Jon82]	Jonker M *et al* 1982 *Nucl. Instrum. Methods* A **200** 183
[Jun01]	Jung C K *et al* 2001 *Annu. Rev. Nucl. Part. Phys.* **51** 451
[Jun02]	Junghans A R *et al* 2002 *Phys. Rev. Lett.* **88** 041101
[Jun03]	Junghans A R *et al* 2003 *Preprint* nucl-ex/0308003
[Jun96]	Jungmann G, Kamionkowski M and Griest K 1996 *Phys. Rep.* **267** 195
[Kac01]	Kachelriess M, Tohas R and Valle J W F 2001 *J. High Energy Phys.* **0101** 030
[Kaf94]	Kafka T *et al* 1994 *Nucl. Phys.* B *(Proc. Suppl.)* **35** 427
[Kah86]	Kahana S 1986 *Proc. Int. Symp. on Weak and Electromagnetic Interaction in Nuclei (WEIN'86)* ed H V Klapdor (Heidelberg: Springer) p 939
[Kaj01]	Kajita T and Totsuka Y 2001 *Rev. Mod. Phys.* **73** 85
[Kam01]	Kampert K H 2001 *Preprint* astro-ph/0102266
[Kan87]	Kane G 1987 *Modern Elementary Particle Physics* (New York: Addison-Wesley)
[Kap00]	Kaplinghat M, Steigman G and Walker T P 2000 *Phys. Rev.* D **62** 043001
[Kar02]	Karle A 2002 *Nucl. Phys.* B *(Proc. Suppl.)* **118**
[Kau98]	Kaulard J *et al* 1998 *Phys. Lett.* B **422** 334
[Kav69]	Kavanagh R W *et al* 1969 *Bull. Am. Phys. Soc.* **14** 1209

[Kaw93]   Kawashima A, Takahashi K and Masuda A 1993 *Phys. Rev.* C **47** 2452
[Kay02]   Kayis-Topasku A *et al* 2002 *Phys. Lett.* B **527** 173
[Kay81]   Kayser B 1981 *Phys. Rev.* D **24** 110
[Kay89]   Kayser B, Gibrat-Debu F and Perrier F 1989 *Physics of Massive Neutrinos* (Singapore: World Scientific)
[Kaz00]   Kazakov D I 2000 *Preprint* hep-ph/0012288, Lecture presented at 2000 European School for High Energy Physics
[Kea02]   Kearns E T 2002 *Preprint* hep-ex/0210019
[Kei03]   Keil M T, Raffelt G G and Janka H T 2003 *Ap. J.* **590** 971
[Kho99]   Khohklov *et al* 1999 *Astrophys. J.* **524** L107
[Kib67]   Kibble T W B 1967 *Phys. Rev.* **155** 1554
[Kie03]   Kiel H, Münstermann D and Zuber K 2003 *Nucl. Phys.* A **723** 499
[Kim93]   Kim C W and Pevsner A 1993 *Massive Neutrinos in Physics and Astrophysics* (Harwood Academic)
[Kin03]   King B 2003 *Nucl. Phys. B (Proc. Suppl.)* **118** 267
[Kip90]   Kippenhahn R and Weigert A 1990 *Stellar Structure and Evolution* (Berlin: Springer)
[Kir03]   Kirsten T A 2003 *Nucl. Phys. B (Proc. Suppl.)* **118**
[Kir67]   Kirsten T, Gentner W and Schaeffer O A 1967 *Z. Phys.* **202** 273
[Kir68]   Kirsten T, Gentner W and Schaeffer O A 1968 *Phys. Rev. Lett.* **20** 1300
[Kir86]   Kirsten T 1986 *Proc. Int. Symp. on Nuclear Beta Decays and Neutrinos (Osaka)* ed T Kotani, H Ejiri and E Takasugi (Singapore: World Scientific) p 81
[Kir99]   Kirsten T A 1999 *Rev. Mod. Phys.* **71** 1213
[Kir99a]   Kirkby J *et al Proposal* PSI R-99-06
[Kit83]   Kitagaki T *et al* 1983 *Phys. Rev.* D **28** 436
[Kla01]   Klapdor-Kleingrothaus H V *et al* 2001 *J. Phys. J.* A **12** 147
[Kla01a]   Klapdor-Kleingrothaus H V (ed) 2001 *Sixty Years of Double Beta Decay* (Singapore: World Scientific)
[Kla02]   Klapdor-Kleingrothaus H V *et al* 2002 *Mod. Phys. Lett.* A **16** 2409
[Kla02a]   Klapdor-Kleingrothaus H V 2002 *Preprint* hep-ph/0205228
[Kla95]   Klapdor-Kleingrothaus H V and Staudt A 1995 *Non Accelerator Particle Physics* (Bristol: IOP)
[Kla95a]   Klapdor-Kleingrothaus H V (ed) 1995 *Proc. ECT Workshop on Double Beta Decay and Related Topics (Trento)* ed H V Klapdor-Kleingrothaus and S Stoica (Singapore: World Scientific)
[Kla97]   Klapdor-Kleingrothaus H V and Zuber K 1997 *Particle Astrophysics* (Bristol: IOP) revised paperback version 1999
[Kla98]   Klapdor-Kleingrothaus H V, Hellmig J and Hirsch M 1998 *J. Phys. G: Nucl. Phys.* **24** 483
[Kla99]   Klapdor-Kleingrothaus H V (ed) 1999 *Baryon and Lepton asymmetry* (Bristol: IOP)
[Kli84]   Klinkhammer D and Manton N 1984 *Phys. Rev.* D **30** 2212
[Kni02]   Kniehl B A and Zwirner L 2002 *Nucl. Phys.* B **637** 311
[Kob73]   Kobayashi M and Maskawa T 1973 *Prog. Theor. Phys.* **49** 652
[Kod01]   Kodama K *et al* 2001 *Phys. Lett.* B **504** 218
[Kog96]   Kogut A *et al* 1996 *Astrophys. J.* **464** L5
[Kol87]   Kolb E W, Stebbins A J and Turner M S 1987 *Phys. Rev.* D **35** 3598

[Kol90]	Kolb E W and Turner M S 1990 *The Early Universe (Frontiers in Physics 69)* (Reading, MA: Addison-Wesley)
[Kol93]	Kolb E W and Turner M S 1993 *The Early Universe* (Reading, MA: Addison-Wesley)
[Kon66]	Konopinski E J 1966 *Beta Decay* (Oxford: Clarendon)
[Kos92]	Koshiba M 1992 *Phys. Rep.* **220** 229
[Kra02]	Kravchenko I *et al* 2002 *Preprint* astro-ph/0206371
[Kra90]	Krakauer D A *et al* 1990 *Phys. Lett.* B **252** 177
[Kra91]	Krakauer D A *et al* 1991 *Phys. Rev.* D **44** R6
[Kun01]	Kuno Y and Okada Y 2001 *Rev. Mod. Phys.* **73** 151
[Kuo02]	Kuo C *et al* 2002 *Preprint* astro-ph/0212289
[Kuo89]	Kuo T K and Pantaleone J 1989 *Rev. Mod. Phys.* **61** 937
[Kuz66]	Kuzmin V A 1966 *Sov. Phys.–JETP* **22** 1050
[Kuz85]	Kuzmin V A, Rubakov V A and Shaposhnikov M E 1985 *Phys. Lett.* B **155** 36
[Lam00]	Lampe B and Reya E 2000 *Phys. Rep.* **332** 1
[Lam97]	Lamoreaux S K 1997 *Phys. Rev. Lett.* **78** 5
[Lan52]	Langer L M and Moffat R D 1952 *Phys. Rev.* **88** 689
[Lan56]	Landau L D 1956 *JETP* **32** 405
	Landau L D 1956 *JETP* **32** 407
[Lan69]	Landolt-Börnstein Bd.I/IV 1969
[Lan75]	Landau L D and Lifschitz E M 1975 *Lehrbuch der Theoretischen Physik* vol 4a (Berlin: Academischer)
[Lan81]	Langacker P 1981 *Phys. Rep.* **72** 185
[Lan86]	Langacker P 1986 *Proc. Int. Symp. Weak and Electromagnetic Interactions in Nuclei, WEIN '86* ed H V Klapdor (Heidelberg: Springer) p 879
[Lan88]	Langacker P 1988 *Neutrinos* ed H V Klapdor (Berlin: Springer)
[Lan91]	Langer N 1991 *Astron. Astrophys.* **243** 155
[Lan93b]	Langacker P and Polonsky N 1993 *Phys. Rev.* D **47** 4028
[Lan94]	Langanke K 1994 *Proc. Solar Modeling Workshop (Seattle, WA)*
[Lat88]	Lattimer J and Cooperstein J 1988 *Phys. Rev. Lett.* **61** 24
[Lea00]	Learned J G and Mannheim K 2000 *Annu. Rev. Nucl. Part. Phys.* **50** 679
[Lea01]	Learned J 2001 *Current Aspects of Neutrino Physics* ed D Caldwell (Berlin: Springer)
[Lea03]	Learned J G 2003 *Nucl. Phys. B (Proc. Suppl.)* **118** 405
[Lea95]	Learned J G and Pakvasa S 1995 *Ann. Phys., Paris* **3** 267
[Lea96]	Leader E and Predazzi E 1996 *An Introduction to Gauge Theories and Modern Particle Physics* (Cambridge: Cambridge University Press)
[Lee01]	Lee A T *et al* 2001 *Astrophys. J.* **561** L1
[Lee56]	Lee T D and Yang C N 1956 *Phys. Rev.* **104** 254
[Lee57]	Lee T D and Yang C N 1957 *Phys. Rev.* **105** 1671
[Lee72]	Lee B W and Zinn-Justin J 1972 *Phys. Rev.* D **5** 3121
[Lee77]	Lee B W and Shrock R E 1977 *Phys. Rev.* D **16** 1444
[Lee77a]	Lee B W and Weinberg S 1977 *Phys. Rev. Lett.* **39** 165
[Lee95]	Lee D G *et al* 1995 *Phys. Rev.* D **51** 229
[Leh02]	Lehtinen N G 2002 *Ann. Phys., Paris* **17** 279
[Lei01]	Leibundgut B 2001 *Astronom. Astrophus. Review* **39** 67
[Lei03]	Leibundgut B and Suntzeff N B 2002 *Preprint* astro-ph/0304112, to be

published in *Supernovae and Gamma Ray Bursters (Lecture Notes in Physics)* Springer

[Len98]   Lenz S *et al* 1998 *Phys. Lett.* B **416** 50

[Let03]   Letessier-Selvon A 2003 *Nucl. Phys. B (Proc. Suppl.)* **118**

[Lew94]   Lewis J R *et al* 1994 *Mon. Not. R. Astron. Soc.* **266** L27

[Lim88]   Lim C S and Marciano W J 1988 *Phys. Rev.* D **37** 1368

[Lin02]   Linde A D 2002 *Preprint* hep-th/0211048

[Lin82]   Linde A D 1982 *Phys. Lett.* B **108** 389

[Lin84]   Linde A D 1984 *Rep. Prog. Phys.* **47** 925

[Lin88]   Lin W J *et al* 1988 *Nucl. Phys.* A **481** 477 and 484

[Lin89]   Lindner M *et al* 1989 *Geochim. Cosmochim. Acta* **53** 1597

[Lip00a]   Lipari P 2000 *Ann. Phys., Paris* **14** 153

[Lip00b]   Lipari P 2000 *Nucl. Phys. B (Proc. Suppl.)* **14** 171

[Lip01]   Lipari P 2001 *Nucl. Phys. B (Proc. Suppl.)* **91** 159

[Lip91]   Lipari P and Stanev T 1991 *Phys. Rev.* D **44** 3543

[Lip93]   Lipari P 1993 *Ann. Phys., Paris* **1** 193

[Lip99]   Lipkin H J 1999 *Preprint* hep-ph/9901399

[Lis97]   Lisi E, Montanino D 1997 *Phys. Rev.* D **56** 1792

[Lle72]   Llewellyn Smith C H 1972 *Phys. Rep.* **3** 261

[Lob85]   Lobashev V M *et al* 1985 *Nucl. Instrum. Methods* A **240** 305

[Lob99]   Lobashev V M *et al* 1999 *Phys. Lett.* B **460** 227

[Lon92, 94] Longair M S 1992, 1994 *High Energy Astrophysics* (Cambridge: Cambridge University Press)

[Lop96]   Lopez J L 1996 1996 *Rep. Prog. Phys.* **59** 819

[Lor02]   Loredo T J and Lamb D Q 2002 *Phys. Rev.* D **65** 063002

[Los97]   Celebrating the neutrino 1997 *Los Alamos Science* vol 25 (Los Alamos: Los Alamos Science)

[Lud01]   Ludovici L 2001 *Nucl. Phys. B (Proc. Suppl.)* **91** 177

[Lue98]   Luescher R *et al* 1998 *Phys. Lett.* B **434** 407

[Lun01]   Lunardini C and Smirnov A Y 2001 *Nucl. Phys.* B **616** 307

[Lun03]   Lunardini C and Smirnov A Y 2003 *Preprint* hep-ph/0302033, *J. CAP* **0306** 009

[Mac84]   McFarland D *et al* 1984 *Z. Phys.* C **26** 1

[Maj37]   Majorana E 1937 *Nuovo Cimento* **14** 171

[Mak62]   Maki Z, Nakagawa M and Sakata S 1962 *Prog. Theor. Phys.* **28** 870

[Mal03]   Malek M *et al* 2003 *Phys. Rev. Lett.* **90** 061101

[Mal93]   Malaney R A and Mathews G J 1993 *Phys. Rep.* **229** 145, 147

[Mal97]   Malaney R A 1997 *Astropart. Phys.* **7** 125

[Man01]   Mann W A 2001 *Nucl. Phys. B (Proc. Suppl.)* **91** 134

[Man01a]   Mannheim K, Protheroe R J and Rachen J P 2001 *Phys. Rev.* D **63** 023003

[Man95]   Mann A K 1995 *Precision Test of the Standard Model* ed P Langacker (Singapore: World Scientific)

[Mar69]   Marshak R E *et al* 1969 *Theory of Weak Interactions in Particle Physics* (New York: Wiley–Interscience)

[Mar77]   Marciano W J and Sanda A I 1977 *Phys. Lett.* B **67** 303

[Mar92]   Marshak R E 1992 *Conceptional Foundations of Modern Particle Physics* (Cambridge: Cambridge University Press)

[Mar97]   Martel H, Shapiro P R and Weinberg S 1997 *Preprint* astro-ph/9701099

[Mar99]   Marciano W J 1999 *Phys. Rev.* D **60** 093006
[Mas95]   Masterson B P and Wiemann C E 1995 *Precision Test of the Standard Model* ed P Langacker (Singapore: World Scientific)
[Mat01]   Matsuno S 2001 *27th Int. Cosmic Ray Conference (ICRC01)* (Hamburg)
[Mat88]   Matz S M *et al* 1988 *Nature* **331** 416
[Mau00]   Mauskopf P D *et al* 2000 *Astrophys. J.* **536** L59
[May87]   Mayle R, Wilson J R and Schramm D N 1987 *Astrophys. J.* **318** 288
[McG93]   McGray R 1993 *Annu. Rev. Astron. Astrophys.* **31** 175
[McN87]   McNaught R M 1987 *IAU Circ. No* **4316**
[Mei30]   Meitner L and Orthmann W 1930 *Z. Phys.* **60** 143
[Mel00]   Melchiorri A *et al* 2000 *Astrophys. J.* **536** L63
[Mel01]   Mele S 2001 *Preprint* hep-ph/0103040
[Mes02]   Meszaros P 2002 *Annu. Rev. Astron. Astrophys.* **40** 137
[Meu98]   Meunier P 1998 *Nucl. Phys. B (Proc. Suppl.)* **66** 207
[Mez01]   Mezzacappa A *et al* 2001 *Phys. Rev. Lett.* **86** 1935
[Mez98]   Mezzacappa A *et al* 1998 *Astrophys. J.* **495** 911
[Mic03]   Michael D 2003 *Nucl. Phys. B (Proc. Suppl.)* **118** 1
[Mik02]   Mikaelyan L A 2002 *Phys. Atom. Nucl.* **65** 1173
[Mik86]   Mikheyev S P and Smirnov A Y 1986 *Nuovo Cimento* C **9** 17
[Mil01]   Mills G F 2001 *Nucl. Phys. B (Proc. Suppl.)* **91** 198
[Min00]   Minakata H and Nunokawa H 2000 *Phys. Lett.* B **495** 369
[Min01]   Minakata H and Nunokawa H 2001 *Phys. Lett.* B **504** 301
[Min02]   Minakata H 2002
[Min98]   Minakata H and Nunokawa H 1998 *Phys. Rev.* D **57** 4403
[Mis73]   Misner C, Thorne K and Wheeler J 1973 *Gravitation* (London: Freeman)
[Mis91]   Mishra S R *et al* 1991 *Phys. Rev. Lett.* **66** 3117
[Mis94]   Missimer J H, Mohapatra R N and Mukhopadhyay N C 1994 *Phys. Rev.* D **50** 2067
[Moe91]   Moe M 1991 *Phys. Rev.* C **44** R931
[Moe91a]   Moe M 1991 *Nucl. Phys. B (Proc. Suppl.)* **19** 158
[Moe94]   Moe M and Vogel P 1994 *Annu. Rev. Nucl. Part. Phys.* **44** 247
[Moh01]   Mohapatra R N 2001 *Current Aspects of Neutrino Physics* (Berlin: Springer)
[Moh80]   Mohapatra R N and Senjanovic G 1980 *Phys. Rev. Lett.* **44** 912
[Moh86, 92] Mohapatra R N 1986 and 1992 *Unification and Supersymmetry* (Heidelberg: Springer)
[Moh86]   Mohapatra R N 1986 *Phys. Rev.* D **34** 3457
[Moh88]   Mohapatra R N and Takasugi E 1988 *Phys. Lett.* B **211** 192
[Moh91]   Mohapatra R N and Pal P B 1991 *Massive Neutrinos in Physics and Astrophysics* (Singapore: World Scientific)
[Moh96a]   Mohapatra R N 1996 *Proc. Int. Workshop on Future Prospects of Baryon Instability Search in p-Decay and n–n̄ Oscillation Experiments (28–30 March, 1996, Oak Ridge)* ed S J Ball and Y A Kamyshkov (US Dept of Energy Publications) p 73
[Mon00]   Agafonova N Y *et al* 2000 *MONOLITH Proposal* LNGS P26/2000, CERN/SPSC 2000-031
[Mon02]   Montaruli T 2002 *Nucl. Phys. B (Proc. Suppl.)* **110** 513
[Mor73]   Morita M 1973 *Beta Decay and Muon Capture* (Reading, MA: Benjamin)

424   References

[Mül95c]   Müller E and Janka H T 1995 *Ann. NY Acad. Sci.* **759** 368
[Mut88]   Muto K and Klapdor H V 1988 *Neutrinos* ed H V Klapdor (Berlin: Springer)
[Mut89]   Muto K 1989
[Mut91]   Muto K, Bender E and Klapdor H V 1991 *Z. Phys.* A **39** 435
[Mut92]   Muto K 1992 *Phys. Lett.* B **277** 13
[Nac86]   Nachtmann O 1986 *Einführung in die Elementarteilchenphysik* (Berlin: Vieweg)
[Nag00]   Nagano M and Watson A A 2000 *Rev. Mod. Phys.* **72** 689
[Nak01]   Nakamura K 2001 *Nucl. Phys. B (Proc. Suppl.)* **91** 203
[Nak02]   Nakamura S *et al* 2002 *Nucl. Phys.* A **707** 561
[Nak03]   Nakaya T 2003 *TAUP 2003 Conference, Seattle*
[Nar93]   Narlikar J V (ed) 1993 *Introduction to Cosmology* 2nd edn (Cambridge: Cambridge University Press)
[Net02]   Netterfield C B *et al* 2002 *Ap. J.* **571** 604
[Nie03]   Niessen P 2003 *Preprint* astro-ph/0306209
[Nil84]   Nilles H P 1984 *Phys. Rep.* **110** 1
[Nil95]   Nilles H P 1995 *Preprint* hep-th/9511313
             Nilles H P 1995 *Conf. on Gauge Theories, Applied Supersymmetry and Quantum Gravity (July 1995, Leuven, Belgium)* ed B de Wit *et al* (Belgium: University Press)
[Nir01]   Nir Y 2001 *Preprint* hep-ph/010909
[Nis03]   Nishikawa K 2003 *Nucl. Phys. B (Proc. Suppl.)* **118** 129
[Noe18]   Noether E 1918 *Kgl. Ges. Wiss. Nachrichten. Math. Phys. Klasse* (Göttingen) p 235
[Noe87]   Nötzold D 1987 *Phys. Lett.* B **196** 315
[Nom84]   Nomoto K 1984 *Astrophys. J.* **277** 791
[Nor84]   Norman E B and DeFaccio M A 1984 *Phys. Lett.* B **148** 31
[Nuf01]   Proc. 3rd International Workshop on neutrino factories based on muon storage rings, 2003 *Nucl. Inst. Meth.* A **503** 1
[Nuf02]   Proc. 4th International Workshop on neutrino factories 2003 *J. Phys.* G **29** 1463
[Nut96]   McNutt J *et al* (NESTOR Collaboration) 1996 *Nucl. Phys. B (Proc. Suppl.)* **48** 469
             McNutt J *et al* (NESTOR Collaboration) 1996 *Proc. 4th Int. Workshop on Theoretical and Phenomenological Aspects of Underground Physics (TAUP'95) (Toledo, Spain)* ed A Morales, J Morales and J A Villar (Amsterdam: North-Holland)
[Obe87]   Oberauer L *et al* 1987 *Phys. Lett.* B **198** 113
[Obe93]   Oberauer L *et al* 1993 *Ann. Phys., Paris* **1** 377
[Ödm03]   Ödman C J *et al* 2003 *Phys. Rev.* D **67** 083511
[Ohs94]   Ohsima T and Kawakami H 1994 *Phys. Lett.* B **235** 337
[Oli00]   Olive K A, Steigman G and Walker T P 2000 *Phys. Rev.* **333** 389
[Oli02]   Olive K A 2002 *Preprint* astro-ph/0202486
[Oli99]   Olive K 1999 *Preprint* hep-ph/9911307
[Ort00]   Ortiz C E *et al* 2000 *Phys. Rev. Lett.* **85** 2909
[Osi01]   Osipowicz A *et al* 2001 *Preprint* hep-ex/0109033
[Ott95]   Otten E W 1995 *Nucl. Phys. B (Proc. Suppl.)* **38** 26

[Oya98]	Oyama Y 1998 *Preprint* hep-ex/9803014
[Pad93]	Padmanabhan T 1993 *Structure Formation in the Universe* (Cambridge: Cambridge University Press)
[Pae99]	Päs H *et al* 1999 *Phys. Lett.* B **453** 194
	Päs H *et al* 2001 *Phys. Lett.* B **498** 35
[Pag97]	Pagel B E J 1997 *Nucleosynthesis and Chemical Evolution of Galaxies* (Cambridge: Cambridge University Press)
[Pak03]	Pakvasa S and Zuber K 2003 *Phys. Lett.* B **566** 207
[Pak03]	Pakvasa S and Valle J W F 2003 *Preprint* hep-ph/0301061
[Pal92]	Pal P B 1992 *Int. J. Mod. Phys.* **A7** 5387
[Pan91]	Panagia N *et al* 1991 *Astrophys. J.* **380** L23
[Pan95]	Panman J 1995 *Precision Tests of the Standard Model* ed P Langacker (Singapore: World Scientific)
[Par01]	Parke S and Weiler T J 2001 *Phys. Lett.* B **501** 106
[Par94]	Parker P D 1994 *Proc. Solar Modeling Workshop (Seattle, WA)* ed A B Balantekin and J N Bahcall (Singapore: World Scientific)
[Par95]	Partridge R B 1995 *3K: The Cosmic Microwave Background Radiation* (Cambridge: Cambridge University Press)
[Pas00]	Pascos E A, Pasquali L and Yu J Y 2000 *Nucl. Phys.* B **588** 263
[Pas02]	Pascoli S, Petcov S T and Rodejohann W 2002 *Phys. Lett.* B **549** 177
[Pas97]	Passalacqua L 1997 *Nucl. Phys. B (Proc. Suppl.)* **55C** 435
[Pat74]	Pati J C and Salam A 1974 *Phys. Rev.* D **10** 275
[Pau77]	Pauli W 1991 On the earlier and more recent history of the neutrino (1957) *Neutrino Physics* ed K Winter (Cambridge: Cambridge University Press)
[PDG00]	Groom D *et al* 2000 Review of particle properties *Eur. Phys. J.* C **3** 1
[PDG02]	Hagiwara K *et al* 2002 Review of particle properties *Phys. Rev.* D **66** 010001
[Pea02a]	Peacock J A *et al* 2002 *Preprint* astro-ph/0204239
[Pea03]	Pearson T J *et al* 2003 *Ap. J.* **591** 556
[Pea98]	Peacock J A 1998 *Cosmological Physics* (Cambridge: Cambridge University Press)
[Pee80]	Peebles P J E 1980 *The Large-Scale Structure of the Universe* (Princeton, NY: Princeton University Press)
[Pee93]	Peebles P J E 1993 *Principles of Physical Cosmology, Princeton Series in Physics* (Princeton, NJ: Princeton University Press)
[Pee95]	Peebles P J E 1995 *Principles of Physical Cosmology, Princeton Series in Physics* (Princeton, NJ: Princeton University Press)
[Pen65]	Penzias A A and Wilson R W 1965 *Astrophys. J.* **142** 419
[Per00]	Perkins D H 2000 *Introduction to High Energy Physics* (New York: Addison-Wesley)
[Per88]	Perkins D H 1988 *Proc. IX Workshop on Grand Unification (Aix-les-Bains)* (Singapore: World Scientific) p 170
[Per94]	Perkins D H 1994 *Ann. Phys., Paris* **2** 249
[Per95a]	Percival J W *et al* 1995 *Astrophys. J.* **446** 832
[Per97]	Perlmutter S *et al* 1997 *Astrophys. J.* **483** 565
[Per99]	Perlmutter S *et al* 1999 *Astrophys. J.* **517** 565
[Pet90]	Petschek A G (ed) 1990 *Supernovae* (Heidelberg: Springer)

[Pet99]   Petrov A A and Torma T 1999 *Phys. Rev.* D **60** 093009
[Pic92]   Picard A *et al* 1992 *Nucl. Instrum. Methods* B **63** 345
[Pie01]   Piepke A 2001 *Nucl. Phys. B (Proc. Suppl.)* **91** 99
[Pil97]   Pilaftsis A 1997 *Phys. Rev.* D **56** 5431
[Pil99]   Pilaftsis A 1999 *Int. J. Mod. Phys.* **A14** 1811
[Pol73]   Politzer H D 1973 *Phys. Rev. Lett.* **30** 1346
[Pon60]   Pontecorvo B 1960 *Sov. J. Phys.* **10** 1256
[Pre85]   Press W H and Spergel D N 1985 *Astrophys. J.* **296** 79
[Pri68]   Primakoff H and Rosen S P 1968 α, β, γ *Spectroscopy* ed K Siegbahn
            (Amsterdam: North-Holland)
[Qia01]   Qian Y Z 2001 *Nucl. Phys. B (Proc. Suppl.)* **91** 345
[Qia93]   Qian Y Z *et al* 1993 *Phys. Rev. Lett.* **71** 1965
[Qia95]   Qian Y Z and Fuller G M 1995 *Phys. Rev.* D **51** 1479
[Qui83]   Quigg C 1983 *Gauge Theories of the Strong, Weak and Electromagnetic
            Interactions (Frontiers in Physics 56)* (New York: Addison-Wesley)
[Qui93]   Quirrenbach A 1993 *Sterne und Weltraum* **32** 98
[Raf01]   Raffelt G G 2001 *Astrophys. J.* **561** 890
[Raf02]   Raffelt G G 2002 *Nucl. Phys. B (Proc. Suppl.)* **110** 254
[Raf85]   Raffelt G G 1985 *Phys. Rev.* D **31** 3002
[Raf90]   Raffelt G G 1990 *Astrophys. J.* **365** 559
[Raf96]   Raffelt G 1996 *Stars as Laboratories for Fundamental Physics* (Chicago,
            IL: University of Chicago Press)
[Raf99]   Raffelt G G 1999 *Annu. Rev. Nucl. Part. Phys.* **49** 163
[Rag94]   Raghavan R S 1994 *Phys. Rev. Lett.* **72** 1411
[Rag97]   Raghavan R S 1997 *Phys. Rev. Lett.* **78** 3618
[Ram00]   Rampp M and Janka H T 2000 *Astrophys. J. Lett.* **539** L33
[Ram03]   Ramachers Y 2003 *Nucl. Phys. B (Proc. Suppl.)* **118**
[Rea92]   Readhead A C S and Lawrence C R 1992 *Ann. Rev. Astron. Astrophys.* **30**
            653
[Ree94]   Reeves H 1994 *Rev. Mod. Phys.* **66** 193
[Rei53]   Reines F and Cowan C L Jr 1953 *Phys. Rev.* **92** 330
[Rei56]   Reines F and Cowan C L Jr 1956 *Nature* **178** 446, 523 (erratum)
            Reines F and Cowan C L Jr 1956 *Science* **124** 103
[Rei58]   Reines F and Cowan C L Jr 1958 *Phys. Rev.* **113** 273
[Rei76]   Reines F *et al* 1976 *Phys. Rev. Lett.* **37** 315
[Rei81]   Rein D and Sehgal L M 1981 *Ann. Phys.* **133** 79
[Rei83]   Rein D and Seghal L M 1983 *Nucl. Phys.* B **223** 29
[Res94]   Resvanis L K (NESTOR Collaboration) 1994 *Nucl. Phys. B (Proc. Suppl.)*
            35 294
            Resvanis L K (NESTOR Collaboration) 1994 *Proc. 3rd Int. Workshop
            on Theoretical and Phenomenological Aspects in Astroparticle and
            Underground Physics (TAUP'93) (Assergi, Italy)* ed C Arpesella,
            E Bellotti and A Bottino (Amsterdam: North-Holland)
[Rho94]   Rhode W *et al* 1994 *Ann. Phys., Paris* **4** 217
[Ric01]   Rich J 2001 *Fundamentals of Cosmology* (Berlin: Springer)
[Rie98]   Riess A G *et al* 1998 *Astronom. J.* **116** 1009
[Rit00]   van Ritbergen T and Stuart R G 2000 *Nucl. Phys.* B **564** 343
[Rod00]   Rodejohann W and Zuber K 2000 *Phys. Rev.* D **62** 094017

[Rod01]   Rodejohann W and Zuber K 2001 *Phys. Rev.* D **63** 054031
[Rod01a]   Rodejohann W 2001 *Nucl. Phys.* B **597** 110
[Rod02]   Rodejohann W 2002 *Phys. Lett.* B **542** 100
[Rod52]   Rodeback G W and Allen J S 1952 *Phys. Rev.* **86** 446
[Rog96]   Rogers F J, Swenson F J and Iglesias C A 1996 *Astrophys. J.* **456** 902
[Rol88]   Rolfs C E and Rodney W S 1988 *Cauldrons in the Cosmos* (Chicago, IL: University of Chicago Press)
[Ros84]   Ross G G 1984 *Grand Unified Theories (Frontiers in Physics 60)* (New York: Addison-Wesley)
[Rou96]   Roulet E, Covi L and Vissani F 1996 *Phys. Lett.* B **384** 169
[Row85a]   Rowley J K, Cleveland B T and Davis R 1985 *Proc. Solar Neutrinos and Neutrino Astronomy (Homestake, USA) (AIP Conf. Proc. 126)* (New York: AIP)
[Rub01]   Rubbia A 2001 *Nucl. Phys. B (Proc. Suppl.)* **91** 223
[Rub96]   Rubbia C 1996 *Nucl. Phys. B (Proc. Suppl.)* **48** 172
[Ruh02]   Ruhl J E *et al* 2002 *Preprint* astro-ph/0212229
[Sac67b]   Sacharov A D 1967 *JETP Lett* **6** 24
[Sah00]   Sahni V and Starobinsky A 2000 *Int. J. Mod. Phys.* **D9** 373
[Sak90]   Sakamoto W *et al* 1990 *Nucl. Instrum. Methods* A **294** 179
[Sal01]   Saltzberg D *et al* 2001 *Phys. Rev. Lett.* **86** 2802
[Sal57]   Salam A 1957 *Nuovo Cimento* **5** 299
[Sal68]   Salam A 1968 *Elementary Particle Theory* ed N Swarthohn, Almquist and Wiksell (Stockholm) p 367
[San00]   Sanuki T *et al* 2000 *Astrophys. J.* **545** 1135
[San88]   Santamaria R and Valle J W F 1988 *Phys. Rev. Lett.* **60** 397
[Sch00]   Schlickeiser R 2000 *Cosmic Ray Astrophysics* (Heidelberg: Springer)
[Sch01]   Scholberg K 2001 *Nucl. Phys. B (Proc. Suppl.)* **91** 331
[Sch03]   Schoenert S 2003 *Nucl. Phys. B (Proc. Suppl.)* **118** 62
[Sch60]   Schwartz M 1960 *Phys. Rev. Lett.* **4** 306
[Sch66]   Schopper H F 1966 *Weak Interactions and Nuclear Beta Decay* (Amsterdam: North-Holland)
[Sch66]   Schopper H 1966 *Weak interactions and Nuclear Beta Decay* (Amsterdam: North-Holland)
[Sch82]   Schechter J and Valle J W F 1982 *Phys. Rev.* D **25** 2591
[Sch97]   Schmitz N 1997 *Neutrinophysik* (Stuttgart: Teubner)
[Sch98]   Schmidt B P *et al* 1998 *Astrophys. J.* **507** 46
[Sco02]   Scott P F *et al* 2002 *Preprint* astro-ph/0205380
[Sec91]   Seckel D, Stanev T and Gaisser T K 1991 *Astrophys. J.* **382** 652
[Sem00]   Semertzidis Y K *et al* 2000 *Preprint* hep-ph/0012087
[Seu01]   Seunarine S *et al* 2001 *Int. J. Mod. Phys.* A **16** 1016
[Sex87]   Sexl R U and Urbantke H K 1987 *Gravitation und Kosmologie* (Mannheim: B I Wissenschaftsverlag) and 1995 (Heidelberg: Spectrum Academischer)
[Sha03]   Shaevitz M H and Link J M 2003 *Preprint* hep-ex/0306031
[Sha83]   Shapiro S L and Teukolsky S A 1983 *Black Holes, White Dwarfs and Neutron Stars* (London: Wiley)
[Sie68]   Siegbahn K (ed) 1968 α, β, γ *Ray Spectroscopy* vol 2
[Sil02]   Silk J *Preprint* astro-ph/0212305

[Sil67]   Silk J 1967 *Nature* **215** 1155
[Sil68]   Silk J 1968 *Ap. J.* **151** 459
[Sim01]   Simkovic F *et al* 2001 *Part. Nucl. Lett.* **104** 40
[Smi01]   Smith P F 1997 *Ann. Phys., Paris* **16** 75
[Smi90]   Smith P F and Lewin J D 1990 *Phys. Rep.* **187** 203
[Smi93b]   Smith M S, Kawano L H and Malaney R A 1993 *Astrophys. J. Suppl.* **85**
            219
[Smi94]   Smirnov A, Spergel D and Bahcall J N 1994 *Phys. Rev.* D **49** 1389
[Smi96]   Smirnov A Y, Vissani F 1996 *Phys. Lett.* B **386** 317
[Smi97]   Smith P F 1997 *Ann. Phys., Paris* **8** 27
[Smo90]   Smoot G F *et al* 1990 *Astrophys. J.* **360** 685
[Smo95]   Smoot G F 1995 *Preprint* astro-ph/9505139
            Smoot G F 1994 *DPF Summer Study on High Energy Physics: Particle
            and Nuclear Astrophysics and Cosmology in the Next Millennium,
            Snowmass* p 547
[Smy03]   Smy M 2003 *Nucl. Phys. B (Proc. Suppl.)* **118**
[Smy03a]   Smy M B *et al* 2003 *Preprint* hep-ex/0309011
[Sne55]   Snell A H and Pleasonton F 1955 *Phys. Rev.* **97** 246
            Snell A H and Pleasonton F 1955 *Phys. Rev.* **100** 1396
[Sob01]   Sobel H 2001 *Nucl. Phys. B (Proc. Suppl.)* **91** 127
[Sok89]   Sokolsky P 1989 *Introduction to Ultrahigh Energy Cosmic Ray Physics
            (Frontiers in Physics 76)* (London: Addison-Wesley)
[Spe03]   Spergel D N *et al* 2003 *Preprint* astro-ph/0302209
[Spi01]   Spiering C 2001 *Nucl. Phys. B (Proc. Suppl.)* **91** 445
[Spi96]   Spiering Ch 1996 *Nucl. Phys. B (Proc. Suppl.)* **48** 463
            Spiering Ch 1996 *Proc. 4th Int. Workshop on Theoretical and
            Phenomenological Aspects of Underground Physics (TAUP'95) (Toledo,
            Spain)* ed A Morales, J Morales and J A Villar (Amsterdam: North-
            Holland)
[Spr87]   Springer P T, Bennett C L and Baisden R A 1987 *Phys. Rev.* A **35** 679
[Sta03]   Stanek K Z *et al* 2003 *Astrophys. J.* **591** L17
[Sta90]   Staudt A, Muto K and Klapdor-Kleingrothaus H V 1990 *Europhys. Lett.*
            13 31
[Ste03]   Steigman G 2003 *Preprint* astro-ph/0307244
[Sti02]   Stix M 2002 *The Sun* 2nd version (Heidelberg: Springer)
[Str03]   Strumia A and Vissani F 2003 *Phys. Lett.* B **564** 42
[Sue03]   Suekane F *et al* 2003 *Preprint* hep-ex/0306029
[Suh93]   Suhonen J and Civitares O 1993 *Phys. Lett.* B **312** 367
[Suh98]   Suhonen J and Civitarese O 1998 *Phys. Rep.* **300** 123
[Sun92]   Suntzeff N B *et al* 1992 *Astrophys. J. Lett.* **384** 33
[Swi97]   Swift A M *et al* 1997 *Proc. Neutrino'96* (Singapore: World Scientific)
            p 278
[Swo02]   Swordy S P *et al* 2002 *Astropart. Phys.* **18** 129
[Tak02]   Takahashi K and Sato K 2002 *Phys. Rev.* D **66** 033006
[Tak84]   Takasugi E 1984 *Phys. Lett.* B **149** 372
[Tam94]   Tammann G, Löffler W and Schröder A 1994 *Astrophys. J. Suppl.* **92** 487
[Tan93]   Tanaka J and Ejiri H 1993 *Phys. Rev.* D **48** 5412

[Tat97]   Tata X 1997 *Preprint* hep-ph/9706307, Lectures given at IX Jorge A
             Swieca Summer School, Campos do Jordao, Brazil, February 1997
[Tay03]   Tayloe R 2003 *Nucl. Phys. B (Proc. Suppl.)* **118** 157
[Tay94]   Taylor J H 1994 *Rev. Mod. Phys.* **66** 711
[Teg95]   Tegmark M 1995 *Preprint* astro-ph/951114
[The58]   Theis W R 1958 *Z. Phys.* **150** 590
[Tho01]   Thompson L E 2001 (ANTARES-coll.) *Nucl. Phys. B (Proc. Suppl.)* **91**
             431
[Tho02]   Thompson T A, Burrows A and Pinto P A 2002 *Ap. J.* **581** L85
[t'Ho72]   't Hooft G and Veltman M 1972 *Nucl. Phys.* B **50** 318
[t'Ho76]   't Hooft G 1976 *Phys. Rev.* D **14** 3432
[Tho95]   Thorne K 1995 *Ann. NY Acad. Sci.* **759** 127
[Thu96]   Thunman M, Ingelman G, and Gondolo P 1996 *Ann. Phys., Paris* **5** 309
[ToI98]   Firestone R B, Chu S Y F and Baglin C M (ed) 1998 *Table of Isotopes*
             (New York: Wiley)
[Tom88]   Tomoda T 1988 *Nucl. Phys.* A **484** 635
[Tom91]   Tomoda T 1991 *Rep. Prog. Phys.* **54** 53
[Ton03]   Tonry J L *et al* 2003 *Preprint* astro-ph/0305008
[Tor99]   Torbet E *et al* 1999 *Astrophys. J.* **521** L79
[Tos01]   Toshito T 2001 *Preprint* hep-ex/0105023
[Tot95]   Totani T and Sato K 1995 *Astropart. Phys.* **3** 367
[Tot96]   Totani T, Sato K and Yoshii Y 1996 *Ap. J.* **460** 303
[Tot98]   Totani T *et al* 1998 *Astrophys. J.* **496** 216
[Tra99]   Trasatti L *et al* 1999 *Nucl. Phys. B (Proc. Suppl.)* **70** 442
[Tre02]   Tretyak V I and Zdesenko Y G 2002 *At. Data Nucl. Data Tables* **80** 83
[Tre79]   Tremaine S and Gunn J E 1979 *Phys. Rev. Lett.* **42** 407
[Tre95]   Tretyak V I and Zdesenko Y G 1995 *At. Data Nucl. Data Tables* **61** 43
[Tri97]   Tripathy S C and Christensen-Dalsgaard J 1997 *Preprint* astro-ph/9709206
[Tse01]   Tserkovnyak V *et al 27th Int. Cosmic Ray Conf. (ICRC01) (Hamburg)*
[Tur02]   Turner M S 2002 *Preprint* astro-ph/0212281
[Tur88]   Turck-Chieze S *et al* 1988 *Astrophys. J.* **335** 415
[Tur92]   Turkevich, A I, Economou T E and Cowen G A 1992 *Phys. Rev. Lett.* **67**
             3211
[Tur92a]   Turner M S 1992 *Trends in Astroparticle Physics* ed D Cline and R Peccei
             (Singapore: World Scientific) p 3
[Tur93a]   Turck-Chieze S *et al* 1993 *Phys. Rep.* **230** 57
             Turck-Chieze S 1995 *Proc. 4th Int. Workshop on Theoretical and
             Phenomenological Aspects of Underground Physics (TAUP'95) (Toledo,
             Spain)* ed A Morales, J Morales and J A Villar (Amsterdam: North-
             Holland)
[Tur93b]   Turck-Chieze S and Lopes I 1993 *Astrophys. J.* **408** 347
[Tur96]   Turck-Chieze S 1996 *Nucl. Phys. B (Proc. Suppl.)* **48** 350
[Tyt00]   Tytler D *et al* 2000 *Preprint* astro-ph/0001318
[Uns92]   Unsöld A and Baschek B 1992 *The New Cosmos* (Heidelberg: Springer)
[Val03]   Valle J W F 2003 *Nucl. Phys. B (Proc. Suppl.)* **118** 255
[Ven01]   McGrew G 2001 *Talk Presented at Neutrino Telescopes (Venice)*
[Vie98]   Vietri M 1998 *Phys. Rev. Lett.* **80** 3690
[Vil94]   Vilain P *et al* 1994 *Phys. Lett.* B **335** 246

[Vil95]   Vilain P *et al* 1995 *Phys. Lett.* B **345** 115
[Vir99]   Viren B 1999 *Preprint* hep-ex/9903029, Proc. DPF Conference 1999
[Vog00]   Vogel P 2001 *Current Topics in Neutrino Physics* (Berlin: Springer)
[Vog86]   Vogel P and Zirnbauer M R 1986 *Phys. Rev. Lett.* **57** 3148
[Vog88]   Vogel P, Ericson M and Vergados J D 1988 *Phys. Lett.* B **212** 259
[Vog95]   Vogel P 1995 *Proc. ECT Workshop on Double Beta Decay and Related Topics (Trento)* ed H V Klapdor-Kleingrothaus and S Stoica (Singapore: World Scientific) p 323
[Vol86]   Voloshin M, Vysotskii M and Okun L B 1986 *Sov. Phys.–JETP* **64** 446
              Voloshin M, Vysotskii M and Okun L B 1986 *Sov. J. Nucl. Phys.* **44** 440
[Wan02]   Wang L *et al* 2002 *Ap. J.* **579** 671
[Wax03]   Waxman E 2003 *Nucl. Phys.* B *(Proc. Suppl.)* **118** 353
[Wax97]   Waxman E and Bahcall J N 1997 *Phys. Rev. Lett.* **78** 2292
[Wax99]   Waxman E and Bahcall J N 1999 *Phys. Rev.* D **59** 023002
[Wea80]   Weaver T A and Woosley S E 1980 *Ann. NY Acad. Sci.* **336** 335
[Wei00]   Weinberg S 2000 *The Quantum Theory of Fields. Vol. 3: Supersymmetry* (Cambridge: Cambridge University Press)
[Wei02]   Weinheimer C 2002 *Preprint* hep-ex/0210050
[Wei03]   Weinheimer C 2003 *Nucl. Phys.* B *(Proc. Suppl.)* **118** 279
[Wei03a]   Weinheimer C 2003 *Preprint* hep-ph/030605
[Wei35]   von Weizsäcker C F 1935 *Z. Phys.* **96** 431
[Wei37]   von Weizsäcker C F 1937 *Z. Phys.* **38** 176
[Wei67]   Weinberg S 1967 *Phys. Rev. Lett.* **19** 1264
              Weinberg S 1971 *Phys. Rev. Lett.* **27** 1688
[Wei72]   Weinberg S 1972 *Gravitation and Cosmology* (New York: Wiley)
[Wei82]   Weiler T J 1982 *Phys. Rev. Lett.* **49** 234
              Weiler T J 1982 *Astrophys. J.* **285** 495
[Wei89]   Weinberg S 1989 *Rev. Mod. Phys.* **61** 1
[Wei99]   Weinheimer C *et al* 1999 *Phys. Lett.* B **460** 219
[Wen01]   Wentz J *27th Int. Cosmic Ray Conf. (ICRC01) (Hamburg)*
[Wes74]   Wess J and Zumino B 1974 *Nucl. Phys.* B **70** 39
[Wes86, 90]   West P 1986, 1990 *Introduction to Supersymmetry* 2nd edn (Singapore: World Scientific)
[Whe02]   Wheeler J C 2003 *Am. J. Phys.* **71** 11
[Whe90]   Wheeler J C 1990 *Supernovae, Jerusalem Winter School for Theoretical Physics, 1989* vol 6, ed J C Wheeler, T Piran and S Weinberg (Singapore: World Scientific) p 1
[Whi94]   White M, Scott D and Silk J 1994 *Ann. Rev. Astron. Astrophys.* **32** 319
[Wie01]   Wieser M E and DeLaeter J R 2001 *Phys. Rev.* C **64** 024308
[Wie96]   Wietfeldt F E and Norman E B 1996 *Phys. Rep.* **273** 149
[Wil01]   Wilkinson J F and Robertson 2001 *Current Aspects of Neutrinophysics* ed D O Caldwell (Springer)
[Wil86]   Wilson J R *et al* 1986 *Ann. NY Acad. Sci.* **470** 267
[Wil93]   Wilson T 1993 *Sterne und Weltraum* **3** 164
[Win00]   Winter K 2000 *Neutrino Physics* ed K Winter (Cambridge: Cambridge University Press)
[Win98]   Wintz P 1998 *Proc. 1st Int. Symp. on Lepton and Baryon Number Violation (Trento)*

[Wis03]	Wischnewski R 2003 *Preprint* astro-ph/0305302, Proc. 28th ICRC Tsukuda, Japan
[Wis99]	Wischnewski R *et al* 1999 *Nucl. Phys. B (Proc. Suppl.)* **75** 412
[Wol78]	Wolfenstein L 1978 *Phys. Rev.* D **17** 2369
[Wol81]	Wolfenstein L 1981 *Phys. Lett.* B **107** 77
[Wol83]	Wolfenstein L 1983 *Phys. Rev. Lett.* **51** 1945
[Won02]	Wong H T and Li J 2002 *Preprint* hep-ex/0201001
[Woo82]	Woosley S E and Weaver T A 1982 *Supernovae: A Survey of Current Research, Proc. NATO Advanced Study Institute, Cambridge* ed M J Rees and R J Sonteham (Dordrecht: Reidel) p 79
[Woo86a]	Woosley S E and Weaver T A 1986 *Ann. Rev. Astron. Astrophys.* **24** 205
[Woo88]	Woosley S E and Phillips M M 1988 *Science* **240** 750
[Woo88]	Woosley S E and Haxton W C 1988 *Nature* **334** 45
[Woo90]	Woosley S E *et al* 1990 *Ast. J.* **356** 272
[Woo92]	Woosley S E and Hoffman R D 1992 *Astrophys. J.* **395** 202
[Woo94]	Woosley S E *et al* 1994 *Astrophys. J.* **433** 229
[Woo95]	Woosley S E and Weaver T A 1995 *Astrophys. J. Suppl.* **101** 81
[Woo95a]	Woosley S E *et al* 1995 *Ann. NY Acad. Sci.* **759** 352
[Woo97]	Woosley S E *et al* 1997 *Preprint* astro-ph/9705146
[Wri03]	Wright E Z 2003 *Preprint* astro-ph/0305591
[Wri94a]	Wright E L *et al* 1994 *Astrophys. J.* **420** 450
[Wu57]	Wu C S *et al* 1957 *Phys. Rev.* **105** 1413
[Wu66]	Wu C S and Moszkowski S A 1966 *Beta Decay* (New York: Wiley)
[Yan54]	Yang C N and Mills R 1954 *Phys. Rev.* **96** 191
[Yan84]	Yang J *et al* 1984 *Astrophys. J.* **281** 493
[Yas94]	Yasumi S *et al* 1994 *Phys. Lett.* B **334** 229
[You91]	You K *et al* 1991 *Phys. Lett.* B **265** 53
[Zac02]	Zach J J *et al* 2002 *Nucl. Instrum. Methods* A **484** 194
[Zal03]	Zaldarriaga M *Preprint* astro-ph/0305272
[Zas92]	Zas E, Halzen F and Stanev T 1992 *Phys. Rev.* D **45** 362
[Zat66]	Zatsepin G T and Kuzmin V A 1966 *JETP* **4** 53
[Zde02]	Zdesenko Y G, Danevich F A and Tretyak V I 2002 *Phys. Lett.* B **546** 206
[Zee80]	Zee A 1980 *Phys. Lett.* B **93** 389
[Zel02]	Zeller G P *et al* 2002 *Phys. Rev. Lett.* **88** 091802
[Zub00]	Zuber K 2000 *Phys. Lett.* B **485** 23
[Zub00a]	Zuber K 2000 *Phys. Lett.* B **479** 33
[Zub01]	Zuber K 2001 *Phys. Lett.* B **519** 1
[Zub02]	Zuber K 2002 *Prog. Nucl. Part. Phys.* **48** 223
[Zub03]	Zuber K 2003 *Phys. Lett.* B **571** 148
[Zub92]	Zuber K 1993 *Ann. NY Acad. Sci* **688** 509 Talk presented at PASCOS'92
[Zub97]	Zuber K 1997 *Phys. Rev.* D **56** 1816
[Zub98]	Zuber K 1998 *Phys. Rep.* **305** 295
[Zuc01]	Zucchelli P 2001 *Preprint* hep-ex/0107006
[Zuc02]	Zucchelli P 2002 *Phys. Lett.* B **532** 166
[Zwe64]	Zweig G 1964 *CERN-Reports* Th-401 and Th-412

Index

ALEPH, 146
AMANDA, ICECUBE, 353
ANTARES, 351
AQUA-RICH, 248
AUGER, 356
axions, 396

Baikal experiment, 349
Baksan underground laboratory,
    267, 308
baryon number, violation, 397
baryons, 109, 110, 375, 378, 395,
    397
BEBC, 74
beta decay, 127
Big Bang, 397
Big Bang model, 361, 368, 370,
    376, 391, 392
Big Bang model, nucleosynthesis
    in, 392
black holes, 308
Borexino, 206
Borexino experiment, 287
bosons, 114, 116, 372
bounce back, of shock wave in
    supernova collapse, 295

Cabibbo angle, 48
Cabibbo–Kobayashi–Maskawa
    matrix, 48, 49
Casimir effect, 365
CCFR, 73, 88, 101
CDHS, 70, 90, 97
CDHSW, 67, 73, 101

Cerenkov detectors, 262, 269, 308,
    309
Cerenkov detectors, in natural water
    sources, NESTOR in
    Mediterranean, 434
Cerenkov detectors, in natural water
    sources, NT-200 in Lake
    Baikal, 349
Chandrasekhar bound, 294
Chandrasekhar mass, 294
charge conjugation, 22
CHARM, 73, 101
CHARM-II, 56
chirality, 21
chlorine experiment, 262, 285
CHOOZ, 200
CHORUS, 72, 96, 97, 211
CNO-cycle, oxygen catalysis of,
    304
COBE, 377
COBRA, 174, 184
collisional damping, 380
Compton wavelength, 373
cosmic microwave background,
    362, 376
cosmic microwave background,
    anisotropies in, 378
cosmic microwave background,
    discovery of, 361, 376
cosmological constant, 362, 365,
    366
cosmological constant, critical
    value, 366

cosmological constant, curvature
        parameter, 366
cosmological principle, 362
CP, eigenstates, 49
CP, violation, 49–51, 397
CP, violation, direct, 50
CP, violation, in early universe,
        397
CP, violation, in K-meson system,
        value of ϵ, 50
creation phase, 24
critical density, 368
critical density parameter Ω, 368,
        382
CVC, 35

dark matter, 382, 385, 386
dark matter, baryonic, 382
dark matter, cold, 384
dark matter, cold, neutrinos as, 385
dark matter, cold, supersymmetric
        particles as, 385
dark matter, experiments for
        detecting, indirect, 386
dark matter, experiments for
        detecting, indirect,
        annihilation in the Sun or
        Earth, 386
dark matter, hot, neutrinos as, 384
de Sitter universe, 368
deceleration parameter, 364
Dirac equation, 38–40
Dirac mass term, 24
DONUT, 69

E734, 80
Einstein's field equations, 361, 362
Einstein–Friedmann–Lemaitre
        equations, 363
electromagnetic interaction, 40
electroweak interaction, 41, 45, 109
electroweak interaction, Weinberg
        angle, 110

electroweak phase transition, in
        early universe, 398
ELEGANT V, 177

Fermi constant, 51
Fermi energy, 294
fermions, 46, 107, 109, 114, 116,
        123, 399
freezing out, of particle species
        from equilibrium, 372
Friedmann equation, 369
Friedmann universe, 376

galaxies, clusters of, 382, 386
galaxies, spiral, rotation curves of
        galaxies, 382
galaxies, superclusters of, 386
galaxies, voids of, 386
GALLEX experiment, 266, 267
gallium experiment, 266
gauge theories, 35–37, 40, 41, 106
gauge theories, Abelian, 40
gauge theories, non-Abelian, 40,
        41, 399
gauge transformation, 39
Gauge transformations, 106
gauge transformations, 105, 107
germane, 266
GNO, 267
Goldhaber experiment, 10
Gran Sasso underground laboratory,
        266, 287
gravitation, 115, 361
gravitational waves, from
        supernovae, 316
GUT, 105, 115, 118, 373, 397
GUT, Pati-Salam model, 113
GUT, SO(10) model, 111, 112, 114,
        123
GUT, SU(5) model, 107, 111, 114
GUT, SU(5) model, generators, 109
GUT, SU(5) model, proton decay
        channels, 110

GUT, SU(5), gauge
    transformations, 109
GWS model, 42
GZK-cutoff, 340

Heidelberg–Moscow experiment,
    173
Heisenberg's uncertainty principle,
    365
helicity, 9, 21
helium flash, 291
HERA, 73, 77, 82, 88, 188
Higgs boson, 47, 61, 115, 398
Higgs field, 47, 109, 123
Higgs field, vacuum expectation
    value, 46
Higgs mechanism, 44
homologous, supernova collapse,
    294
Hubble, 361
Hubble constant, 363, 383, 396
Hubble Space Telescope, 304, 305
Hubble space telescope, 306
Hubble time, 365
Hyper-Kamiokande, 248

ICARUS, 245, 315
IGEX, 174
IMB experiment, 308, 309
inflationary models, 368
instantons, 399
iron core, in supernova collapse,
    disruption of, 296
IUE satellite, 307

K2K, 81, 242
Kamiokande experiments, 264,
    308, 309
KamLAND, 204, 301
KARMEN, 102, 209
KATRIN, 384
KEK, *B* factory, 50

least action, principle, 38
Lemaitre universe, 366

LENS, 289
lepton flavour violation, 30
leptons, 109, 110, 113, 397
LIGO experiment, for gravitational
    waves, 316
LSND, 56, 104, 207
LUNA, 255
LVD, 301

MACRO, 239
main sequence, 254
main sequence, Sun as member of,
    259
Mainz experiment, 137
Majorana mass term, 25
Majorana particle, 24
Majorana phase, 126, 185
Majoron, 396
Maxwell equations, 36
mesons, 375
MIBETA, 176
microwave background radiation,
    375
MiniBooNE, 209
MINOS, 244
MNS matrix, 50, 124
MONOLITH, 247
Mont Blanc experiment, 308
MOON, 288
MUNU, 150

narrow-band beam, 66
NEMOIII, 177
NESTOR experiment, 350
Neutrino Factory, 220
neutrino sphere, in supernova
    collapse, 295, 301
neutrinos, 51, 123, 384, 388,
    390–392
neutrinos, birth of neutrino
    astrophysics, 302
neutrinos, decay, 312
neutrinos, decouple in Big bang,
    375

neutrinos, Dirac, 383
neutrinos, from SN 1987A, 308
neutrinos, from SN 1987A, in IMB experiment, 308, 309
neutrinos, from SN 1987A, in Kamiokande experiments, 308, 309
neutrinos, from SN 1987A, lifetime of, 311
neutrinos, from supernovae, 299
neutrinos, from supernovae, detection of, 301
neutrinos, heavy, as dark matter, 385
neutrinos, hep, 252
neutrinos, light, as candidates for hot dark matter, 384
neutrinos, limit on electric charge of, 314
neutrinos, Majorana, 123, 284, 383
neutrinos, number of, 16
neutrinos, number of flavours from ^{4}He abundance, 391
neutrinos, oscillations, 272, 276
neutrinos, oscillations, MSW effect, 272
neutrinos, oscillations, seesaw mechanism, 123
neutrinos, solar, 250, 259, 264, 267, 286, 289
neutrinos, solar, ^{7}Be, 287
neutrinos, solar, ^{8}B, 265
neutrinos, solar, correlation with sun-spot activity, 286
neutrinos, solar, definition of SNU, 262
neutrinos, solar, experiments, 260
neutrinos, solar, experiments, radiochemical, 260, 266
neutrinos, solar, experiments, radiochemical, chlorine experiment, 262, 285
neutrinos, solar, experiments, radiochemical, GALLEX, 266, 267
neutrinos, solar, experiments, radiochemical, SAGE, 266, 267
neutrinos, solar, experiments, real time, 264
neutrinos, solar, experiments, real time, Borexino experiment, 287
neutrinos, solar, experiments, real time, SNO experiment, 269
neutrinos, solar, experiments, real-time, 260
neutrinos, solar, from pp cycle, 259, 263, 266
neutrinos, solar, high-energy, 386
neutrinos, solar, Kamiokande experiments, 264
neutrinos, solar, problem, 264, 271, 281, 283
neutrinos, solar, spectrum of, 255
neutrinos, solar, Super-Kamiokande, 264
neutron star, 299, 308, 313
neutron star, binding energy of, 310
neutron–antineutron oscillations, 114
Noether's Theorem, 39
NOMAD, 71, 96, 98, 211
NT-200 experiment, 349
NuTeV, 97

opacity, neutrino, in supernova collapse, 295
opacity, of stars, 304
opacity, of Sun, 258
OPERA, 246

Palo Verde, 202
parity transformation, 23
parity violation, 8
Pauli exclusion principle, role in supernova formation, 292
Pauli spin matrix, 40

PCAC, 35
photino, 116, 386
Planck scale, 373
primordial nucleosynthesis, 375,
     391
primordial nucleosynthesis, ^{4}He
     abundance, 392
primordial nucleosynthesis, ^{4}He
     abundance, parameters
     controlling, 395
primordial nucleosynthesis, ^{4}He
     abundance,influence of
     number of relativistic
     degrees of freedom, 396
primordial nucleosynthesis, ^{7}Li
     abundance, 393
primordial nucleosynthesis,
     baryonic density, 395
prompt explosion, in supernovae,
     297
proton decay, 110

QCD, 109
QCD, lambda, 111, 373
QCD, quark–gluon plasma, 375
QED, 39
quarks, 48, 109, 373
quarks, free, 373

redshifts, 361, 363
relativity, general theory of, 361
relativity, special theory, 362, 363,
     365
renormalization group equations,
     106, 110
resonant spin flavour precession,
     284, 325
Robertson–Walker metric, 362

SAGE, 266, 267
Schrödinger equation, 38
Schwarzschild radius, 373
seesaw, 28
shock wave in supernova collapse,
     295, 296

Silk-damping, 380
SLAC, *B* factory, 50
Sloan Digital Sky Survey (SDSS),
     388
solar maximum satellite, 304, 312
sonic point, in supernova collapse,
     294
Soudan-2, 239
sphaleron, 399
spontaneous symmetry breaking,
     44, 45
squark, 116
standard model, 41
standard model, of cosmology, 361,
     368, 370, 376
standard model, of particle physics,
     41, 396, 399
standard solar model, 250, 259
standard solar model, CNO cycle,
     252, 253
standard solar model, D production,
     252
standard solar model, Gamow peak,
     254
standard solar model, He
     production, 252
standard solar model, hep
     neutrinos, 252
standard solar model, pp cycle, 252,
     254, 263, 266
standard solar model, pp cycle,
     neutrinos from, 259
standard solar model, pp I process,
     252
standard solar model, pp II process,
     253
standard solar model, pp III
     process, 253
standard solar model, Sommerfeld
     parameter, 254
steady-state model, 361
Stefan–Boltzmann law, 376, 390
strong interaction, 41, 113, 373
strong mixing, in supernovae, 298

Sudbury Neutrino Observatory (SNO), 269, 281, 301
Sun-spots, 285
Super-Kamiokande, 111, 120, 281, 301
Superbeams, 220
supernovae, 290, 315, 316, 391
supernovae, bounce back of shock wave in collapse, 295
supernovae, computer simulation of, 298
supernovae, frequency, 315
supernovae, gravitational waves from, 316
supernovae, homologous collapse in, 294
supernovae, iron core in collapse, disruption of, 296
supernovae, Kepler's, 302
supernovae, neutrino opacity in collapse, 295
supernovae, neutrino sphere in collapse, 295, 301
supernovae, neutrinos from, 290
supernovae, numbering scheme, 302
supernovae, prompt explosion of, 297
supernovae, shock wave in collapse, 295, 296
supernovae, SN 1987A, 298, 302, 304, 305, 307, 308, 315
supernovae, SN 1987A, γ radiation from, 304
supernovae, SN 1987A, characteristics of, 304
supernovae, SN 1987A, distance to, 307
supernovae, SN 1987A, light curve, 304, 305
supernovae, SN 1987A, neutrino properties from, 310
supernovae, SN 1987A, neutrinos from, 308

supernovae, SN 1987A, precursor star of, 304
supernovae, SN 1987A, ring from, 307
supernovae, SN 1987A, Sanduleak as precursor star, 304
supernovae, SN 1987A, search for pulsar from, 305
supernovae, sonic point in collapse, 294
supernovae, strong mixing in, 298
supernovae, type II, 290, 304, 315
supersymmetry, 114, 118
supersymmetry, as dark matter, 385
supersymmetry, local, 115
supersymmetry, LSP, 117
supersymmetry, MSSM, 115
supersymmetry, Weinberg angle in, 118
symmetries, continuous, 38
symmetry breaking, spontaneous, 45

Thompson scattering, 376
Thomson scattering, 375
Troitsk experiment, 137

universe, large scale structure in, 376, 378
universe, large scale structures in, power spectrum, 387
UNO, 248

(V–A) structure, 34
VIRGO experiment, for gravitational waves, 316
virial theorem, 291

W, Z bosons, 109
weak interaction, 48, 386, 392
Wein's law, 376
Weinberg angle, 44, 58
Weyl spinor, 22
white dwarf, 290
wide band beam, 67

WIMPs, 381, 384
WIMPs, cross sections for, 385
WMAP, 379
Wu experiment, 7

Yang–Mills theories, 40
Yukawa couplings, 46